D1175985

Magnetic Information Storage Technology

ACADEMIC PRESS SERIES IN ELECTROMAGNETISM

....

Electromagnetism is a classical area of physics and engineering which still plays a very important role in the development of new technology. Electromagnetism often serves as a link between electrical engineers, material scientists, and applied physicists. This series presents volumes on those aspects of applied and theoretical electromagnetism that are becoming increasingly important in modern and rapidly developing technology. Its objective is to meet the needs of researchers, students, and practicing engineers.

This is a volume in
ELECTROMAGNETISM

....

ISAAK MAYERGOYZ, SERIES EDITOR
UNIVERSITY OF MARYLAND, COLLEGE PARK, MARYLAND

Magnetic Information Storage Technology

Shan X. Wang

*Department of Materials Science and Engineering
(and of Electrical Engineering)
Stanford University
Stanford, California*

Alexander M. Taratorin

*IBM
Almaden Research Center
San Jose, California*

ACADEMIC PRESS

San Diego London Boston
New York Sydney Tokyo Toronto

The cover page illustrates a hard disk drive used in modern computers. Hard disk drives are the most widely used information storage devices. The block diagram shows the principle of the state-of-the-art disk drives. Any information such as text files or images are first translated into binary user data, which are then encoded into binary channel data. The binary data are recorded in magnetic disks by a write head. For example, the magnetization pointing to the right (or left) represents "1" (or "0"). The magnetization pattern generates a voltage waveform when passing underneath a read head, which could be integrated with the write head and located near the tip of the stainless suspension. The voltage waveform is then equalized, i.e., reshaped into a proper form. This equalization step, along with a so-called maximum likelihood detection algorithm, allows us to detect the binary channel data with a high reliability. The trellis diagram shown with 16 circles and 16 arrows is the foundation of maximum likelihood detection. The detected binary channel data are then decoded back to the binary user data, which are finally translated back to the original information.

This book is printed on acid-free paper. ⊗

Copyright © 1999 by Academic Press

ACADEMIC PRESS
a division of Harcourt Brace & Company
525 B Street, Suite 1900, San Diego, California 92101-4495, USA
http://www.apnet.com

Academic Press
24-28 Oval Road, London NW1 7DX, UK
http://www.hbuk.co.uk/ap/

Library of Congress Cataloging-in-Publication Data

Wang, Shan X.
      Magnetic information storage technology / Shan
   X. Wang, Alex Taratorin.
         p. cm. — (Electromagnetism)
      Includes index.
      ISBN 0-12-734570-1
      1. Magnetic recorders and recording.   I. Wang, Shan X.
   II. Taratorin, A. M.   III. Title.   IV. Series.
   TK7881.6.W26   1998
   621.39'7—dc21   98-24797
                                                                    CIP

Printed in the United States of America
99 00 01 02 03   MB   9 8 7 6 5 4 3 2 1

Academic Press Series in Electromagnetism

Edited by: Isaak Mayergoyz, University of Maryland, College Park, Maryland

To Tsing, Natasha, Winston, Melinda,
Roman and Nellie.

Contents

Foreword

This volume in the Academic Press Electromagnetism series presents a modern, comprehensive and self-contained treatment of magnetic information storage technology. This technology evolves at a remarkable pace in the directions of increasing areal density of data storage, increasing rate of data transfer and decreasing cost. The phenomenal and explosive growth and expansion of magnetic storage technology has been achieved as a result of continuous and coordinated progress in such diverse areas as physics and processing of novel magnetic thin film materials and structures, mechanics of positioning and flying with extremely small tolerances and signal coding, detection and processing techniques. Thus, it is becoming increasingly important and challenging to present comprehensive and up-to-date exposition of magnetic storage technology with the emphasis on its interdisciplinary nature and the latest technical achievements. This book responds to this challenge.

Magnetic Information Storage Technology is written by Professor Shan Wang and Dr. Alexander Taratorin who are young, very active and productive researchers in the area of magnetic storage technology with intimate and firsthand knowledge of the latest technological innovations. The authors have managed to put together in one volume an extraordinary amount of technical information. The salient and unique feature of the book is its interdisciplinary nature. It covers with equal depth the topics related to magnetic heads, media, write and read-back processes as well as system aspects of magnetic recording such as coding, signal detection and processing. This book is probably the first attempt to present these diverse topics in the same volume.

I maintain that this book will be a valuable source of ideas, concepts, insights, and facts for students, practicing engineers, material scientists and physicists involved in the fundamental study and development of modern magnetic storage technology.

Isaak Mayergoyz, Series Editor

Preface

Most of us now use magnetic information storage technology, one way or another, on a daily basis. Billions of bytes of digital information storage space can be accessed at a touch of our fingertips, and at a dirt-cheap price! This feat is made possible by the ingenious creativity and hard work of many scientists and engineers who devoted themselves to magnetic information technology over the years. The technology involves many scientific disciplines, and is progressing at a lightning speed. The world-wide economy derived from magnetic and nonmagnetic information storage products is truly global and gigantic. As the new millennium approaches and information revolution deepens, the information storage technology will have unlimited promises and challenges.

There are many excellent books written on this exciting subject, most of which will be listed in the end of Section 1.2. For example, Profs. Hoagland and Monson's classical textbook, *Digital Magnetic Recording*, seems to be more suitable for an undergraduate course nowadays. Drs. Mee and Daniel's *Magnetic Recording Technology* and *Magnetic Storage Handbook* are essential references for practicing engineers, scientists, and specializing students, but somewhat intimidating for beginners. Prof. Bertram's *Theory of Magnetic Recording* is an excellent book for advanced graduate students and researchers, but indeed heavily emphasizes theoretical approaches, as its title suggests. In our humble opinions, there is a need for a textbook on magnetic information storage technology that is reasonably self-contained, emphasizes both experimental and theoretical concepts, and contains the important developments of the 1990s. In addition, the number of books on the subject does not nearly match the size of the information storage industry. Therefore, we invested a great deal of time and energy to write this new book and aimed for the following readers:

1. Graduate and undergraduate students who may or may not be directly involved with the research of magnetic information storage technology, but are interested in the subject at the materials, device, and system levels
2. Professors and instructors who teach courses related to information storage

3. Seasoned researchers who desire an all-around understanding of the fundamentals of the magnetic information storage and a comparison of alternative storage technologies
4. Entry-level engineers at data storage industry who wish to quickly acquire relevant working knowledge
5. Professionals who want to develop new information storage technologies and need to know the competition

We have tried to make the book friendly to beginners, yet at the same time reasonably rigorous and useful to advanced readers. Thus, it is written assuming that the readers have had some basic knowledge of *electromagnetism, Fourier transform, digital circuit,* and *probability and statistics.* Generally speaking, the level of difficulty of this book is between that of Haogland and Monson's and that of Bertram's textbook. However, it contains many materials not covered in either book.

This book evolved from the lecture notes developed for a two-quarter course on magnetic recording and information storage systems at Stanford University, California. The course is designed for graduate students or upper-level undergraduate students in the departments of materials science and engineering, electrical engineering, mechanical engineering, applied physics, and other related areas. It aims to help students gain an appreciation of the fascinating science and technology of magnetic information storage, and more importantly, to develop an ability to solve problems in magnetic recording using the basic techniques and models introduced in the course. We try to select the models and techniques that are always useful even if the technology changes from generation to generation. In keeping with this spirit, the book emphasizes basic concepts and most fundamental aspects of magnetic information storage. It is relatively self-contained and allows the readers to follow the contents without resorting to technical papers. Technical papers are cited for relatively new and specialized topics and should be studied to enhance the depth and breadth of your understanding. We try to be fair in citing the right papers for original contributions, but sometimes we cite papers for convenience and easy access to students. For the fundamental chapters, we cite fewer papers to force us to make the presentations as readable as possible to beginners and experts alike.

"A picture is worth a thousand words," so we have created many illustrations and figures to demonstrate basic concepts, models and results. We hope to convey to the readers the simplicity and beauty of the fundamental concepts through pictures that will stick in the readers' minds for a long time. Many of the figures and tables are taken or modified from our own research work, published or unpublished. Some of the figures are modified from other books and papers, and their original authors are gratefully acknowledged.

The organization of this book is very straightforward. A brief introduction to digital magnetic information storage and basic magnetics is given in Chapter 1. The latter part is designed for the readers who do not have a good exposure to magnetism, and should be skipped by advanced readers. The two building blocks of magnetic recording—the magnetic recording head and medium—are introduced in Chapter 2, with just enought detail to allow serious discussions of magnetic recording processes. Chapters 3 and 4 deal with write and read processes in magnetic recording, which are the foundation of the whole book. The next three chapters (Chapters 5, 6, and 7) are devoted to the detailed discussions of inductive magnetic heads, magnetic recording media, and magnetoresistive (MR) heads. Device performance, fabrication, and materials issues are covered here. Some sections in these three chapters can be skipped without affecting the reading of later chapters. Chapter 8 covers channel coding and error corrections. This chapter is essential for understanding how digital information storage is actually done, and why. The noise formulation and mechanism in magnetic recording is presented in Chapter 9, with an emphasis of signal-to-noise ratio considerations in magnetic recording. Signal-to-noise ratio is the starting point for understanding magnetic information storage at the system level. Other than noise, magnetic storage is also disturbed by nonlinear distortion and off-track interference, which are discussed in Chapters 10 and 14, respectively. These studies are necessary to make magnetic storage actually work. Chapters 11, 12, and 13 are devoted to three common data detection channels: peak detection, partial-response maximum likelihood (PRML), and decision feedback equalization (DFE). This book is probably the first attempt to cover these signal detection and processing methods in the same context. Chapters 9–13 are intended for advanced readers or students in an advanced course on megnatic information storage, so they can be largely skipped by beginners. Chapter 15 deals with servo and position error signal generation. The coverage does not do justice to the importance of the subject, but it is unintentionally short because of the authors' limited expertise. Similar situations happened in the treatment of other topics such as tribology, and we ask for readers' tolerance. Chapter 16 touches on the fundamental limitations of magnetic recording, including superparamagnetic effects and dynamic effects. These subjects are the focus of intense investigations as we speak. Finally, Chapter 17 is devoted to a brief examination of "alternative," or nonmagnetic information storage technologies, such as optic disk recording (including magneto-optic recording), holographic recording, semiconductor flash memory, and magnetic random access memory (RAM). This chapter is worth studying on its own right, but it also serves to show why magnetic information storage technology, particularly hard disk drive, is called the "once and future king."

Although great efforts have been made to write and edit the book, it is far from being perfect. There are inevitably some errors, typos, and even misconceptions undetected by the authors. If you notice any short-comings, or simply have a comment, please contact the authors at sxwang@ee.stanford.edu, who will be very grateful.

This book is probably suitable as a textbook for a one-semester or two-quarter course on magnetic information storage technology. It has worked quite well at Stanford. Obviously, an instructor may choose to use some chapters and add his or her own supplemental materials. The authors do have a limited number of problem sets that can be shared on an informal basis. It is very important to do *homework*. To paraphrase a cliché, you forget after hearing, you remember after seeing, but you re-member and understand only after doing.

The writing of this book is a substantial endeavor, which will not be possible without the help and support of many individuals. S. X. W. wishes to thank his colleagues at Stanford University, who have created an atmosphere conducive to productive work. In particular, Professors Robert White and Bruce Clemens have provided vital intellectual support over the years. The students at Stanford have also been very helpful and stimulating. A. T. wishes to thank his former colleagues at Guzik Technical Enterprises for their essential technical expertise and engineering support, and his fellow researchers and the management at IBM for providing an excellent work environment and for allowing him to take on this endeavor. Both authors wish to thank Professor Isaak Mayergoyz for his encourage-ment, and Zvi Ruder, Diane Grossman, and Michael Granger for their professionalism during the production of this book. Finally, we wish to thank our families, to whom this book is dedicated, for their loving sup-port and sacrifices.

List of Acronyms

2D	two-dimensional
3D	three-dimensional
a.u.	arbitrary unit
ABS	air-bearing surface
AC	alternating current
ADC	analog-to-digital converter
AF	antiferromagnetic
AFM	antiferromagnet
AGC	automatic gain circuit
Al	aluminum
Al-Mg	aluminum magnesium
AMR	anisotropic magnetoresistive
Ar	Argon
AWGN	additive white Gaussian noise
BER	bit error rate
BLS	baseline shift
CCD	charge-coupled device
CD	compact disk
CD-ROM	compact disk read-only memory
CGS	centimeter, gram, and second (unit system)
CMOS	complementary metal oxide semiconductor
CNR	carrier-to-noise ratio
CTF	continuos time filter
Cu	copper
DASD	direct access storage devices
DC	direct current
DFE	decision feedback equalization
DRAM	dynamic random access memory
DVD	digital video disk or digital versatile disk
DVD-RAM	digital versatile disk random access memory (rewritable)
DVD-ROM	digital versatile disk read-only memory
E/H	easy/hard (transitions)
E^2PROM	electrically erasable programmable read-only memory
EA	easy axis
ECC	error-correction code
EFM	eight-to-fourteen modulation (code)
emu	electromagnetic unit
EPR4	class 4 extended partial response (channel)

EPROM	erasable programmable read-only memory
ESD	electrostatic discharge
esu	electrostatic unit
ETOM	electron trapping optical memory
FCI	flux change per inch
FDTS/DF	fixed-depth tree search with decision feedback
FE	ferroelectric
FFT	fast Fourier transform
FIR	finite impulse response
FM	frequency modulation
FM	ferromagnet
FT	Fourier transform
GB	gigabyte
Gb	gigabit
GMR	giant magnetoresistance
H/E	hard/easy (transitions)
HA	hard axis
HGA	head-gimbal assembly
HREM	high-resolution (transmission) electron microscopy
HTS	hard transition shift
IC	integrated circuit
IF	intermediate frequency
ISI	intersymbol interference
KB	kilobyte
Kb	kilobit
KBPI	1000 bit per millimeter
KFCI	1000 flux change per inch
LAN	local area network
MB	megabyte
Mb	megabit
MDFE	multilevel decision feedback equalization (channel)
MFM	magnetic force microscope
MFM	modified frequency modulation (code)
M-H	M (magnetization) versus H (magnetic field)
MIG	metal-in-gap (recording head)
MKSA	meter, kilogram, second, and ampere (unit system)
ML	maximum likelihood
M-O or MO	magneto-optic
MOKE	magneto-optical Kerr effect
MPEG	Moving Pictures Experts Group
MR	magnetoresistive or magnetoresistance
MRAM	magnetic random access memory
MSD	mean squared distance
MSE	mean squared error
NEP	noise equivalent (optical) power
Ni	nickel
NLTS	nonlinear transition shift
NM	nonmagnetic metal

NRZ	non-return-to-zero (code)
NRZI	modified non-return-to-zero (code)
OD	outer diameter (of disk)
OTC	off-track capability
PBSC	polarizing beam-splitter cube
PC	personal computer
PDF	probability density function
PE	partial erasure
PES	position error signal
PFPE	perfluoropolyethers
PLL	phase-locked loop
PPM	peak-to-peak modulation
PR4	class 4 partial response (channel)
PRML	partial-response maximum likelihood (channel)
PROM	programmable read-only memory
PW	pulse width
PWM	pulse width modulation
QWP	quarter-wave plate
RAM	random access memory
RAMAC	random access method of accounting and control
RE-TM	rare earth-transition metal (alloy)
RLL	run-length limited (code)
rms	root-mean-square
RPM	revolutions per minute
SAL	soft adjacent layer
SAM	sequenced amplitude margin
SDT	spin-dependent tunneling
SI	International System (of units)
SLM	spatial light modulator
SNR	signal-to-noise ratio
SR	shift register
SRAM	static random access memory
TAA	track-averaged amplitude
TCR	temperature coefficient of resistivity
TEM	transmission electron microscopy
TMR	tunneling magnetoresistance
TMR	track misregistration
UHV	ultrahigh vacuum
UV	ultraviolet (light)
VGA	variable-gain amplifier
VLSI	very-large-scale integration
VSM	vibrating sample magnetometer
XOR	exclusive OR (gate)

List of Symbols

$\vec{A}$	Vector form of any variable A
$\langle\,\rangle$	Average or expected value
$*$	Conjugate of a function or variable, variable after differentiation
∇	Gradient
$\nabla\times$	Curl
$\nabla\cdot$	Divergence
α	Direction cosine, angle between field and easy axis, flux leakage factor, temperature coefficient of resistivity, amplitude loss factor due to partial erasure, convergence parameter to minimize MSE, convergence parameter in clock recovery, attenuation factor to cancel precursor ISI in DFE, damping constant
β	Convergence parameter in gain recovery, factor in head field rise time
γ	Domain wall energy (per unit area), gyromagnetic ratio
γ'	Gyromagnetic ratio in the Landau-Lifshitz equation
γ_w	Domain wall energy per unit wall area
$\varDelta$	Hard transition shift or nonlinear transition shift, parameter related to reading adjacent track
$\varDelta_k,\ \varDelta(k)$	NLTS associated with the kth transition in a series of transitions
Δ	Change of a variable
$\Delta\rho_{\max}$	Maximum change in resistivity
$\Delta\rho_{\max}/\rho_0$	Magnetoresistance ratio
ΔE	Energy barrier to thermal switching
Δf	Frequency bandwidth, resolution bandwidth
Δl	Change in l
δ	Medium magnetic layer thickness, delta-function, skin depth
δ_n	Thickness of the nth particle
ε	Relative dielectric constant
ε_0	Absolute dielectric constant of vacuum
η	White noise spectral density, photodetector quantum efficiency
η_k	Ellipticity
θ	Angle between magnetization and easy axis, angle between magnetization and sense current
θ_1	Angle between spin valve free layer magnetization and sense current
θ_a	Analyzer orientation angle
θ_k	Kerr rotation angle
Λ	Grating period
λ	Wavelength, relaxation frequency, characteristic length of magnetic flux leakage

xxiv

λ_s	Saturation magnetostriction constant
μ	Relative permeability, absolute permeability when noted
μ_0	Absolute permeability of vacuum
μ_c	Core relative permeability
μ_{eff}	Effective relative permeability
μ_n	Horizontal magnetization of the nth particle
ρ	Resistivity, electric charge density, information capacity per unit volume
$\rho_\parallel$	Resistivity parallel to magnetization
$\rho_\perp$	Resistivity perpendicular to magnetization
ρ_0	Minimum resistivity
ρ_j	Resistivity along current direction
ρ_m	Magnetic charge volume density
σ	Stress, electric conductivity
σ_Σ	Standard deviation of the total noise at the slicer of MDFE detector
σ_e	Stress along easy axis
σ_h	Stress along hard axis
σ_m	Magnetic charge surface density
σ_n	Effective mean transition jitter
σ_p	Standard deviation of $p(x,y,z)$
σ_{PW50}	Standard deviation of half-amplitude pulse width
σ_t	Standard deviation of t_s
σ_V	Standard deviation of zero-to-peak voltage amplitude
σ_x	Standard deviation of transition center
τ	Access time, head field rise time constant
τ_c	Current rise time
τ_f	Field or flux rise time
τ_h	Intrinsic head field rise time
τ_l	Rotational latency
$\tau_{\min}$	Switching time at critical damping
τ_s	Head settling time
τ_{sk}	Seek time
Φ	Magnetic flux
$\Phi_{\text{sig}}(x)$	Signal flux injected into MR element from ABS
$\Phi_{\text{sig}}(x,y')$	Signal flux propagating in MR element
ϕ	Magnetic potential
$\phi(k,y)$	Spatial Fourier transform of $\phi(x,y)$
$\phi_s(x)$	Surface potential at $y = 0$
χ	Magnetic susceptibility, electric susceptibility, magnetic susceptibility of minor loops
ψ	Attenuation factor of HTS due to the proximity effect, spatial transition shift due to head field rise time
ω	Angular frequency
ω_0	Fundamental angular frequency
ω_1	Low angular frequency during overwrite
ω_2	High angular frequency during overwrite
ω_c	Cut-off angular frequency of a low-pass filter
A	Area, distance between the (normalized) sample values in PRML

	channels, variable threshold in the error filter method of SAM plot
A_c	Magnetic core cross sectional area
A_g	Head gap cross sectional area
a	Transition parameter
$a(t)$	Average pulse of $e(t)$ and $h(t)$
a_d	Transition parameter limited by demagnetization
a_f	Final transition parameter after relaxation
a_{im}	Transition parameter with head imaging
$a_k, a(k)$	The kth bit of in an NRZ data sequence
B	Magnetic induction, distance between magnetic transitions, flux change period, difference between the correct path metric and the wrong path metric
B_0	Required distance between magnetic transitions
b	Ratio of disk inner-diameter over outer-diameter
C	Formatted box capacity, data capacity of a single disk surface, Curie constant, variable threshold of the difference between the correct path metric and the wrong path metric
c	Speed of light
CDR	Channel data rate
CNR	Ratio of rms carrier signal power over slot noise power near the carrier frequency
D	Electric displacement, MR element stripe height, delay operator
$D(\omega)$	Fourier spectrum of $d(t)$
$D(k)$	Difference metric at the kth step
$D(\varepsilon)$	Density of states at an energy level of ε
D_{50}	Linear density at 50% resolution
D_a	Areal data density
D_{ch}	Channel density
D_f	Flux density
D_l	Linear data density
D_r	Data rate
D_t	Track density
D_u	User density
d	Magnetic spacing, differential, coil width, linear grain size, distance, hologram thickness
$d(\mathbf{A}, \mathbf{B})$	Hamming distance between codeword $\mathbf{A}$ and $\mathbf{B}$
d_{min}	Minimum distance, minimum Hamming distance
d^2	Squared distance between two trajectories in a trellis
d_{min}^2	Minimum squared distance in a trellis
(d, k)	Minimum and maximum channel bit periods inserted between two magnetic transitions in RLL coding
$d(k)$	Shape distortion noise in sample values at the kth step
$d(t)$	Error pulse of $e(t)$ and $h(t)$
E	Electric field, electric field of transmitted wave, energy, energy per unit volume, head efficiency, mean-squared error (MSE) of equalizer
$\mathbf{E}$	Error vector
$E(\omega)$	Fourier spectrum of $e(t)$
E_{ex}	Exchange energy

E_{GMR}	GMR head efficiency
E_j	Electric field along current direction
E_{mag}	Magnetostatic energy per unit volume
E_{MR}	MR head efficiency
E_p, E_s	Electric fields of two orthogonal polarization states that are parallel and perpendicular to the incident plane of light respectively
E_w	Energy due to domain wall
E_x, E_y	Electric fields of two orthogonal polarization states of light beam
e	Base of natural logarithms, electron charge Erase band
$e(t)$	Voltage pulse due to an easy transition
$e(X)$	Error polynomial
F	Factor due to head imaging
FT	Fourier transform
f	Any function, frequency, probability of thermal switching per unit time, noise filter expression in PRML detector, fraction of p-polarization being transmitted through a leaky PBSC
$f * g$	Convolution of function f and g
$f(t)$	Function f in time domain
$f(x)$	Function f in spatial domain
f_0	Fundamental harmonic frequency, attempt frequency
f_1	Old signal frequency which was overwritten, low frequency
f_2	New signal frequency used to overwrite, high frequency
f_{old}	Old voltage before overwrite
f_{ow}	New voltage after overwrite
G	Photodetector built-in current gain
$\mathbf{G}$	Generator matrix
$G(A/2)$	Total probability of error in ML detector, including both random noise and shape distortion
$G(f)$	Temporal Fourier transform of $g(t)$
$G(k)$	Spatial Fourier transform of $g(x)$
G_{ap}	Conductance at antiparallel magnetization configuration
G_p	Conductance at parallel magnetization configuration
g	Head gap length
$g(f)$	Probability density function of distortion filter expression
$g(nT)$	Sample values of $g(t)$ with a sampling period of T
$g(t)$	Any analog or continuous time signal in time domain
$g(X)$	Generator polynomial
gb	Guard band
H	Magnetic field
$\mathbf{H}$	Parity-check matrix
H_0	Intrinsic coercivity or anisotropy field
H_{app}	Applied magnetic field
H_{bias}	Bias field (transverse except when noted)
H_c	Coercivity, magnetic field inside magnetic core
$H_c(t)$	Effective coercivity
H_{cr}	Remanence coercivity
H_d	Demagnetizing field
H_{df}	Final demagnetizing field after relaxation

H_{dy}	Demagnetizing field along y-direction
H_g	Head deep gap field
H_h	Head field
H_{in}	Magnetic field inside a material
H_J	Field due to sense current density J
H_k	Anisotropy field
$H_s(k)$	Spatial Fourier transform of head field at surface $y = 0$
$H_s(x)$	Surface magnetic field at $y = 0$
$H_x(k,y)$	Spatial Fourier transform of $H_x(x,y)$
$H_x(x,y)$	Head field in x-direction (longitudinal), demagnetizing field in x-direciton
$H_y(k,y)$	Spatial Fourier transform of $H_y(x,y)$
$H_y(x,y)$	Head field in y-direction (perpendicular), demagnetizing field in y-direction
H_z	Field along z-direction (crosstrack or transverse)
H_x^{MR}	MR head field (due to imaginary coil) along x-direction
$H(\omega)$	Transfer function of equalizer
$H(D)$	Partial response polynomial
$H(f)$	Fourier transform of channel impulse response
$H(t)$	Step response function in write heads
$H(t_k)$	Histogram of bit shift
$h(k)$	Sampled version of $h(t)$, k is an integer
$h(t)$	Voltage pulse due to a hard transition, sinc function corresponding a sample of "1", impulse response of equalizer
I	Electric current, photodetector electric current
I_c	rms noise current due to preamplifier
I_e	rms noise current due to Johnson thermal noise
I_s	Sense current or bias current of MR head
I_{shot}	rms photon shot noise current
I_w	Zero-to-peak write current
i	Imaginary number, electric current, any integer
J	Current density
J_{ij}	Exchange integral
j	Any integer
K	Anisotropy constant (uniaxial when unnoted)
K_1, K_2	Cubic anisotropy constants
K_u	Uniaxial anisotropy constant
$K_{u,eff}$	Effective uniaxial anisotropy constant
k	Any integer, constant in the Biot-Savart law, Boltzmann constant, wave vector or spatial frequency, step in PRML detection, fraction of read track width allowed to be out of the written track width
k'	Noise-to-signal voltage ratio
$k\omega_0$	Angular frequency of the kth harmonic
k_0	Fundamental wave vector, fraction of read track width allowed to be out of the written track width in the absence of erase band
$ke^{i\kappa}$	Absolute ratio of r_{ps} over reflection coefficient r
k_F	Fermi wave vector
L	Inductance, length, observation length of signal

l	Length
l_2	Throat height of thin film inductive head
l_c	Magnetic core length, cross-track correlation length
l_n	Length of the nth particle
M	Magnetization, path metric or mean-squared distance
M_f	Final magnetization after relaxation
M_s^f	Saturation magnetization of the free layer in spin valve
M_r	Remanence magnetization
M_s	Saturation magnetization
MSD	Mean-squared distance
m	Magnetic moment
$\mathbf{m}$	Message vector
$m(k)$	Number of levels between correct samples and wrong samples
$m(X)$	Message polynomial
$m(x,y)$	Normalized recorded magnetization
m/n	Code rate of RLL code
N	Demagnetizing tensor, number of coil turns, number of magnetic particles per flux change period, wide-band noise power, ratio of the high frequency over the low frequency during overwrite
N^*	Wide-band noise power after differentiation
N_a, N_b	Demagnetizing factors along the main axes of an ellipsoid
N_d	Demagnetizing factor along y-direction
NEP	Noise-equivalent (optical) power
n	Any integer, number of data zones, track number, particle number density, refractive index
$n(k)$	Random noise in sample values at the kth step
(n, k)	ECC cyclic code encoding k message bits into n code bits
n^*	Noise voltage after differentiation
n_e	Electronics noise voltage
n_m	Medium noise voltage
O	Electric field of object wave
$O(\omega)$	Fourier spectrum of $o(t)$
OTC	Off-track capability
OW	Overwrite ratio
$o(t)$	Voltage associated with residual old signal
P	Electric polarization
$P(\omega)$	Fourier spectrum of an isolated pulse $p(t)$
$P(k)$	Slot noise power at a spatial frequency of k
$P_{0/1}$	Bit error rate for mistaking 0 for 1
$P_{1/0}$	Bit error rate for mistaking 1 for 0
P_e	Bit error rate
$P_{e,\text{bs}}$	Bit error rate due to bit shift at zero-crossing detector
$P_{e,\text{th}}$	Bit error rate at threshold detector
PES	Position error signal
P_l	Optical power of laser
P_{ML}	ML detector error rate
P_n	Slot noise power
$P_{\text{p-p}}$	Peak-to-peak optical power

P_s	Average power of signal s
$\langle PSD \rangle$	Average power spectral density
PSD_n	Noise power spectral density
$PSD_s(f)$	Power spectral density in frequency domain of signal s
$PSD_s(k)$	Power spectral density in spatial frequency domain of signal s
PSD_{tot}	Total noise power spectral density in the presence of multiple noise sources
PW_{50}	Half-amplitude pulse width
p	Magnetic pole tip thickness, spin polarization, average packing fraction of magnetic particles, number of bit periods between two unipolar pulses in the spectral elimination method, write precompensation parameter
$p(f)$	Probability density function of noise filter expression f
$p(k)$	Sampled version of $p(t)$, k is an integer
$p(t)$	Single pulse voltage in time domain (due to an isolated magnetic transition), target pulse of equalizer
$p(x,y,z)$	Normalized packing fraction of magnetic particles
p_k	Write precompensation parameter for the kth transition
Q	Q-factor of head field gradient, hologram Q-parameter
$Q(A/2)$	Probability of error in ML detector, i.e., when the noise filter expression f excceds $A/2$
$Q(C)$	Probability of error in ML detector, i.e., when the difference between the correct path metric and the wrong path metric exceeds C
$Q(x)$	Probability for a Gaussian noise (unit-variance, zero-mean) exceeding x
q_m	Magnetic charge
R	Reluctance, resistance, readback resolution, horizontal size of the write bubble, reflectivity, electric field of reference wave
$\mathbf{R}$	Received code vector
$R(\omega)$	Fourier spectrum of $r(t)$
$R(X)$	Received code polynomial
R_{ap}	Resistance at antiparallel magnetization configuration
R_c	Magnetic core reluctance
R_g	Gap reluctance
R_p	Resistance at parallel magnetization configuration, cross-correlation coefficient of packing fraction
R_s	Sheet resistance
$R_s(\zeta)$	Autocorrelation function of signal $s(t)$
$R_s(\eta)$	Autocorrelation function of signal $s(x)$
R^{st}	Statistical autocorrelation function
r	Radius, ratio of H_c over H_{cr}, reflection coefficient
r, r'	Position
$r(\tau)$	Cross-correlation coefficient of voltage signals in time domain
$r(t)$	Read voltage from square-wave recording
r_c	Critical radius of magnetic particle
r_{ID}	Disk radius at inner-diameter
r_{OD}	Disk radius at outer-diameter
$r_{pp}, r_{ps}, r_{sp}, r_{ss}$	Fresnel coefficients
S	Remanence squareness, zero-to-peak signal power, magnetic viscosity

	coefficient
S	Syndrome vector
$S(t)$	Square wave of alternating ± 1 amplitude
$S(X)$	Syndrome polynomial
S^*	Coercive squareness
S_A	Signal due to A-burst servo pattern
S_B	Signal due to B-burst servo pattern
$S_L(k)$	Spatial Fourier transform of $S_L(x)$
$S_L(x)$	Truncated version of signal $s(x)$ over length L
SNR	Ratio of signal power over noise power
$SNR(\text{dB})$	SNR in the unit of decibel (dB)
SNR^*	SNR after differentiation
SNR_{rms}	Ratio of rms signal power over noise power
SNR_{slot}	Ratio of signal power over slot noise power
SNR_{sp}	Ratio of single pulse peak signal power over wide-band noise power
SNR_{wb}	Ratio of signal power over wide-band noise power
$S_T(f)$	Temporal Fourier transform of $S_T(t)$
$S_T(t)$	Truncated version of signal $s(t)$ over time period T
$s(k)$	Sampled version of $s(t)$, k is an integer
$s(t)$	Signal s in time domain, voltage signal (including noise) in time domain, analog isolated pulse in partial-response channels, source pulse of equalizer
$s(x)$	Signal s in real space (position domain)
s_i	Spin angular momentum of the ith electron
$s_k, s(k)$	The kth sample value in a detection channel
T	Temperature, bit period, sampling period, observation time period of signal
T_a	Amplitude transmittance of hologram
T_c	Curie temperature
T_{ch}	Channel bit period
T_m	Melting point
$\text{TMR}_{\text{w,r}}$	Write-to-read track misregistration
T_u	User bit period
T_w	Timing window (or detection window or clock window), channel bit period
T_x	Crystallization temperature
t	Time, magnetic yoke thickness, MR element thickness
t_k	Thickness of keeper layer, bit shift
t_s	Zero-crossing shift
U	Code vector
$U(X)$	Code polynomial
UDR	User data rate
V	Voltage, magnetic grain volume
$V(k)$	Spatial Fourier transform of $V(x)$
$V(t)$	On-track readback voltage (excluding noise) in time domain
$V(x)$	Readback voltage
V^*	Voltage in MR head imaginary coil, voltage in zero-crossing detector
$V_{0\text{-p}}$	Zero-to-peak signal voltage amplitude

V_{3f}	Voltage from square-wave recording at a frequency of $3f$
V_a	Interference signal voltage from adjacent track
V_f	Voltage from square-wave recording at a frequency of f
V_G	Applied voltage at the select gate
V_{MR}	MR element or head voltage
V_n	Noise voltage
V_{oi}	Old information noise voltage
V^{peak}	Peak voltage (zero-to-peak) of $V(x)$, voltage amplitude
$V_{\text{p-p}}$	Peak-to-peak voltage (twice zero-to-peak voltage)
V_{rms}	Root-mean-square voltage
$V_{\text{rms,n}}$	rms noise voltage
V_s	Signal voltage
$V_{\text{sp}}(f)$	Temporal Fourier transform of an isolated pulse
$V_{\text{sp}}(k)$	Spatial Fourier transform of $V_{\text{sp}}(x)$
$V_{\text{sp}}(x)$	Single pulse or voltage from an isolated transition
V_T	Transistor turn-on threshold voltage
v	Linear velocity between head and disk
v_n	Volume of the nth particle
W	Physical trackwidth of head or medium, information channel signal levels
W_n	Width of the nth particle
w	Transition center fluctuation
w_r	Read track width
w_w	Written track width, written track width plus erase band
x modulo y	Remainder obtained from dividing x by y
x	Position (often along the downtrack direction), radial position of head
$\bar{x}$	Relative displacement of medium with respect to head
x_0	Transition center
x_n	Longitudinal position of the nth magnetic transition, radial position of the nth track,
y	Position (often perpendicular to disk plane), Magnetic spacing
z	Position (often along the crosstrack direction)
z_0	Side write distance

CHAPTER 1

Introduction

We are in an information age, thus, the demand for high performance, low cost, and nonvolatile information storage systems is ever-increasing. There are a variety of information storage systems with varying degrees of development and commercialization, including magnetic tape drives, magnetic hard disk drives, magnetic floppy disk drives, magneto-optic (MO) disk drives, phase-change optic disk drives, semiconductor flash memory, magnetic random access memory (RAM), and holographic optical storage. To date magnetic information storage technology (hard disk, floppy disk, and tape drives) is most widely used. Therefore, it is the main subject of this book.

The first part of this chapter will give an overview of magnetic information storage systems, particularly magnetic hard disk drives. The second part gives a brief review or introduction to basic magnetics to prepare for further reading of this book.

1.1 OVERVIEW OF DIGITAL MAGNETIC INFORMATION STORAGE

1.1.1 Information storage hierarchy

We are all familiar with computers, particularly personal computers (PCs), which are now ubiquitous and indispensable tools to improve our productivity and communication. As an example, a notebook computer is shown in Fig. 1.1, which contains a microprocessor, a keyboard, a flat panel display monitor, a semiconductor dynamic random access memory (DRAM), a hard disk drive, and a floppy disk drive, etc. The microprocessor and the information storage devices (DRAM, hard disk, floppy disk),

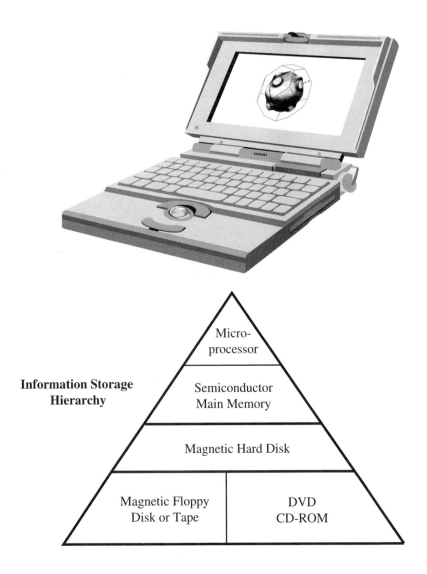

FIGURE 1.1. A notebook personal computer (top) and information storage hierarchy (bottom). Displayed by the notebook is the Fermi surface of copper.

along with appropriate software, constitute the "brain" of the computer, which remembers and processes all kinds of information.

The computer can be further expanded to a printer that provides a hard copy of information if desired. Paper, the traditional form of information storage device which was invented thousands of years ago in China, still holds the vast majority of the information on the Earth. It is believed that eventually digital information storage devices will surpass paper, but nobody knows when. The computer can also be connected to a network such as a local area network (LAN) and the Internet, through which one computer can access and communicate with virtually any other computers in the same network. The explosion of the World Wide Web or "cyberspace," the popular names of the Internet, is one of the driving forces behind the rapidly increasing demand for low-cost and high-performance digital information storage devices.

A typical hierarchical architecture of information storage in a PC is shown in Fig. 1.1. It is most efficient for the microprocessor to access and store the information in the semiconductor main memory (DRAM) alone because the DRAM access time is <100 ns. However, DRAM is volatile, any information in the DRAM will be lost at a power down. Therefore, lower levels of information storage devices are absolutely necessary. The lowest level of storage devices may include optical compact disk-read only memory (CD-ROM) in most desktop PCs, and magnetic tape drives for many workstations and servers. Digital versatile disk (DVD) is an emerging optical storage device targeted as a replacement of CD-ROM, both to be discussed in Chapter 17. Generally speaking, the higher levels of the memory hierarchy are superior in performance (e.g., data access time) but inferior in cost [$S/megabyte (MB)]. Furthermore, the lower levels of the memory hierarchy are nonvolatile. The economics and the nonvolatile requirements dictate that the majority of digital information is now stored in the lower levels of the memory hierarchy, including magnetic hard/floppy disks, magnetic tapes, and CD-ROM.

1.1.2 Magnetic and optical storage devices

A *nonvolatile* digital information storage device is constructed from a physical system that can accommodate two distinctive states. These two states can be altered back and forth by a transducer (called write head or writer) at a localized storage site. The two states will also produce distinguishable signals in a transducer (called read head or reader). Without external transducers the two states should be stable indefinitely, i.e., nonvolatile. Here we will briefly examine how magnetic and optical infor-

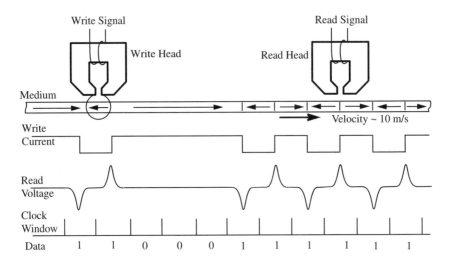

FIGURE 1.2.   Schematic principle of magnetic recording.

mation storage devices work. Today's digital information is mostly stored by one of the two storage technologies.

Magnetic hard disk drives, floppy disk drives, and tape drives are all based on the same fundamental principle of magnetic recording which involves an inductive recording head[1] and a recording medium, as shown in Fig. 1.2. The former is a transducer essentially consisting of some coils wound around a horseshoe-shaped soft magnetic material (with a low coercivity and high permeability) which has an air gap (called the head gap), while the latter is made of a hard magnetic material with a large coercivity.

There are possibly three modes of magnetic recording. One is the *longitudinal recording,* in which the medium magnetization is parallel to the disk plane, as shown in Figure 1.2. The second mode is *perpendicular recording,* in which the medium magnetization is normal to the disk plane. The third mode is *transverse recording,* in which the medium magnetization is normal to the page. This book largely focuses on longitudinal recording because it is adopted in all commercial magnetic recording systems now. Perpendicular recording is still under research, while transverse recording is not being actively pursued because of various technical constraints.

During a write process, a write current is passed through the head coils to produce a magnetic write field in the medium near the head gap.

---

[1]Magnetoresistive (MR) read heads are also being used nowadays, which will be discussed in Chapter 6.

The write field must be larger than the medium coercivity to magnetize the medium along the field direction. By switching the direction of the write field or the write current, magnetization transitions can be written in the medium. The magnetic transition length is finite in reality, but for simplicity it is shown to be zero in Fig. 1.2.

During a read process, the medium with transitions is passed underneath the head gap. The relative motion between the head and the magnetostatic stray fields produced by the magnetic transitions results in an induced voltage pulses in the coils due to Faraday's law. Each transition corresponds to a voltage pulse that has finite amplitude and finite half-amplitude pulse width PW_{50}, which limits how close the voltage pulses can be spaced and still be resolved.

In a simple coding scheme of digital magnetic recording called non-return-to-zero modified (NRZI) coding, the presence of a transition or a voltage pulse in a given clock window (timing window) represents a "1", while the absence of a transition or pulse represents a "0". Therefore, the linear data density is fundamentally limited by the value of PW_{50}.

Obviously, the above is a much simplified version of magnetic recording. A detailed understanding will emerge as we go through this book.

An optical data storage device is achieved in a very similar manner except that the two distinct states are now represented by what will lead to a difference in optical read signal. The examples include the following, which will be discussed in Chapter 17:

1. CD-ROM—small "bumps" in a compact disk cause destructive interference in reflected beams.
2. Phase-change optical storage: two different phases (crystalline/amorphous) cause a large difference in optical reflectivity.
3. Magnetic-optic storage: magnetic domains cause magneto-optical Kerr effect (MOKE), leading to a modulation of reflected light intensity.
4. Holographic optical storage: trapped electrical charge patterns cause a modulation in the phase of the diffracted light beam.

1.1.3 Overview of magnetic hard disk drives

Among the magnetic storage systems, the magnetic hard disk drive is the dominant secondary mass storage device for computers ranging from notebooks to mainframes because of its large storage capacity [>1 gigabyte (GB)], low cost per megabyte (<$0.20/MB), reasonably fast access time

(~10 ms), and a relatively mature manufacturing infrastructure. The worldwide hard disk drive revenue is ~$33 billion in 1997, and is growing at an annual rate of 10–20%.[2] Therefore, the magnetic hard disk drive will be the primary information storage system to be analyzed in this book.

A magnetic disk drive is mainly composed of four parts:

1. *Magnetic read/write heads and magnetic disks (platters).* Each write/read head (too small to be visible to the eye) is located on the trailing edge of a *slider* (or *head-slider*). A slider is mounted to the end of a stainless steel gimbal-suspension, forming a so-called *head-gimbal assembly* (HGA). A pair of a write/read head and a recording disk surface is often called a *head-disk assembly.*
2. *Data detection electronics and write circuit,* mostly located on a print-circuit board with many very-large-scale integration (VLSI) chips.
3. *Mechanical servo and control system,* including spindles, actuators, suspensions, and control chips.
4. *Interface* to microprocessor, located at one edge of the print-circuit board and through which the microprocessor input information from or output information to the disk drive.

A photograph of two hard disk drives is shown in Fig. 1.3. One drive contains five disks (10 head-disk assemblies), the other contains nine disks (18 head-disk assemblies). Note that the print-circuit board at the bottom of the drive and the sliders between disks are invisible in the photo. As mentioned in the Preface, magnetic heads and disks will be the focus through Chapters 2–7; data detection electronics will be discussed extensively in Chapter 8 and through Chapters 11–13; and mechanical servo will be discussed relatively briefly in Chapters 14–15. Interface will not be covered in this book.

The operation of the head-disk assembly is based on a self-pressurized *air bearing* between the slider and the *spinning* disk, which maintains a constant separation (called the *fly height*) between them, as shown in Fig. 1.4. By positioning the head-slider along the radial direction, different data *tracks* can be written on the disk. Each data track is divided into many data *sectors*. The state-of-the-art fly height is on the order of 1 micro-inch (μin.) or 25 nm, while the relative speed between slider and disk is extremely high (~10 m/s or higher). A good analogy is that the flying slider is like a Boeing 747 jetliner flying near the ground constantly. This

---

[2]P. Devin, M. Bourdon, and F. Yale, "Outlook '97: let the good times roll," *Data Storage,* January, 1997, p. 22.

FIGURE 1.3. Magnetic hard disk drives.

amazing feat is made possible by the wealth of knowledge from tribology. Tribology is the science which studies friction, wear, and lubrication between surfaces. In the context of magnetic recording, it is the science of head-medium interface. Tribological issues are touched superficially in Chapters 5 and 7, and readers are directed to other books and technical journals for more appropriate study of this very important subject.

The performance of hard disk drives is specified by the following parameters:

1. *Formatted box capacity C* (in MB) is the data storage capacity available to users. It is defined as

$$C = \frac{\text{User bytes}}{\text{Track}} \cdot \frac{\text{Tracks}}{\text{Data band}} \cdot \frac{\text{Data bands}}{\text{Head-disk assembly}} \cdot \frac{\text{Head-disk assemblies}}{\text{Box}}, \quad (1.1)$$

where

User bytes = Total data bytes − Overhead bytes (formatting),
Data bands = Zones with same data rate

A major advantage of magnetic hard disk drives over optical disk drives is that the former can have more head-disk assemblies per box because magnetic read/write transducers are much smaller than optical heads.

(a)

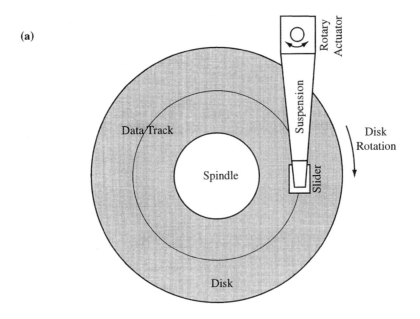

(b)

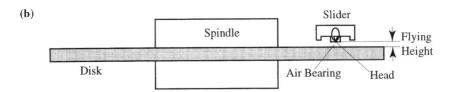

(c)

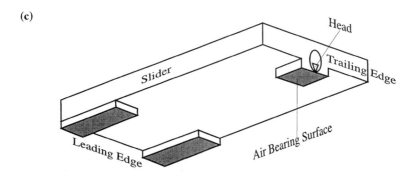

FIGURE 1.4. Schematic hard-disk assembly: (a) top view, (b) side view; and magnified view of slider (c).

2. *Rotational latency* τ_l is the time that the read head takes to rotate from an arbitrary data sector to the desired data sector on the same data track. The average rotational latency is defined as

$$\bar{\tau_l} = \text{Time of } \tfrac{1}{2} \text{ revolution of disk} \tag{1.2}$$
$$= \tfrac{1}{2} \div \text{Disk angular speed [revolutions per minute (RPM)].}$$

3. *Access time* τ is the time of read head moving from the current track to reading target data sector. It is the sum of the *seek time* τ_{sk} (time needed to move from an arbitrary track to the target track, but not yet ready to read), the *head setting time* τ_s (time needed to let the head stop vibration at the end of seek), and rotational latency:

$$\tau = \tau_{sk} + \tau_s + \tau_l \tag{1.3}$$

The head starts to read after $\tau_{sk} + \tau_s$, and it reaches the target data sector with an additional time of rotational latency τ_l.

4. *Linear data density* D_l is defined as the inverse of the smallest *bit length*, often given in the units of byte per millimeter, bit per millimeter, or (1000 bits per inch) (KBPI).

Another common unit for linear data density is flux change per inch (FCI). In the simplest data coding scheme like NRZI coding, one bit is equivalent to one magnetic transition or one flux change, so 1 KBPI is the same as 1000 FCI (KFCI). There is a difference between KBPI and KFCI in most practical magnetic recording systems, which will be discussed in Chapter 8.

5. *Track density* D_t is defined as the inverse of the *track pitch* (the distance between adjacent data track centers). Its unit is track/mm or KTPI (1000 tracks per inch).

6. *Data rate* D_r (byte/s, bit/s, MB/s, or Mb/s) is the number of bits per unit time that the write/read head can deliver. It is related to linear density and disk linear velocity v as follows:

$$D_r \text{ (bits per second)} = D_l \text{ (bits per inch)} \times v \text{ (inches per second).} \tag{1.4}$$

7. *Areal data density* D_a is defined as the inverse of bit area (bit length times track pitch), and is often given in the units of bit/mm^2, bit/in.^{2}, Mb/in.^{2}, Gb/in.^{2}, etc. Note that by convention "b" stands for bit, "B" stands for byte (except BPI and KBPI), so 1 MB = 8 × 10^6 bits. Obviously, areal data density is the product of linear data density and track density:

$$D_a \text{ (bit/in.}^2\text{)} = D_l \text{ (bit/in.)} \times D_t \text{ (track per inch).} \tag{1.5}$$

The ultimate goal in the magnetic storage research and development is to raise both linear density and track density to achieve the largest feasible areal recording density, which will usually translate into larger capacity, lower cost per megabyte and faster access time.

1.1.4 History and trends in digital magnetic recording

Hard disk drives, also called rigid disk files, or direct access storage devices (DASD), were first developed by IBM, San Jose, California in the late 1950s. The very first hard disk drive introduced in 1956 was called the random access method of accounting and control (RAMAC) or IBM 350. It contained 50 disks with a diameter of 24 in. and provided a data capacity of 5 MB and a data rate of 8.8 kilobytes (KB)/s. The areal density was about 2 Kb/in^2. This was no doubt an epoch-making invention. An interesting and extentive review of the history of magnetic recording can be found in the *Magnetic Recording Handbook* by Camras, or in the *Magnetic Recording: The First 100 Years*, edited by Daniel, Mee and Clark.

RAMAC used a hydrostatic air bearing, but later generations of hard disk drives have employed hydrodynamic air bearing resembling that shown in Fig. 1.4. In the beginning a slider in hard disk drives is formed by attaching a separate write/read head to a mechanical air bearing design. One generation of hard disk drive introduced in 1973—IBM 3340, also named Winchester after the famous rifle—is worth special mention, for it is the first disk drive to use low-mass ferrite sliders and lubricated disks. The Winchester slider design was fabricated entirely of ferrite. The Winchester design contained two (or four) disks with a diameter of 14 in., provided a data capacity of 35 (or 70) MB, a data rate of 0.8 MB/s, and an areal density of 1.69 Mb/in^2. In contrast, the highest areal density of any hard disk drive in 1997 was 2.6 Gb/in^2 found in IBM Travelstar 4GT. It had a data capacity of 4 GB and a data rate of 10.4 MB/s, and contained four disks with a diameter of 2.5 in. Today's hard disk drives, in spite of much larger data capacities, are sometimes still called Winchester drives.

Since the invention of the hard disk drive, it has undergone both evolutionary and revolutionary changes at a tremendous pace. The progress of the areal data density is shown in Fig. 1.5, which indicates an annual compound growth rate of ~30% in the 1970s and 1980s, but an even more astonishing rate of ~60% in the 1990s. The faster pace in the early 1990s is mainly due to the introduction of magnetoresistive (MR) heads and partial-response maximum likelihood (PRML) detection channels. A compound growth rate of 60% means that *areal density of hard disk drives doubles every 18 months*. This is analogous to the Moore's law that microprocessor performance doubles every 18 months.

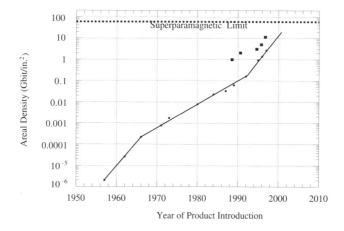

FIGURE 1.5. Leading-edge magnetic hard disk drive areal density vs. year of product introduction. The black dots represent commercial products, while the squares represent laboratory demonstrations of 1–10 Gb/in.^{2} densities. The line is for visual guidance only.

Magnetic data storage is currently the dominant form of storage for most computers. They are widely used in today's computers including laptop (notebook) PC, desktop PC, and mainframe computer. Magnetic disk recording up to 10 Gb/in.^{2} areal densities have been demonstrated by 1997, and there is still much room until it approaches the so-called *superparamagnetic limit* in the 40–70 Gb/in.^{2} range when the bits are so small that they become thermally unstable. This topic will be discussed in Chapters 7 and 16. Many ideas have been proposed in the literature to extend magnetic storage beyond the perceived superparamagnetic limit.

Optical data storage products remain to be marginal except for the CD-ROM (and possibly the DVD in the future). The success of optical storage technology will depend upon their performance/cost improvement as well as the dynamics of the marketplace. No convincing evidence exists to indicate that optical storage can replace magnetic storage in the foreseeable future. Flash memory (see Chapter 17) is a type of nonvolatile semiconductor memory, but is much more expensive than magnetic storage systems, and is not likely to replace magnetic storage either. It is anticipated that magnetic storage will continue to be the primary type of computer storage while other types play secondary roles in the next 10–20 years. Nobody can certainly predict what is far ahead of us, and you may judge for yourself after reading this book.

1.2 REVIEW OF BASIC MAGNETICS

1.2.1 Magnetic Field

Magnetic field is a form of matter produced by electrical currents. The most fundamental quality describing magnetic field is magnetic induction or magnetic flux density, $\vec{B}$, which is related to electrical current by Biot-Savart's law:

$$\vec{B} = k\iiint \frac{\vec{J}(\vec{r}') \times (\vec{r} - \vec{r}')}{|\vec{r} - \vec{r}'|^3} d^3\vec{r}', \qquad (1.6)$$

where $\vec{J}$ is the current density at position $\vec{r}'$, $\vec{B}$ is the magnetic flux density at position $\vec{r}$, $k = \mu_0 / 4\pi = 1 \times 10^{-7}$ H/m in the International System (SI), or $k = 1$ in the Gaussian system. The unit systems will be reviewed in the later part of this section.

Magnetic flux through a cross-sectional area A that is normal to magnetic induction $\vec{B}$ is defined as

$$\Phi = BA. \qquad (1.7)$$

Magnetic dipole is an electrical current flowing in a loop. The magnetic moment of a dipole is (in SI)

$$m = IA, \qquad (1.8)$$

where I is the current and A is the loop area. The direction of the vector $\vec{m}$ is defined by the right-hand rule. The magnetic induction at position $\vec{r}$ due to a magnetic dipole at the origin is given by

$$\vec{B} = \frac{\mu_0}{4\pi}\left[\frac{3(\vec{m}\cdot\vec{r})}{r^5}\vec{r} - \frac{\vec{m}}{r^3}\right]. \qquad (1.9)$$

For example, orbital motion and spin of electrons in molecules can be equivalent to a magnetic dipole. The magnetic moment per unit volume in a material is called magnetization, $\vec{M}$. The spontaneous magnetization (with no applied field) of a nonmagnetic material such as air or vacuum is usually zero. The magnetic field, $\vec{H}$, in a material is generally (in SI)

$$\vec{H} = \frac{\vec{B}}{\mu_0} - \vec{M}.$$

For an isotropic uniform material, the constitutive relations that relate magnetic field, magnetic induction, and magnetization are (in SI)

$$\vec{B} = \mu\mu_0\vec{H}, \; \vec{M} = \chi\vec{H}, \; \mu = 1 + \chi, \qquad (1.10)$$

where μ is the (relative) permeability, and χ is the susceptibility. Similarly, the constitutive relations that relate electric field $\vec{E}$, electric displacement $\vec{D}$, and electric polarization $\vec{P}$ are

$$\vec{D} = \varepsilon\varepsilon_0\vec{E}, \quad \vec{P} = \chi\varepsilon_0\vec{E}, \quad \varepsilon = 1 + \chi,$$

where ε is the (relative) dielectric constant, and χ is the electric susceptibility.

Electromagnetic field is fundamentally described by the Maxwell's equations that are as follows (in SI):

$$\nabla \cdot \vec{D} = \rho, \qquad \nabla \times \vec{E} = -\frac{\partial \vec{B}}{\partial t},$$

$$\nabla \cdot \vec{B} = 0, \qquad \nabla \times \vec{H} = \vec{J} + \frac{\partial \vec{D}}{\partial t}.$$

For static magnetic field, the relevant equations are

$$\nabla \cdot \vec{B} = 0, \qquad \nabla \times \vec{H} = \vec{J}. \tag{1.11}$$

They can be written in an integral form:

$$\oint \vec{B} \cdot d\vec{S} = 0, \qquad \oint \vec{H} \cdot d\vec{l} = I, \tag{1.12}$$

where I is the net current enclosed by the integral loop. The former is called Gauss's law, and the latter Ampere's law. Gauss's law means that there are no magnetic monopoles like electric charges, or *magnetic flux is always conserved*. However, Gauss's law can be rewritten as:

$$\nabla \cdot \vec{H} = -\nabla \cdot \vec{M} \equiv \rho_m. \tag{1.13}$$

If we draw an analogy between $\vec{H}$ and $\vec{E}$, then ρ_m can be regarded as the "magnetic charge" density corresponding to the electric charge density. At an interface, magnetic charge density per unit surface is

$$\sigma_m \equiv -\vec{n}_{12} \cdot (\vec{M}_2 - \vec{M}_1), \tag{1.14}$$

where $\vec{n}_{12}$ is the unit vector normal to the interface and pointing from medium 1 to medium 2. For example, as shown in Fig. 1.6, a uniformly

FIGURE 1.6.   Magnetization and magnetic charge.

magnetized bar has positive or negative magnetic charges at either end surface.
If we define a magnetic potential ϕ such that

$$\vec{H} = -\nabla\phi, \tag{1.15}$$

then the magnetic potential obeys the Poisson's equation:

$$\nabla^2\phi = -\rho_m. \tag{1.16}$$

In the space free of magnetic charges, we can find magnetic potential and
magnetic field by solving Laplace's equation:

$$\nabla^2\phi = 0. \tag{1.17}$$

1.2.2 Unit system

So far we have mostly adopted the SI unit system in the book. Readers may
be aware that the equations and numerical values of magnetic field change
with the system used! Therefore, it is necessary to review the basics of unit
systems here. The units in magnetics have always been somewhat confus-
ing, and both SI and Gaussian unit systems are widely used in the literature.
This situation can be understood from the following observations:

1. If one always uses the same set of units and certain sets of equations,
   then systems of units can be introduced with much freedom of
   choice.
2. One tends to use the units that most closely describe the phenomenon being observed. Therefore, astronomers use the light-year ($=$
   9.46×10^{15} m) as atomic physicists use angstrom ($\mathring{A} = 10^{-10}$ m)
   to measure length. It is not surprising that the SI unit ampere (A)
   is used to measure current instead of Gaussian unit statampere ($=$
   $\frac{1}{3} \times 10^{-9}$ A), the latter being too small. On the other hand, the
   Gaussian unit gauss (G) may be used to measure the earth's magnetic field ($\sim$0.4 G) instead of the SI unit tesla (1 T $= 10^4$ G), the
   latter being too large.

A unit system is established based on a set of *basic units*. Once the
basic units are chosen, all other units *(derived units)* can be derived from
the basic ones. The SI unit system is based on the basic units of meter,
kilogram, second, and ampere, so it is also called the MKSA system. The
SI system is the most widely used unit system in the world today. The
Gaussian unit system is based on the basic units of centimeter, gram, and
second, so it is sometimes called the CGS system. However, additional

basic units are needed to uniquely define the Gaussian system. On one hand, it uses CGS electrostatic units (esu) for electrical quantities. For example, the Gaussian units for electric current, charge, and voltage, are statampere, statcoulomb, and statvolt, respectively. On the other hand, it uses CGS electromagnetic units (emu) for magnetic quantities. For example, the units for magnetic induction, field, and dipole moment are gauss, oersted, and emu, respectively. In this sense, Gaussian system is a "mixed" unit system in which no basic units or derived units are clearly defined for electric and magnetic quantities. A more complete discussion of the unit systems in magnetism can found in the literature.

If unspecified explicitly, the SI unit system is used hereafter in this book. However, one should be able to comprehend both SI and Gaussian systems and convert between them. When you solve a magnetics problem, stick with only a single unit system to avoid errors. A conversion table of selected formulas and physical quantities are listed in Table 1.1, where $c = 3 \times 10^{10}$ cm/s is the speed of light.

1.2.3 Demagnetizing field

It has been long recognized that the magnetostatic field inside a magnetic material is often opposite to the magnetization such that it tends to "demagnetize" the latter. This can be understood by superposing the magnetic

TABLE 1.1. Comparison of the SI and Gaussian Unit Systems

SI (MKSA)	Gaussian (CGS)
$\vec{H} = \dfrac{\vec{B}}{\mu_0} - \vec{M}$	$\vec{H} = \vec{B} - 4\pi\vec{M}$
$m = I\,A$	$m = I\,A/c$
$\nabla \cdot \vec{H} = -\nabla \cdot \vec{M} \equiv \rho_m$	$\nabla \cdot \vec{H} = -4\pi\nabla \cdot \vec{M} \equiv 4\pi\rho_m$
$\nabla \times \vec{H} = \vec{J}$	$\nabla \times \vec{H} = \dfrac{4\pi}{c}\vec{J}$
B: tesla (T)	gauss (G), 1 G $= 10^{-4}$ T
H: ampere/meter (A/m)	oersted (Oe), 1 Oe = 1 G = 80 A/m
M: ampere/meter (A/m)	emu/cm^3, 1 emu/cm^3 = 1 kA/m
I: ampere (A)	statamp, 1 A $= 3 \times 10^9$ statamp

field due to point magnetic charges. If a point charge q_m is at position $\vec{r}'$, then its magnetostatic field at position $\vec{r}$, is

$$\vec{H} = \frac{q_m}{4\pi} \frac{\vec{r} - \vec{r}'}{|\vec{r} - \vec{r}'|^3}.$$

(1.18)

By the superposition principle we obtain that the magnetostatic field produced by a distribution of volume charge ρ_m and surface charge σ_m is

$$\vec{H}(\vec{r}) = \frac{1}{4\pi} \iiint \rho_m \frac{\vec{r} - \vec{r}'}{|\vec{r} - \vec{r}'|^3} d^3\vec{r}' + \frac{1}{4\pi} \iint \sigma_m \frac{\vec{r} - \vec{r}'}{|\vec{r} - \vec{r}'|^3} d^2\vec{r}'$$

or

$$\vec{H}(\vec{r}) = -\frac{1}{4\pi} \iiint (\nabla \cdot \vec{M}(\vec{r}')) \frac{\vec{r} - \vec{r}'}{|\vec{r} - \vec{r}'|^3} d^3\vec{r}'$$

$$+ \frac{1}{4\pi} \iint (\vec{n}' \cdot \vec{M}(\vec{r}')) \frac{\vec{r} - \vec{r}'}{|\vec{r} - \vec{r}'|^3} d^2\vec{r}'.$$

(1.19)

These are equivalent to $\nabla \cdot \vec{H} = \rho_m$. The magnetostatic field produced by the magnetization itself is called the demagnetizing field. It can be shown that for a uniformly magnetized material, the demagnetization field is related to the magnetization by a demagnetizing factor tensor:

$$\vec{H}(\vec{r}) = -\overleftrightarrow{N}(\vec{r}) \cdot \vec{M},$$

(1.20)

where the demagnetizing factor $\overleftrightarrow{N}(\vec{r})$ is a tensor function of position independent of the magnetization. In many symmetrical materials such as any ellipsoids of revolution, the demagnetizing factor tensor only has three principal components, i.e.,

$$\begin{pmatrix} H_1 \\ H_2 \\ H_3 \end{pmatrix} = - \begin{pmatrix} N_1 & 0 & 0 \\ 0 & N_2 & 0 \\ 0 & 0 & N_3 \end{pmatrix} \begin{pmatrix} M_1 \\ M_2 \\ M_3 \end{pmatrix},$$

where

$$N_1 + N_2 + N_3 = 1 \quad \text{(SI)},$$
$$N_1 + N_2 + N_3 = 4\pi \quad \text{(Gaussian)}.$$

(1.21)

TABLE 1.2. Demagnetizing Factors of Selected Shapes

Shape	N_1	N_2	N_3
Sphere	1/3	1/3	1/3
Long cylinder along z-axis	1/2	1/2	0
Infinite film normal to z-axis	0	0	1
Strip film normal to z-axis[a]	0	$\sim\dfrac{2}{\pi}\dfrac{t}{W}$	~ 1
(t, thickness; W, width; L, length; $t \ll W \ll L$)			

[a] The demagnetizing factors listed are at the center of the strip film.

Selected examples are given in Table 1.2, and the demagnetizing factors for ellipsoids of revolution can be found in the literature (e.g., Cullity, *Introduction to Magnetic Materials*, Appendix 6).

1.2.4 Ferromagnetism and magnetic hysteresis

Many materials such as iron and steel display a spontaneous magnetization, so they are called ferromagnetic materials. The spontaneous magnetization arises from magnetic dipoles due to *itinerant* electrons and unclosed atomic electron orbitals. These magnetic dipoles are coupled in parallel by the exchange interaction between electrons. The *exchange interaction* was the origin of the "molecular field" hypothesized by Weiss, long before the discovery of spin and quantum physics. Weiss found that a molecular field on the order of 10^3 T is necessary to align magnetic dipoles in parallel against thermal energy ($k_B T$, k_B is the Boltzmann's constant) that randomizes the orientation of magnetic dipoles.

The exchange interaction energy originates from the Pauli exclusion principle in quantum physics and Coulomb electrostatic energy. For two arbitrary electrons (denoted as i and j) the exchange interaction energy is proportional to the dot product of the two spins:

$$E_{ex} = -2J_{ij}\vec{s}_i \cdot \vec{s}_j, \tag{1.22}$$

where J_{ij} is the exchange integral, which is most significant for nearest-neighbor electrons. If $J_{ij} > 0$, the spins tend to orient parallel to minimize the exchange interaction energy, leading to ferromagnetism. If $J_{ij} < 0$, the spins tend to orient antiparallel, leading to antiferromagnetism.

In the absence of the exchange interaction, thermal energy aligns magnetic dipoles randomly, resulting in paramagnetic materials that do not have a spontaneous magnetization. A paramagnet has a small positive susceptibility described by Curie's law:

$$\chi = C/T, \tag{1.23}$$

where C is the Curie constant. A ferromagnet will become a paramagnet above a certain transition temperature called Curie temperature, T_c, when the thermal energy overcomes the exchange interaction. Above the Curie temperature, the ferromagnetic susceptibility obeys Curie-Weiss's law:

$$\chi = C/(T - T_c). \tag{1.24}$$

Below the Curie temperature, a ferromagnet usually has a temperature-dependent spontaneous magnetization. The ferromagnet-paramagnet transition is shown schematically in Fig. 1.7. The Curie temperatures of transition metal ferromagnets iron (Fe), cobalt (Co), and nickel (Ni) are much higher than room temperature, so their room temperature magnetization is very close to that at 0 K.

The ferromagnetic susceptibility below the Curie temperature is a complex function because of magnetic hysteresis. In other words, the

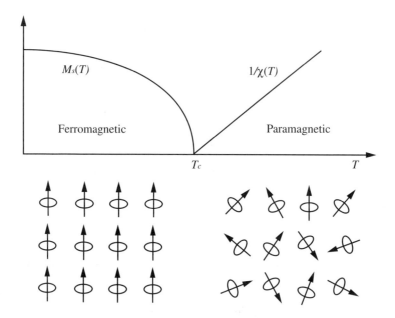

FIGURE 1.7. Ferromagnet-paramagnet transition. The arrows represent magnetic dipole moments.

magnetization of a ferromagnetic material as a function of applied field is dependent on the history of the applied field. A typical magnetic hysteresis loop, also called M-H loop, is shown in Fig. 1.8, which is a plot of magnetization M vs. applied field H. Usually, the applied field is the same as the internal field of the magnetic material with no demagnetizing field, and the M-H loop is an intrinsic property of magnetic material, independent of sample geometry (if it is much larger than magnetic domain size). If the demagnetizing factor N_d is not zero, the intrinsic M-H loop will be sheared by an amount determined by the demagnetizing factor, for the applied field H_{app} is related to the internal field H_{in} by the following equation:

$$H_{in} = H_{app} - N_d M. \tag{1.25}$$

Therefore, an intrinsic hysteresis loop $M = M(H_{in})$ will change into a sheared loop:

$$M = M(H_{app} - N_d M).$$

An originally square loop will have a slope of $1/N_d$ (with a minimum of 1) at coercive field. In any case, the hysteresis loop are described by several important quantities:

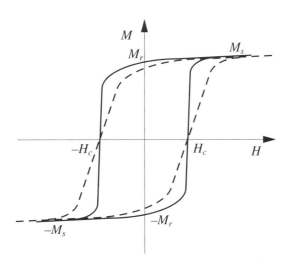

FIGURE 1.8. Magnetic hysteresis loop. The dashed loop indicates shearing due to demagnetizing field.

1. *Coercivity* H_c: the magnetic field needed to switch the direction of magnetization. Its unit is A/m in SI and Oe in the Gaussian system, respectively (1 Oe = $1000/4\pi$ A/m $\cong$ 80 A/m).

2. *Saturation magnetization* M_s: the maximum magnetization of a magnetic material when fully saturated ($H \gg H_c$). Its unit is A/m in SI and emu/cm^3 in the Gaussian system. The saturation induction or saturation magnetic flux density is given by $B_s = \mu_0(H + M_s)$ in SI and $B_s = H + 4\pi M_s$ in the Gaussian system. Its unit is tesla (T) in SI and gauss (G) in the Gaussian system, respectively. It is easy to remember that 1 T = 10^4 G, and 1 G = 1 Oe = 4π emu/cm^3, but one should not be confused that in the literature M_s is often quoted in emu/cm^3, while $4\pi M_s$ in G.

3. *Remanence magnetization* M_r: the magnetization after the applied field is removed.

4. M-H loop shape is often specified by *remanence squareness* $S = M_r/M_s$ and *coercive squareness* S^*. The latter is related to the slope at coercivity:

$$\left.\frac{dM(H)}{dH}\right|_{H_c} = \frac{M_r}{H_c(1 - S^*)}. \tag{1.26}$$

Generally the hysteresis loop of a magnetic material is symmetrical, i.e., $M(-H) = -M(H)$. Magnetic materials with $H_c < 10$ Oe are usually called *soft magnetic materials*, while those with $H_c > 100$ Oe are called *hard magnetic materials*. The coercivity is generally not affected by demagnetization because the magnetization is zero at coercive field. The hard magnetic materials are naturally used as magnetic recording media since their high coercivity value prevents them from demagnetization if the demagnetizing field H_d is smaller than H_c. Usually large S and S^* are also required for magnetic media, and $S = S^* = 1$ for a perfect square hysteresis loop. The commonly used magnetic recording media include γ-Fe$_2$O$_3$ particles (in plastic binder), CoCrTa thin films and CoCrPt thin films, the thin-film disks being the main stream products. In contrast, inductive recording head materials are magnetically soft, typically have a coercivity ≤ 1 Oe. The typical M-H loops of hard and soft magnetic materials are illustrated in Fig. 1.9.

1.2.5 Magnetic anisotropy

In magnetic materials it is often found that the magnetization has certain favorable directions, called the *easy axes* (EAs) and unfavorable directions,

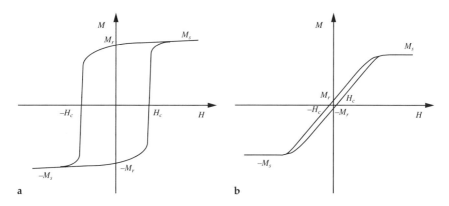

FIGURE 1.9. Hysteresis loops of (a) hard and (b) soft magnetic materials.

called the *hard axes* (HAs). In other words, the energy of a magnetic material displays an anisotropy in orientations, often termed magnetic anisotropy. Magnetic anisotropy energy is due to a number of mechanisms which will be discussed in the following.

Magnetocrystalline Anisotropy.

Due to the crystal symmetry, certain electron orbitals are special. In the presence of a weak spin-orbit coupling, magnetization (antiparallel to spin) will be coupled to electron orbitals and thus have a lowest or highest energy along certain symmetry axes.

For cubic crystals, the magnetocrystalline energy per unit volume can be generally expressed as

$$E = K_0 + K_1(\alpha_1^2\alpha_2^2 + \alpha_2^2\alpha_3^2 + \alpha_3^2\alpha_1^2) + K_2(\alpha_1^2\alpha_2^2\alpha_3^2) + \cdots, \quad (1.27)$$

where α_i are the direction cosines of the magnetization with respect to crystal axes. K_0 is a constant which can be ignored, K_2 is often much smaller than K_1. Examples of cubic anisotropic materials are given in Table 1.3.

TABLE 1.3. Magnetic Anisotropy of Selected Magnetic Materials (at Room Temperature)

Crystal	Structure	K_1 or K_u (10^6 J/m^3)	K_2 or K_{u2} (10^6 J/m^3)	Easy axes	Hard axes
Fe	BCC	4.8	±0.5	<100>	<111>
Ni	FCC	−0.5	−0.2	<111>	<100>
Co	HCP	45	15	[0001]	
$BaO \cdot 6Fe_2O_3$	HCP	33		[0001]	

For hexagonal crystals, the magnetocrystalline energy per unit volume can be generally expressed as

$$E = K_0 + K_u \sin^2\theta + K_{u2} \sin^4\theta + \cdots , \qquad (1.28)$$

where θ is the angle between the magnetization and the [0001] c-axis, K_0 is a constant which can be ignored, K_{u2} is often much smaller than K_u. This form of magnetic anisotropy is often called uniaxial anisotropy. Examples are given in Table 1.3.

Shape Anisotropy.   Due to the magnetostatic interaction between magnetization $\vec{M}$ and its own demagnetization field $\vec{H}_d$, a magnetic material possesses so-called demagnetizing energy (per unit volume):

$$E = -\frac{1}{2}\mu_0 \vec{M} \cdot \vec{H}_d. \qquad (1.29)$$

The demagnetizing energy is always positive or zero, and its magnitude depends on the shape of specimen. Thus shape anisotropy arises. For example, consider a very long cylinder along the z-axis, the demagnetizing factors are $N_1 = N_2 = 1/2$, $N_3 = 0$. The shape anisotropy energy per unit volume is

$$E = \frac{1}{4}\mu_0 M^2 \sin^2\theta, \qquad (1.30)$$

where θ is the angle between the magnetization and the z-axis. Thus the cylindrical axis is the EA. For a thin magnetic film perpendicular to the z-axis, the demagnetizing factors are $N_1 = N_2 = 0$, $N_3 = 1$. Thus the shape anisotropy energy per unit volume is

$$E = \frac{1}{2}\mu_0 M^2 \cos^2\theta, \qquad (1.31)$$

where θ is the angle between the magnetization and the z-axis. Thus the film normal axis is the HA. The magnetization energy is isotropic in the film plane.

Stress Anisotropy.   When a magnetic material is saturated by a magnetic field, its dimension will change accordingly. The saturation magnetostriction constant is defined as the fractional change in length along the magnetization:

$$\lambda_s = \frac{\Delta l}{l}. \qquad (1.32)$$

The magnetostriction constant is usually a small number ($<10^{-4}$).

If the magnetostriction constant is isotropic, but the magnetic material is subject to a uniaxial stress σ, then the stress anisotropy energy per unit volume is

$$E = \frac{3}{2}\lambda_s\sigma\sin^2\theta. \tag{1.33}$$

If the material is positively magnetostrictive ($\lambda_s > 0$), a uniaxial tensile stress will produce an EA along the same direction, causing the magnetization to align along the stress direction. This is often called the inverse magnetostriction effect.

Induced Anisotropy. If a soft magnetic material is deposited or postannealed in a magnetic field, a uniaxial anisotropy can be induced along the applied field direction, mainly due to atomic directional ordering mechanism and/or inverse magnetostriction effect. Any uniaxial anisotropy energy per unit volume can be generally expressed as

$$E = K_0 + K_u \sin^2\theta + K_{u2}\sin^4\theta + \cdots, \tag{1.28}$$

where the terms other than the second term can usually be neglected.

It should be noted that any combination of the above anisotropy mechanisms can exist in a magnetic material. The total magnetic anisotropy energy is the sum of all the anisotropy energies present.

1.2.6 Magnetic domains

The presence of a strong exchange interaction in a ferromagnetic material such as iron does not necessarily lead to a large spontaneous magnetization. Moreover, a ferromagnet with a large spontaneous magnetization can be *demagnetized*, i.e., made into one having no spontaneous magnetization. The reason was unknown for a long time until Weiss solved the puzzle by proposing that a ferromagnet is usually divided into magnetic domains. Each domain is uniformly magnetized and possesses the same magnetization, but they orient along different directions. Therefore, a ferromagnet could have a spontaneous magnetization much smaller than the saturation magnetization. The saturation magnetization of a ferromagnet, equal to the magnetization of each domain, is realized when the multidomain state is transformed into a single-domain state, as shown in Fig. 1.10.*

The boundary between two neighboring domains is called a domain wall or magnetic transition. Domain walls have finite width within which

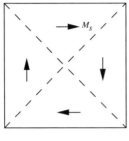
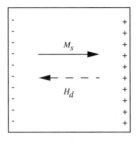

Multidomain State   Single-Domain State

FIGURE 1.10.   Magnetic domains. Note that the single domain state has magnetic charges while the multidomain state has none.

spins rotate gradually from one direction to the other. Since the spins have to overcome the exchange interaction and magnetic anisotropy to rotate, a domain wall has certain energy.

The origin of magnetic domains is nature's tendency of occupying the state of the lowest possible energy. A single domain state is usually unfavorable because of its large demagnetizing energy. When a single domain is broken into a multidomain state, demagnetizing energy diminishes along with magnetic charges. (This is not always true, a multidomain state could have magnetic charges as well.) The decrease in demagnetizing energy, however, is traded by an increase in domain wall energy. The balance of demagnetizing energy, domain wall energy, magnetic anisotropy energy, etc., determines whether a domain configuration is stable.

1.2.7 Magnetization process

When a multidomain ferromagnet is subjected to a magnetic field, the domains with a lower energy will expand while the domains with a higher energy will shrink, resulting in a domain wall displacement. In addition, magnetization in domains can also rotate toward the direction of the applied field without displacing domain walls. Two most commonly observed magnetization processes: domain wall motion and coherent magnetization rotation, are shown in Fig. 1.11 and will be discussed briefly. It should be noted that a magnetization process is generally more complicated than the simple picture presented here, and the readers should consult the literature for more in-depth insights.

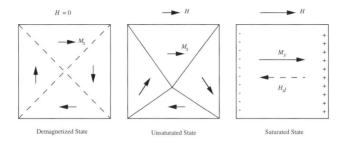

FIGURE 1.11. Magnetization process: domain wall motion and coherent magnetization rotation.

Domain Wall Motion and Barkhausen Noise. A ferromagnetic material is usually not perfect; it contains grain boundaries, defects, impurity inclusions, etc. These are so-called domain wall pinning sites. Therefore, the domain wall energy (per unit wall area) γ_w is generally a function of location x as shown in Fig. 1.12a, where each local minimum is a pinning

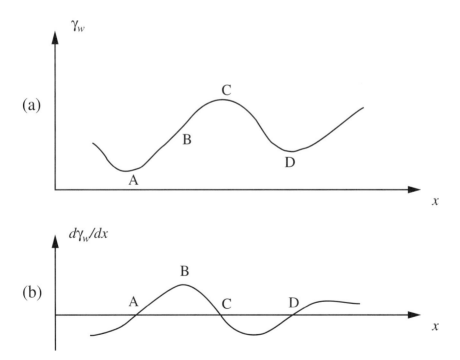

FIGURE 1.12. Domain wall energy (a) and restoring force (b) vs. wall location.

site (e.g., points A and D). The restoring force (per unit area) which pins a domain wall in a particular site is given in Fig. 1.12b. When a domain wall is subjected to a magnetic field H, the external magnetostatic force on the wall is (in SI)

$$\frac{dE_{mag}}{dx} = 2\mu_0 M_s HA,$$

where A is the domain wall area. If the external force exceeds the maximum restoring force (at point B), the domain wall will be swept through the specimen and the specimen will be magnetized along the applied field. Therefore, the domain wall coercivity H_c is given by

$$2\mu_0 M_s H_c = \left.\frac{d\gamma_w}{dx}\right|_{max}. \tag{1.34}$$

If the external force exceeds the local restoring force, the domain wall will move to another pinning site when the applied field is removed. In general, such domain wall motion is *irreversible.* In contrast, if the external force is less than the local restoring force, the domain wall will return to the original pinning site, such domain wall motion is then *reversible.*

Irreversible domain wall motion tends to be discontinuous and irregular. In 1919, Barkhausen performed an experiment by connecting a solenoid (with a magnetic core) to an amplifier and a loudspeaker. When the magnet is subjected to a full hysteresis, he heard a crackling noise in the speaker. If the speaker is replaced with an oscilloscope, irregular voltage spikes were seen. These noises and spikes were later named Barkhausen noise; they are due to discontinuous irreversible domain wall motion, often called Barkhausen jumps. A hysteresis loop, if magnified, usually contains discontinuous steps due to the Barkhausen jumps.

Coherent Rotation.   Coherent rotation of magnetic dipoles is another common mode in magnetization process, which simply means all the spins rotate together. Assume that the magnet has a uniaxial anisotropy, and it is subjected to a magnetic field with an angle of α from the EA, as shown in Fig. 1.13, then the total energy per unit volume of the magnet is

$$E = K_u \sin^2 \theta - \mu_0 M_s H \cos(\alpha - \theta), \tag{1.35}$$

where θ is the angle between the magnetization and the EA. The direction of magnetization is given by local minimums of E:

$$\frac{\partial E}{\partial \theta} = 2K_u \sin\theta\cos\theta - \mu_0 M_s H \sin(\alpha - \theta) = 0,$$

$$\frac{\partial^2 E}{\partial \theta^2} = 2K_u \cos2\theta + \mu_0 M_s H\cos(\alpha - \theta) > 0.$$

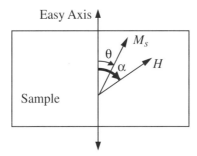

FIGURE 1.13. Coherent rotation.

Setting the second derivative to zero, we obtain the condition for the magnetization to rotate (switch) from one energy minimum orientation to another. Such rotation is usually *irreversible*. A small rotation near an energy minimum tends to be *reversible*.

Now consider two special cases of coherent rotation. Along the EA, $\alpha = 0$, $\theta = 0$ or π. The EA loop is shown in Fig. 1.14a. The coherent rotation is irreversible. Along the HA, $\alpha = \pi/2$, $\sin \theta = \mu_0 M_s H / 2K_u$. The HA loop is shown in Fig. 1.14b. The coherent rotation under an applied field along the HA is reversible. The magnetic material saturates at $H_k = 2K_u/\mu_0 M_s$ (in SI) or $H_k = 2K_u/M_s$ (in the Gaussian system). The saturation field is often called the anisotropy field because it is proportional to the anisotropy constant. The initial permeability μ along the HA, or the HA

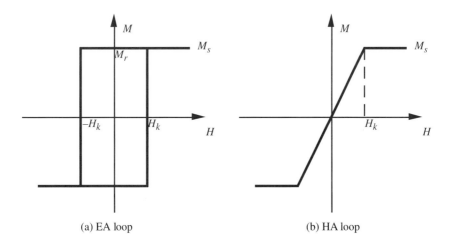

(a) EA loop   (b) HA loop

FIGURE 1.14. Magnetic hysteresis loops due to coherent rotation: (a) along easy axis (EA), (b) along hard axis (HA).

loop slope before saturation, is approximately equal to magnetic susceptibility χ because magnetization is usually much greater than the applied
field:

$$\mu \approx \chi = \frac{M_s}{H_k}\text{(SI)}, \quad \text{or} \quad \mu \approx 4\pi\chi = \frac{4\pi M_s}{H_k}\text{(Gaussian)}. \qquad (1.36)$$

Very soft magnetic materials ($H_c < 1$ Oe) with a steep HA loop slope
are used in magnetic recording heads. They behave like a "conductor"
of magnetic flux. Their ability to conduct magnetic flux is represented by
the permeability which is often >1000. The magnetic flux and applied
field along the HA are related by

$$M = \mu H, \qquad (1.37)$$

where μ is the initial slope of the hard axis M-H loop of the soft magnetic
material in Fig. 1.14b. Similar to Ohm's law in electric circuits: electric
current density is related to applied electric field as

$$J = \sigma E,$$

where σ is the conductivity. Therefore, one can draw an analogy between
magnetic flux density and electric current density, permeability and conductivity, magnetic field and electric field, magnetic flux and electric
current, etc. This analogy has proven to be quite useful when analyzing
magnetic circuits made of soft magnetic materials and permanent magnets.

Bibliography

Magnetic Information Storage Technology

T. C. Arnoldussen and L. L. Nunnelley, *Noise in Digital Magnetic Recording*, (Singa-
pore: World Scientific, 1992).

K. Ashar, *Magnetic Disk Drive Technology*, (Piscataway, NJ: IEEE Press, 1997).

H. N. Bertram, *Theory of Magnetic Recording*, (Cambridge, UK: Cambridge University Press, 1994).

B. Bhushan, *Tribology and Mechanics of Magnetic Storage Devices*, (New York, NY:
Springer, 1996).

M. Camras, *Magnetic Recording Handbook*, (New York, NY: Van Nostrand, 1988).

P. Ciureanu, *Magnetic Heads for Digital Recording*, (Amsterdam, Holland: Elsevier,
1990).

A. S. Hoagland and J. E. Monson, *Digital Magnetic Recording*, (New York, NY: John
Wiley & Sons, 1991).

S. Iwasaki and J. Hokkyo, *Perpendicular Magnetic Recording*, IOS Press, 1991.

F. Jorgensen, *The Complete Handbook of Magnetic Recording*, (Blue Ridge Summit,
PA: TAB Books, 1995).

J. C. Mallinson, *The Foundation of Magnetic Recording* (San Diego, CA: Academic Press, 1993).

C. D. Mee and E. D. Daniel, *Magnetic Storage Handbook,* (New York, NY: McGraw-Hill, 1996).

C. D. Mee and E. D. Daniel, *Magnetic Recording Technology,* (New York, NY: McGraw-Hill, 1996).

E. D. Daniel, C. D. Mee, and M. Clark, *Magnetic Recording: The First 100 Years,* (New York, NY: IEEE Press, 1998).

J. J. M. Ruigrok, *Short Wavelength Magnetic Recording: New Methods and Analyses,* (Amsterdam, Holland: Elsevier, 1990).

R. M. White, *Introduction to Magnetic Recording,* (New York, NY: IEEE Press, 1985).

Magnetics and Magnetic Materials

A. Aharoni, *Introduction to the Theory of Ferromagnetism,* (New York, NY: Oxford Univ. Press, 1996).

J. A. C. Bland and B. Heinrich, *Ultrathin Magnetic Structures I & II,* (Berlin, Germany: Springer-Verlag, 1994).

W. F. Brown, *Micromagnetics,* (Malabar, FL: Krieger, 1978).

R. M. Bozorth, *Ferromagnetism,* (New York, NY: Van Nostrand, 1951).

S. Chikazumi and S. H. Charap, *Physics of Magnetism,* (Malabar, FL: Krieger, 1978)

B. D. Cullity, *Introduction to Magnetic Materials,* (Reading, MA: Addison-Wesley, 1972).

J. D. Jackson, "Appendix on units and dimensions," in *Classical Electrodynamics,* (New York, NY: John Wiley & Sons, 1975).

D. Jiles, *Introduction to Magnetism and Magnetic Materials,* (London, UK: Chapman & Hall, 1991).

E. P. Wohlfarth, *Ferromagnetic Materials,* (Amsterdam: North-Holland, I–IV, 1980–2).

Optical and Other Information Storage Technologies

W. Gerhartz, *Imaging and Information Storage Technology,* (Weinheim, Germany: VCH (press); 1992).

C. Hu, *Nonvolatile Semiconductor Memories,* (Piscataway, NJ: IEEE Press, 1991).

M. Mansuripur, *Physical Principles of Magneto-optical Recording,* (Cambridge, UK: Cambridge University Press, 1995).

A. B. Marchant, *Optical Recording,* (Reading, MA: Addison-Wesley, 1990).

T. W. McDaniel and R. H. Victora, *Handbook of Magneto-optical Data Recording,* (Park Ridge, NJ: Noyes Publications, 1997).

A. K. Sharma, *Semiconductor Memories,* (Piscataway, NJ: IEEE Press, 1996).

K. K. Shvarts, *The Physics of Optical Recording,* (Berlin, Germany: Springer-Verlag, 1993).

H. M. Smith, *Holographic Recording Materials,* (Berlin, Germany: Springer-Verlag, 1977).

J. Watkinson, *Coding for Digital Recording,* (Kent, UK: Focal Press, 1990).

E. W. Williams, *The CD-ROM and Optical Disc Recording Systems,* (Oxford, UK: Oxford University Press, 1994).

CHAPTER 2

Fundamentals of Inductive Recording Head and Magnetic Medium

In this chapter we will discuss the magnetic fields generated by an idealized magnetic recording head and magnetic medium to pave the way for the presentation of read and write processes in the next two chapters. Magnetic recording is mostly concerned with very-high-frequency signals, and the high-frequency components of a magnetic field are of vital importance. Therefore, Fourier transforms are employed extensively to analyze magnetic fields and readback signal waveforms, and a brief review of Fourier transform and notations used in this book will be given first.

2.1 FOURIER TRANSFORMS OF MAGNETIC FIELDS IN FREE SPACE

Fourier transform is widely used in science and engineering, but its definition is not unique. Waveforms are most often expressed in time (t) or displacement (x) domain, $f(t)$ or $f(x)$, respectively, where $x = vt$ and v is the velocity. The *temporal* Fourier transform of waveform $f(t)$ is defined as

$$F(f) = \int_{-\infty}^{\infty} dt\, f(t)e^{-i2\pi ft} \tag{2.1}$$

where the variable f is the frequency. The corresponding inverse Fourier transform is

$$f(t) = \int_{-\infty}^{\infty} df\, F(f)e^{i2\pi ft}. \tag{2.2}$$

The *spatial* Fourier transform of $f(x)$ is defined as

$$F(k) = \int_{-\infty}^{\infty} dx\, f(x)e^{-ikx}, \tag{2.3}$$

where variable k is the wavevector. The corresponding inverse Fourier transform is

$$f(x) = \frac{1}{2\pi}\int_{-\infty}^{\infty} dk\, F(k)e^{ikx}, \tag{2.4}$$

Note that there are other forms of definition for Fourier transforms, and one should be careful when using a mathematics handbook to look up the Fourier transforms of a given function. In magnetic recording it usually holds that $x = vt$, $k = 2\pi/\lambda = 2\pi f/v$, where x is the displacement along the head-medium relative motion direction, v is the relative velocity between recording head and medium, t is the time, and λ is the wavelength. Magnetic information is recorded along a linear track spatially, but it is often captured on an oscilloscope in time domain. It is important to remember that the spatial and temporal waveforms have the following relations:

$$\begin{aligned} f(x) &\equiv f(t), \\ F(k) &= v \cdot F(f). \end{aligned} \tag{2.5}$$

Hereafter we often use spatial Fourier transform in analyzing magnetic recording. *The results must be translated into experiments in time domain (oscilloscope) or frequency domain (spectrum analyzer) by the above equations.*

The Fourier transform is a linear operation. When it is combined with some other operations, there are useful relations as listed in Table 2.1. In particular, the Fourier transform of the *convolution* of two functions, defined as

$$f(x) * g(x) \equiv \int_{-\infty}^{\infty} dx'\, f(x')g(x - x'), \tag{2.6}$$

is the product of their Fourier transforms.

There is an important relation between a function and its Fourier transform called Parseval's theorem, indicating that the total energy of a

TABLE 2.1.   Fourier Transforms[1]

	Function	Fourier transform
Differentiation	$\dfrac{\partial f(x)}{\partial x}$	$ikF(k)$
Translation	$f(x - a)$	$e^{-ika}F(k)$
Modulation	$e^{ik_0 x}f(x)$	$F(k - k_0)$
Convolution	$f(x) * g(x)$	$F(k)G(k)$
Multiplication	$f(x)g(x)$	$F(k) * G(k)$

function is the same whether it is expressed in the real space (x) or in the Fourier space (k):

$$\int_{-\infty}^{\infty} dx |f(x)|^2 = \frac{1}{2\pi} \int_{-\infty}^{\infty} dk |F(k)|^2. \tag{2.7}$$

Now let us consider the free space magnetostatic potential, which obeys Laplace's equation:

$$\nabla^2 \phi = 0. \tag{1.17}$$

The magnetic field is, of course,

$$H = -\nabla\phi. \tag{1.15}$$

In a two-dimensional (2D) problem (x–y plane) Laplace's equation is simply

$$\frac{\partial^2 \phi(x, y)}{\partial x^2} + \frac{\partial^2 \phi(x, y)}{\partial y^2} = 0. \tag{2.8}$$

Performing Fourier transform to Equation (2.8) with respect to x, we obtain

$$(ik)^2 \phi(k, y) + \frac{\partial^2 \phi(k, y)}{\partial y^2} = \frac{\partial^2 \phi(k, y)}{\partial y^2} - k^2 \phi(k, y) = 0. \tag{2.9}$$

where $\phi(k, y) \equiv \int_{-\infty}^{\infty} dx \phi(x,y) e^{-ikx}$. A general solution of $\phi(k, y)$ is

$$\phi(k, y) = A_+(k)e^{ky} + A_-(k)e^{-ky}. \tag{2.10}$$

The potential must approach zero when it is infinitely distant from the field sources. If the free space in question is completely at one side of the field sources such as magnetic charges or permeable materials, i.e., the field sources are below a certain plane ($y < 0$), then the potential will vanish at $y = +\infty$, and the potential in the free space is

$$\phi(k, y) = \phi(k,0)e^{-|k|y} \quad (y > 0). \tag{2.11}$$

This means that the Fourier transform (FT) of a magnetostatic potential with respect to x would decay exponentially in y direction if the free space is at one side of field sources! It follows that *the Fourier transform of the magnetostatic field with respect to x will decay exponentially in the y direction* as well:

$$H_x(k, y) \equiv FT\left(-\frac{\partial\phi}{\partial x}\right) = -ik\phi(k, y) = H_x(k,0)e^{-|k|y} \quad (y > 0), \tag{2.12}$$

$$H_y(k, y) \equiv FT\left(-\frac{\partial\phi}{\partial y}\right) = -\frac{\partial\phi(k, y)}{\partial y} = |k|\phi(k, y) = H_y(k,0)e^{-|k|y}$$

where $H_x(k,0) = -ik\phi(k,0)$, $H_y(k,0) = |k|\phi(k,0)$. It can be shown that

$$H_y(k, y) = i\,\mathrm{sgn}(k)H_x(k, y), \quad \mathrm{sgn}(k) \equiv \begin{cases} 1 \text{ for } k > 0, \\ 0 \text{ for } k = 0, \\ -1 \text{ for } k < 0. \end{cases} \quad (2.13)$$

This states that *in 2D free space, the Fourier transforms of the two field components are identical in magnitude and only differ in phase by* ± 90°. Taking a logarithmic of the field components, we obtain that

$$20\log H_x(k, y) = 20\log H_x(k,0) - 54.6\frac{y}{\lambda} \quad \text{(dB)},$$
$$20\log H_y(k, y) = 20\log H_y(k,0) - 54.6\frac{y}{\lambda} \quad \text{(dB)},$$

$$(2.14)$$

where $\lambda = 2\pi/k$ is the wavelength. This means that *the Fourier transform of a magnetic field component with respect to x will drop "linearly" in the y direction at a rate of 54.6 dB per wavelength.* This is an equivalent way of describing the exponential decay. The exponential decay is often called the *Wallace spacing loss* in magnetic recording, and is only valid for the Fourier transforms rather than a magnetic field itself! Spacing loss is one of the most important factors to be considered in magnetic recording, and will be discussed more in Chapter 3.

Now let us derive another important relation for a free space potential, $\phi(x,y)$, based on the Fourier transform of convolution. It is the inverse spatial Fourier transform of $\phi(k, y)$:

$$\phi(x, y) = FT^{-1}[\phi(k, y)] = FT^{-1}[\phi(k,0)e^{-|k|y}]$$
$$= \phi(x,0) * FT^{-1}[e^{-|k|y}]$$
$$= \phi_s(x) * \frac{y}{\pi(x^2 + y^2)}$$

where $\phi_s(x) \equiv \phi(x,0)$ is the surface magnetic potential at $y = 0$. Thus

$$\phi(x, y) = \frac{y}{\pi}\int_{-\infty}^{\infty} dx' \frac{\phi_s(x')}{(x - x')^2 + y^2}. \quad (2.15)$$

This means that *the magnetic potential (and thus magnetic field) can be expressed by the surface magnetic potential at y = 0 alone without any other information.* With the above relations in hand, we now turn to the fundamentals of inductive recording heads.

2.2 INDUCTIVE RECORDING HEADS

As introduced in Chapter 1, an inductive recording head is essentially a horseshoe-shaped soft magnet with an air gap and electric coils. Of course, the real structure of inductive recording heads is actually more complicated. Most audio and video recorders, and low-end hard disk drives employ so-called ferrite ring heads that consist of magnetically soft bulk ferrites machined into head structures. A more advanced version of the ring head is metal-in-gap (MIG) ring head, in which a metal layer with a higher saturation such as Sendust (FeSiAl alloy) is deposited in the head gap such that a larger write field can be achieved. The ring heads are not the focus of this book, and their details can be found in the reference books.[2]

Nowadays most hard disk drives being shipped employ thin film heads that evolve from IBM 3370 thin-film heads, first introduced in the 1970s. The cross section of an inductive thin-film head is shown in Fig. 2.1a. The bottom and top magnetic poles are electroplated permalloy ($Ni_{80}Fe_{20}$) layers, typically 2–4 μm thick. The four-layer, 54-turn coils are made from electroplated copper. The insulation layers separating coils and magnetic poles are formed by hard-cured photoresists. The head gap region is enlarged and shown in Fig. 2.1b. The gap layer is a sputtered alumina (Al_2O_3) layer in submicron thickness ($\sim$0.9 μm in original IBM 3370, now commonly <0.3 μm). The pole tip width of two magnetic poles (P1 and P2), along the direction perpendicular to the cross section, is approximately the data trackwidth written to and read from the magnetic disk. Therefore, they are made very narrow, ranging from >10 μm in the old days to submicron width in the state-of-the-art heads. Thin-film heads are batch fabricated in a fashion similar to that in VLSI technology, and will be discussed in detail in Chapter 5.

2.2.1 Karlqvist head equation

The magnetic field of an inductive head is complex because the head geometry is very complicated. Head fields can be solved numerically by finite element modeling or boundary element modeling, and sometimes analytically by conformal mapping, Green functions, or transmission line model, some of which will be described in Chapter 5. Here we describe the simplest yet most useful head model. In a classical paper published in 1954, Karlqvist idealized an inductive head with the following assumptions[3]:

1. The permeability of magnetic core or film is infinite.
2. The region of interest is small compared with the size of the magnetic core that the head can be approximated geometrically as

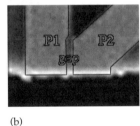

(a)   (b)

FIGURE 2.1.   Inductive thin-film head. (a) A cross section showing the alumina-TiC substrate, magnetic layers (P1 and P2), coils, insulation layers, alumina under-coat, and overcoat. (b) Enlarged head gap region.

shown in Fig. 2.2a. This is traditionally called a Karlqvist head. The magnetic core is much wider than the head gap, and the whole head is infinite in the z direction. Therefore, the problem becomes two-dimensional in the x–y plane, i.e., the paper plane.

3. The magnetic potential across the head gap varies linearly with x, as shown in Fig. 2.2b. This means that the magnetic field in head gap, H_g, is constant down to $y = 0$, and the magnetic potential in the magnetic core is zero because of its infinite permeability. The surface magnetic potential of a Karlqvist head at $y = 0$ is the same as the magnetic potential deep across the head gap:

$$\phi_s(x) = \begin{cases} -H_g x & \text{for} \quad |x| \leq g/2, \\ -gH_g/2 & \text{for} \quad x \geq g/2, \\ gH_g/2 & \text{for} \quad x \leq -g/2. \end{cases} \tag{2.16}$$

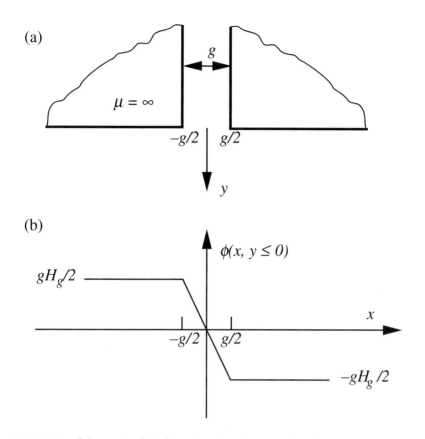

FIGURE 2.2. Schematic of Karlqvist head and magnetic potential.

Consequently, the magnetic potential of the Karlqvist head at $y > 0$ is

$$\phi(x, y) = \frac{y}{\pi}\int_{-\infty}^{\infty} dx' \frac{\phi_s(x')}{(x - x')^2 + y^2}$$

$$= -\frac{H_g}{\pi}\left[(x + g/2)\tan^{-1}\left(\frac{x + g/2}{y}\right) - (x - g/2)\tan^{-1}\left(\frac{x - g/2}{y}\right)\right.$$

$$\left. -\frac{y}{2}\ln\frac{(x + g/2)^2 + y^2}{(x - g/2)^2 + y^2}\right].$$

Then the x and y components of the Karlqvist field are given by

$$H_x(x, y) = -\frac{\partial \phi}{\partial x} = \frac{H_g}{\pi}\left[\tan^{-1}\left(\frac{x + g/2}{y}\right) - \tan^{-1}\left(\frac{x - g/2}{y}\right)\right]$$

$$= \frac{H_g}{\pi}\tan^{-1}\left[\frac{gy}{x^2 + y^2 - (g/2)^2}\right],$$

$$H_y(x, y) = -\frac{\partial \phi}{\partial y} = -\frac{H_g}{2\pi}\ln\frac{(x + g/2)^2 + y^2}{(x - g/2)^2 + y^2}.$$

(2.17)

It can be readily shown that, *for a Karlqvist head, the contours of constant H_x are circles going through the head gap corners.*

At the small gap limit, $g \to 0$, then

$$H_x(x, y) = \frac{H_g g}{\pi}\frac{y}{x^2 + y^2},$$

$$H_y(x, y) = -\frac{H_g g}{\pi}\frac{x}{x^2 + y^2}.$$

(2.18)

This is the magnetic field produced by a line current located at the head gap edge.

The Karlqvist field components as a function of x (displacement along data track) at a constant y (spacing between magnetic medium and head air-bearing surface) are plotted in Fig. 2.3. The field amplitudes drop and the pulses widen with increasing y. Both the amplitude and shape of the head field are important in magnetic write process, while the shape determines the sensitivity of an inductive read head. These concepts will become clear in Chapters 3 and 4.

In this text we will mostly be concerned with longitudinal magnetic recording in which magnetic medium is magnetized along the data track direction, so the x component of head field is very relevant. The Fourier transform of the x component of the Karlqvist field is

$$H_x(k, y \geq 0) = gH_g\frac{\sin(kg/2)}{kg/2}e^{-|k|y}$$

$$= gH_g\frac{\sin(kg/2)}{kg/2}e^{-ky} \quad \text{for} \quad k > 0.$$

(2.19)

We often concentrate on $k > 0$ in Fourier transforms, particularly in experiments. The constant coefficient above is the Fourier transform of the surface magnetic field, which is in the form of a single square pulse:

$$H_x(x, y = 0) = \begin{cases} H_g & \text{for} \quad |x| \leq g/2, \\ 0 & \text{for} \quad |x| > g/2. \end{cases}$$

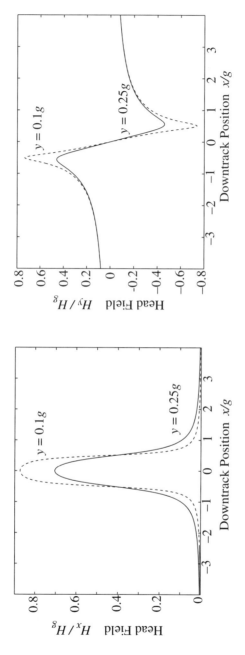

FIGURE 2.3. Karlqvist longitudinal and perpendicular fields *vs.* downtrack position *x*.

This square wave nature leads to the vanishing of the field transform when the wavevector satisfies $k = 2n\pi/g$, or when the wavelengths are $\lambda = 2\pi/k = g/n$, where $n = 1, 2, 3. \ldots$ These vanishing points are called the *gap nulls*.

The assumptions made by Karlqvist about the surface potential is not very accurate near the gap corners (or air-bearing surface), a conformal mapping technique is used by Westmijze to solve the head field accurately.[4] The accurate solution of the Karlqvist head without using assumption #3 (on p. 36) is very close to the Karlqvist expression far away from the gap corners, but drastically different near the gap corners. One consequence is that the gap nulls are slightly modified. For example, the first gap null actually occurs at a wavelength of $\sim1.136g$ instead of g!

2.2.2 Head magnetic circuit

In the Karlqvist model, the deep gap field H_g must be known to calculate the head field in magnetic medium, but only write current is directly controlled. Therefore, we need to understand the relation between the write current and the deep gap field. As a simplest example, an inductive recording head is modeled as a soft magnet core of constant cross section area A_c with an air gap of constant area A_g, as shown in Fig. 2.4a. If the magnetic flux due to the fringe field near the head gap is neglected, then the magnetic flux Φ is constant throughout the inductive head because of flux conservation:

$$\Phi = B_c A_c = B_g A_g , \tag{1.7}$$

where B_c and B_g are the magnetic flux densities in core and gap, respectively. From Ampere's law, we have

$$NI = H_c l_c + H_g g , \tag{1.12}$$

where N is the number of coil turns, I is the write current, H_c is magnetic field in the magnetic core (not to be confused with coercivity), and l_c and g are the length of the magnetic core and head gap, respectively. Substituting magnetic field by magnetic flux density using the constitutive relations [Equation (1.10)], Ampere's law becomes

$$NI = \frac{B_c}{\mu_c \mu_0} l_c + \frac{B_g}{\mu_0} g = \Phi\left(\frac{l_c}{\mu_c \mu_0 A_c} + \frac{g}{\mu_0 A_c}\right).$$

where μ_c is the permeability of magnetic core, while the air permeability is 1.

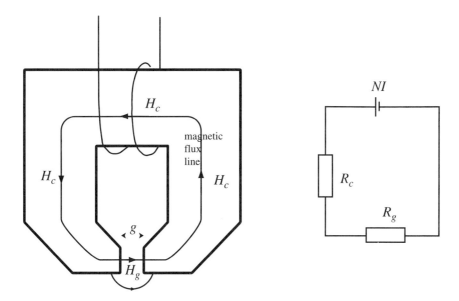

FIGURE 2.4. (a) Simplest inductive recording head model, and (b) its equivalent magnetic circuit.

Similar to the concept of electrical resistance, we define the magnetic *reluctance* of a section of soft magnet of length l, cross section area A, and absolute permeability μ as

$$R \equiv \frac{l}{\mu A}. \qquad (2.20)$$

Then, Ampere's law can be rewritten as

$$NI = \Phi(R_c + R_g), \quad R_c \equiv \frac{l_c}{\mu_c \mu_0 A_c}, \quad R_g \equiv \frac{g}{\mu_0 A_g}. \qquad (2.21)$$

In other words, the recording head model in Fig. 2.4a can be represented by a magnetic circuit as shown in Fig. 2.4b. The magnetomotive force is NI, the magnetic flux is Φ, and the core and gap reluctances are in series, corresponding to voltage, current, and two resistors in series in an electric circuit. This type of magnetic circuit model is very useful in analyzing magnetic recording heads and other magnetic devices.

The head efficiency E is defined as the percentage of magnetomotive force across the head gap out of the total magnetomotive force of the head coils: (We used the same symbol for efficiency, electric field, energy, etc.; the meaning is distinguishable from the context.)

$$E \equiv \frac{H_g g}{NI} = \frac{R_s}{R_g + R_c} \geq 1. \tag{2.22}$$

The head efficiency is determined by the head geometry and core material property. If $E = 1$, the inductive head transfers all the coil magnetomotive force NI to the gap region. If the head efficiency is known, the head deep gap field H_g can be easily calculated from write current.

2.3 MAGNETIC RECORDING MEDIUM

Magnetic recording media in hard disk drives contain either magnetic particles in organic binders (particulate disks) or continuous magnetic thin films (thin-film disks), which must have coercivities large enough to support the magnetic transitions to be written. Thin-film disks, presently the dominant magnetic disks, are actually complex multilayer structures as shown schematically in Fig. 2.5. The aluminum substrate is plated with an amorphous NiP undercoat to make the disk rigid, smooth, and properly textured. A chromium underlayer is often used to control magnetic prop-

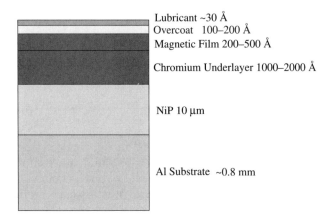

Lubricant ~30 Å
Overcoat   100–200 Å
Magnetic Film 200–500 Å

Chromium Underlayer 1000–2000 Å

NiP 10 μm

Al Substrate ~0.8 mm

FIGURE 2.5.   A schematic cross section of thin film hard disk.

erties and microstructures of the magnetic recording layer. The magnetic layer (typically Co-based alloy) is covered by a carbon overcoat layer and lubricant. The last two layers are necessary for the tribological perform-ance of the head-disk interface and for the protection of magnetic layer. The disk must rotate underneath a flying head at a very high speed (>5000 RPM), with a fly height as small as ~1.5 μin. (~38 nm). The *fly height* is the distance between the head air bearing surface to the disk top surface, while the *magnetic spacing* is the distance between head pole tips and the magnetic layer. The fly height is only a fraction of the magnetic spacing, and the latter is most relevant in the write and read processes. The non-magnetic layers in thin film disks are treated just like air when calculating magnetic fields.

2.3.1 Magnetic fields of step transition

Magnetic field of magnetized recording medium is produced by bound currents (in current picture), or equivalently by magnetic charges (in charge picture) within the magnetic medium. The latter picture is often more convenient and is thus widely used. In this case, the magnetic field can be computed from the volume charges inside the medium and the surface charges at the interfaces (Chapter 1):

$$\vec{H}(\vec{r}) = -\frac{1}{4\pi}\iiint \left(\nabla \cdot \vec{M}(\vec{r}')\right)\frac{\vec{r} - \vec{r}'}{|\vec{r} - \vec{r}'|^3}d^3\vec{r}' \qquad (1.19)$$

$$+ \frac{1}{4\pi}\iint (\hat{n}' \cdot \vec{M}(\vec{r}'))\frac{\vec{r} - \vec{r}'}{|\vec{r} - \vec{r}'|^3}d^2\vec{r}' \ .$$

A magnetic medium typically extends infinitely along the x-axis and has a finite thickness. Similar to the case of recording heads, the z direction (across track) can often be assumed to be uniform, and the problem can be simplified into a 2D geometry in the x–y plane.

Consider a single, perfectly sharp step transition written on a magnetic medium:

$$M(x) = \begin{cases} -M_r & \text{for} \quad x < 0, \\ M_r & \text{for} \quad x > 0. \end{cases} \qquad (2.23)$$

Then there is a surface charge density of $\sigma_m = -2M_r$ at the transition center $x = 0$ extending from $-\partial/2 < y < \partial/2$, where ∂ is the magnetic medium thickness, M_r is the remanence magnetization of the magnetic medium, and the center plane of the magnetic medium is $y = 0$. (Note that the y-origin is different from that in Section 2.2.) Using the second term of Equation (1.19), we can easily show that the horizontal and vertical magnetic fields produced by the transition are

$$H_x(x, y) = -\frac{M_r}{\pi}\left[\tan^{-1}\left(\frac{y + \delta/2}{x}\right) - \tan^{-1}\left(\frac{y - \delta/2}{x}\right)\right], \quad (2.24)$$

$$H_y(x, y) = -\frac{M_r}{2\pi}\ln\left[\frac{(y + \delta/2)^2 + x^2}{(y - \delta/2)^2 + x^2}\right].$$

These equations should be valid both inside and outside the magnetic layer. The magnetic field of a unit charge step transition at $x = 0$ is

$$H_x^{\text{ustep}}(x, y) = \frac{1}{2\pi}\left[\tan^{-1}\left(\frac{y + \delta/2}{x}\right) - \tan^{-1}\left(\frac{y - \delta/2}{x}\right)\right], \quad (2.25)$$

$$H_y^{\text{ustep}}(x, y) = \frac{1}{4\pi}\ln\left[\frac{(y + \delta/2)^2 + x^2}{(y - \delta/2)^2 + x^2}\right].$$

2.3.2 Magnetic fields of finite transition

If a magnetic transition is not sharp, then the field patterns produced are broadened compared with the single-step transition. Assuming that the finite transition does not change shape in the y and z direction, then the magnetic charge at each location x' is equivalent to a step transition with a surface charge density of $-\partial M(x')/\partial x'$. The magnetic field at x produced by such a step transition (located at $x = x'$) is

$$H_x^{\text{step}}(x - x', y) = \frac{-\partial M_r(x')/\partial x'}{2\pi}\left[\tan^{-1}\left(\frac{y + \delta/2}{x - x'}\right) - \tan^{-1}\left(\frac{y - \delta/2}{x - x'}\right)\right],$$

$$\quad (2.26)$$

$$H_y^{\text{step}}(x - x', y) = \frac{-\partial M_r(x')/\partial x'}{4\pi}\ln\left[\frac{(y + \delta/2)^2 + (x - x')^2}{(y - \delta/2)^2 + (x - x')^2}\right].$$

Based on the principle of superposition, the net magnetic field produced by a finite transition is the integration of the fields due to all the step transition distributed from $-\infty$ to $+\infty$:

$$H_{x \text{ or } y}(x, y) = \int_{-\infty}^{\infty} dx' H_{x \text{ or } y}^{\text{step}}(x - x', y) = -\frac{\partial M(x)}{\partial x} * H_{x \text{ or } y}^{\text{ustep}}(x, y). \quad (2.27)$$

That is to say, *the magnetic field of a finite transition is the convolution of the longitudinal gradient of the magnetization and the magnetic field of a unit charge step transition.* We must know the shape of the magnetization transition to calculate its magnetic field.

A real magnetization transition in magnetic media can be modeled by an analytical function such as *arctangent function, hyperbolic tangent function,* and *error function,* of which the arctangent function is the most convenient.[5] In this case, a magnetic transition centered at $x = 0$ can be modeled as

$$M(x) = \frac{2M_r}{\pi} \tan^{-1}\left(\frac{x}{a}\right),$$ (2.28)

where a is the transition parameter. The bigger it is, the wider the transition. If $a = 0$, the transition is an ideal sharp step transition. Both the step transition and arctangent transition are shown in Fig. 2.6.

The longitudinal gradient of the arctangent transition is

$$\frac{\partial M(x)}{\partial x} = \frac{2M_r}{\pi} \frac{a}{x^2 + a^2},$$ (2.29)

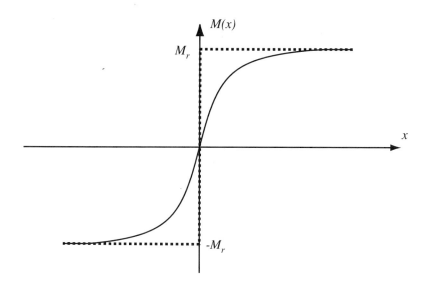

FIGURE 2.6. Step (dashed) and arctangent (solid) transitions of magnetization.

so the magnetic fields of a finite transition can now be expressed by the following convolution [Equation (2.27)]:

$$H_{x\text{ or }y}(x, y) = -\frac{2M_r}{\pi}\frac{a}{x^2 + a^2} * H_{x\text{ or }y}^{\text{ustep}}(x, y).$$

Now the magnetic fields can be solved analytically.[6] Inside the medium ($|y| \leq \partial/2$),

$$H_x(x, y) = -\frac{M_r}{\pi}\left[\tan^{-1}\left(\frac{a + y + \partial/2}{x}\right) + \tan^{-1}\left(\frac{a - y + \partial/2}{x}\right) - 2\tan^{-1}\left(\frac{a}{x}\right)\right],$$

$$H_y(x, y) = \frac{M_r}{2\pi}\ln\left[\frac{(a - y + \partial/2)^2 + x^2}{(a + y + \partial/2)^2 + x^2}\right]. \qquad (2.30)$$

Outside the medium ($|y| \geq \partial/2$),

$$H_x(x, y) = -\frac{M_r}{\pi}\left[\tan^{-1}\left(\frac{a + |y| + \partial/2}{x}\right) - \tan^{-1}\left(\frac{a + |y| - \partial/2}{x}\right)\right], \qquad (2.31)$$

$$H_y(x, y) = \pm\frac{M_r}{2\pi}\ln\left[\frac{(a + |y| - \partial/2)^2 + x^2}{(a + |y| + \partial/2)^2 + x^2}\right] (+ \text{ for } y > \partial/2, - \text{ for } y < -\partial/2).$$

Note that Equations (2.30) and (2.31) become the same as those of the step transition [Equation (2.24)] if $a = 0$. Furthermore, *all the fields exterior to the medium are identical to those from the step transition when y is replaced by y ± a (plus for y > 0, minus for y < 0), as if the field were produced by a step transition displaced further by a distance of the transition parameter.*

The x-component of magnetic field along the medium centerline, with an arctangent magnetization transition and the medium thickness $\partial = a$ or $2a$, are shown in Fig. 2.7a. The magnetic fields inside the medium are always opposite in direction to the magnetization as they should. The demagnetizing field diminishes at the center of the transition ($x = 0$) or far away. (The perpendicular demagnetizing field at $y = 0$ is zero because of symmetry.) The x- and y-component of magnetic field above the magnetic medium is shown in Fig. 2.7b. This field is the field detected by the magnetic read head.

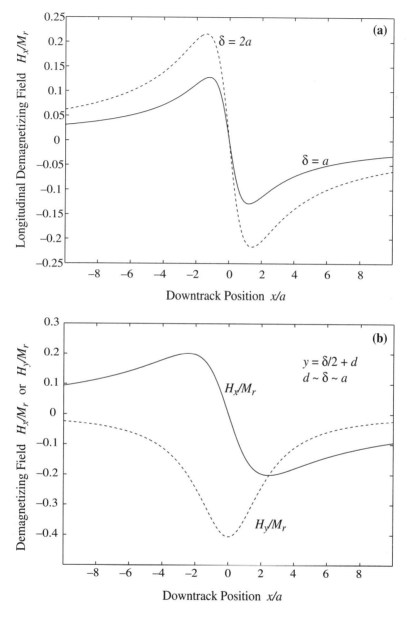

FIGURE 2.7. Longitudinal demagnetizing field (a) along the centerline in the medium ($y = 0$), (b) above the medium ($y = \delta/2 + d, d \sim \delta \sim a$).

References

1. R. A. Witte, *Spectrum and Network Measurements*, (Engelwood Cliffs, NJ: Prentice Hall, 1991). H. N. Bertram, *Theory of Magnetic Recording*, (Cambridge, UK: Cambridge University Press, 1994, pp. 38–46).

2. R. M. White, *Introduction to Magnetic Recording*, (New York, NY: IEEE Press, 1985, Chapter 3).

3. O. Karlqvist, "Calculation of the magnetic field in the ferromagnetic layer of a magnetic drum," *Trans. Roy. Inst. Technol. Stockholm*, vol. 86, p. 3, 1954.

4. W. K. Westmijze, "Studies on magnetic recording," *Philips Res. Rep.*, Part II, 8(3), 161, 1953.

5. H. N. Bertram, *Theory of Magnetic Recording*, (Cambridge, UK: Cambridge University Press, 1994, Chapter 4).

6. R. I. Potter, "Analysis of saturation magnetic recording based on arctangent magnetization transitions," *J. Appl. Phys.*, 4(4), 1647, 1970.

CHAPTER 3

Read Process in Magnetic Recording

Important aspects of the readback process in magnetic recording will be studied in this chapter. We will first discuss the reciprocity principle for calculating readback signal, then derive the readback voltage pulses of various types of magnetic transitions. The chapter ends with the discussion of linear superposition of multiple transitions and spectrum analysis of readback signals.

When a magnetic recording medium with written magnetic transitions passes underneath an inductive recording head, the magnetic flux produced by each transition is picked up by the head. The variation of the magnetic flux produces an "induced" voltage in the head coil, i.e., a readback signal. Therefore, the readback signal can be calculated from Faraday's law. However, the magnetic flux actually going through the head coil is very difficult to calculate *directly*. Instead of using the cumbersome direct method, a powerful theorem called reciprocity principle, commonly used to calculate the magnetic flux through the head coil, will be introduced first.

3.1 RECIPROCITY PRINCIPLE

The reciprocity principle is based on the fact that the mutual inductance between any two objects #1 and #2 is one quantity and the same: $M_{12} = M_{21}$. Consider the two objects to be a recording head and a magnetic element of recording medium located at (x, y, z) with a volume of $dxdydz$, as shown in Fig. 3.1, where d is the magnetic spacing between the head pole tips and the top of magnetic recording layer. Assume that the magnetization in the medium is horizontal along the x-axis, then we only need to consider the longitudinal component of the magnetic field. This is often called *longitudinal recording* as opposed to *perpendicular recording* in which

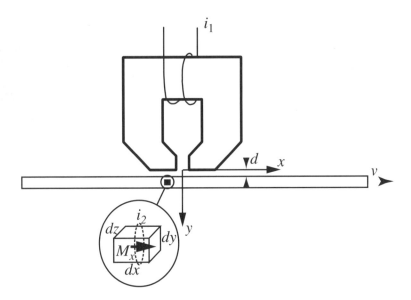

FIGURE 3.1.   Reciprocity between recording head and magnetic medium.

the magnetization is vertical (along the y-axis). If the coil (object #1) has an imaginary write current i_1, which produces a head field H_x, then the magnetic flux through the medium element $dxdydz$ at location (x, y, z) (object #2) due to current i_1 is

$$d\Phi_{21} = \mu_0 H_x(x, y, z)dydz.$$

Because the equivalent surface current density is $M_x(x - \bar{x}, y, z)$, in the unit of A/m, the equivalent current at the four surfaces (parallel to the x-axis) of the magnetic element $dxdydz$ is

$$i_2 = M_x(x - \bar{x}, y, z)dx,$$

where $\bar{x} = vt$ is the moving distance of the medium relative to the head. Then, the magnetic flux through the head coil due to the magnetic element $dxdydz$ (or equivalent current i_2), $d\Phi_{12}$, is related to $d\Phi_{21}$ by the following relation:

$$\frac{d\Phi_{12}}{i_2} = \frac{d\Phi_{21}}{i_1}.$$

The magnetic flux through the head coil due to the magnetic element $dxdydz$ is then

$$d\Phi_{12} = \frac{\mu_0 H_x(x, y, z)dydz}{i_1} M_x(x - \bar{x}, y, z)dx$$

$$= \frac{\mu_0 H_x(x, y, z)}{i_1} M_x(x - \bar{x}, y, z)dxdydz.$$

Therefore, the total magnetic flux through the head coil is expressed by head field per unit head current and the medium magnetization:

$$\Phi = \mu_0 \int_{-\infty}^{\infty} dx \int_{d}^{d+\delta} dy \int_{-\infty}^{\infty} dz \frac{H_x(x, y, z)}{i} M_x(x - \bar{x}, y, z), \tag{3.1}$$

where the subscripts of Φ_{12} and i_1 have been dropped, δ is the magnetic medium thickness, and d is the magnetic spacing. This expression is called the *reciprocity formula* or *reciprocity principle*. Note that the head coil current i is the imaginary write current rather than the readback current. Even though the head coil current is contained in the reciprocity formula, the head field is normally proportional to the coil current. Therefore, the expression H_x/i and the magnetic flux Φ are independent of the imaginary coil current i. Assume that both the magnetization and head field are *approximately* uniform over the data track (the data trackwidth is W) along the z-axis and that the magnetization is uniform through the medium thickness along the y-axis, then the reciprocity formula can be simplified as

$$\Phi = \mu_0 W \int_{-\infty}^{\infty} dx \int_{d}^{d+\delta} dy \frac{H_x(x, y)}{i} M_x(x - \bar{x}). \tag{3.2}$$

The readback voltage is given by Faraday's law:

$$V_x(\bar{x}) = -\frac{d\Phi}{dt} = -\frac{d\bar{x}}{dt}\frac{d\Phi}{d\bar{x}} = -\mu_0 v W \int_{-\infty}^{\infty} dx \int_{d}^{d+\delta} dy \frac{H_x(x, y)}{i} \frac{dM_x(x - \bar{x})}{d\bar{x}}.$$

$$\tag{3.3}$$

Since the derivative of magnetization is the magnetic charge density, the above expression reaffirms the fact that the head essentially senses the moving magnetic charge. If the magnetization derivative is zero, or the velocity is zero, then there is no readback signal.

The reciprocity formula can be easily extended to the three-dimensional (3D) vector form.[1] It must be cautioned that the *validity* of the reciprocity principle is based on a number of subtle but important assumptions: (1) The imaginary write current is so small that the head is in a *linear* write mode, i.e., the head material permeability is independent of drive field amplitude. In this case, the write head efficiency and read head efficiency are identical. (2) The imaginary write current and the read excitation must be either both direct current (DC) or both quasi-static. (3) The reciprocity principle can be extended to high frequencies. It remains valid for sinusoidal write current and read excitation at a single high frequency if the head designs and materials have adequately flat permeability spectra over the bandwidth concerned.[2,3]

3.2 READBACK FROM SINGLE TRANSITION

With the reciprocity formula in hand, we can now readily derive the readback voltage from various magnetic recording patterns. We first consider a single magnetic transition.

Infinitely Sharp Step Transition.   A single infinitely sharp step transition with a moving center $\bar{x} = vt$ is expressed as

$$M_x(x - \bar{x}) = \begin{cases} -M_r & \text{for} \quad x < \bar{x}, \\ M_r & \text{for} \quad x > \bar{x} \end{cases} \tag{2.23}$$

where M_r is the remanence magnetization of the medium. Then

$$\frac{dM_x(x - \bar{x})}{d\bar{x}} = -\frac{dM_x(x - \bar{x})}{dx} = -2M_r\delta(x - \bar{x}),$$

where $\delta(x)$ is the δ-function (not to be confused with medium thickness). From the reciprocity formula, we obtain

$$V_x(\bar{x}) = 2\mu_0 v W M_r \int_d^{d+\delta} dy \, \frac{H_x(\bar{x}, y)}{i} \tag{3.4}$$

$$\cong 2\mu_0 v W M_r \delta \frac{H_x(\bar{x}, d)}{i} \quad \text{for thin medium } (\delta \ll d).$$

This means that the readback signal of a thin magnetic medium with an infinitely sharp magnetic transition is in the shape of the head field profile at the medium. Therefore, head field expression is often called head sensitivity function.

Arctangent Transition. The real magnetic transition is not a sharp step transition, but rather a transition with a finite width. It is often approximated by an arctangent function:

$$M_x(x - \bar{x}) = \frac{2M_r}{\pi} \tan^{-1}\left(\frac{x - \bar{x}}{a}\right),\qquad(2.28)$$

where a is the transition parameter. As mentioned before, the arctangent transition becomes a step transition if $a = 0$.

From the reciprocity formula, Karlqvist equation, and the following identities:

$$\frac{d}{dx}\tan^{-1}x = \frac{1}{1 + x^2},$$

$$\int_{-\infty}^{\infty} dx \frac{1}{a^2 + (\bar{x} - x)^2}$$

$$\left[\tan^{-1}\left(\frac{x + c}{y}\right) - \tan^{-1}\left(\frac{x - c}{y}\right)\right] = \frac{\pi}{a}\left[\tan^{-1}\left(\frac{\bar{x} + c}{y + a}\right) - \tan^{-1}\left(\frac{\bar{x} - c}{y + a}\right)\right],$$

it can be readily derived that the readback voltage from an arctangent transition is

$$V_x(\bar{x}) = -\mu_0 v W \int_{-\infty}^{\infty} dx \int_d^{d+\delta} dy \frac{H_x(x, y)}{i} \frac{dM_x(x - \bar{x})}{d\bar{x}}$$

$$= \mu_0 v W \int_{-\infty}^{\infty} dx \int_d^{d+\delta} dy \frac{H_g}{\pi i}\left[\tan^{-1}\left(\frac{x + g/2}{y}\right) - \tan^{-1}\left(\frac{x - g/2}{y}\right)\right]\frac{2M_r}{\pi}\frac{a}{a^2 + (x - \bar{x})^2}$$

$$= \frac{2\mu_0 v W H_g M_r}{\pi i}\int_d^{d+\delta} dy\left[\tan^{-1}\left(\frac{\bar{x} + g/2}{y + a}\right) - \tan^{-1}\left(\frac{\bar{x} - g/2}{y + a}\right)\right].$$

or

$$V_x(\bar{x}) = \frac{2\mu_0 v W M_r}{i}\int_d^{d+\delta} dy H_x(\bar{x}, y + a) = \frac{2\mu_0 v W M_r}{i}\int_{d+a}^{d+a+\delta} dy H_x(\bar{x}, y).\qquad(3.5)$$

In other words, *an arctangent transition at a magnetic spacing of d is equivalent to a step transition at an effective magnetic spacing of d + a.* This is consistent with the results of the demagnetizing field of an arctangent transition derived in Chapter 2. If the magnetic medium is thin enough, then

$$V_x(\bar{x}) \cong 2\mu_0 v W M_r \delta \frac{H_x(\bar{x}, d + a)}{i}\quad \text{for thin medium } (\delta \ll d).\quad(3.6)$$

Its peak voltage is

$$V^{\text{peak}} = V_x(\bar{x} = 0) \cong 4\mu_0 v W M_r \delta \frac{H_g}{\pi i} \tan^{-1}\left(\frac{g/2}{d+a}\right) \quad (\delta << d). \quad (3.7)$$

For thin magnetic medium, the readback voltage pulse from an arctangent transition is in the shape of head field profile with an effective magnetic spacing of $d + a$.

Now consider a thin medium read by a small gap Karlqvist head. The readback voltage pulse is

$$V_x(\bar{x}) = 2\mu_0 v W M_r \delta \frac{H_g}{\pi i} \tan^{-1}\left[\frac{g(d+a)}{\bar{x}^2 + (d+a)^2 - (g/2)^2}\right]$$

$$\approx 2\mu_0 v W M_r \delta \frac{H_g g}{\pi i} \frac{d+a}{\bar{x}^2 + (d+a)^2}.$$

The single pulse is a Lorentzian pulse, which can be written in the form of

$$V_{\text{sp}}(x) = \frac{V^{\text{peak}}(PW_{50}/2)^2}{x^2 + (PW_{50}/2)^2}, \quad (3.8)$$

where $V^{\text{peak}} = 2\mu_0 v W M_r \delta H_g g / \pi i (d + a)$ is the pulse amplitude, and $PW_{50} = 2(d + a)$ is the pulse width at half-amplitude. This Lorentzian pulse is so simple that it is often used in magnetic recording channel modeling.

More generally, the readback pulse width at half-amplitude can be expressed as

$$PW_{50}\begin{cases} = \sqrt{g^2 + 4(a+d)^2} & \text{for} \quad \delta << d, \\ = \sqrt{4(a+d)(a+d+\delta)} & \text{for} \quad g << d, \\ = \sqrt{g^2 + 4(a+d)(a+d+\delta)} & \text{in general.} \end{cases} \quad (3.9)$$

The second equation is derived for a small-gap Karlqvist head ($g \sim 0$), while the last equation is the extrapolation of the first and second equations. This is a good approximation in many recording environments where the Karlqvist head and arctangent transition models are valid. The expression breaks down when the magnetic medium used is relatively thick (such as tape medium). Small gap length, transition parameter, medium thickness, and magnetic spacing are required to achieve a narrow pulse. An efficient inductive head design to minimize PW_{50} is given by

$$g^2 \approx 4(a+d)(a+d+\delta). \quad (3.10)$$

The Fourier transform with respect to $\bar{x}$ of Equation (3.5), which is the voltage pulse of a Karlqvist head reading an arctangent transition, gives

$$V_x(k) = \frac{2\mu_0 v WM_r}{i}\int_{d+a}^{d+a+\delta} dy\ FT[H_x(\bar{x}, y)]$$

$$= \frac{2\mu_0 v WM_r}{i}\int_{d+a}^{d+a+\delta} dy\ gH_g\ \frac{\sin(kg/2)}{kg/2}e^{-ky},$$

which leads to

$$V_x(k) = 2\mu_0 v WM_r\delta\frac{H_g g}{i}e^{-k(d+a)}\ \frac{1-e^{-k\delta}}{k\delta}\ \frac{\sin(kg/2)}{kg/2}. \qquad (3.11)$$

The Fourier component of the readback voltage of an arctangent magnetization is a function of wavevector (or wavelength), as plotted in Fig. 3.2.
At low frequencies the voltage is simply due to head differentiation, so
it is proportional to velocity v, head efficiency, and number of coil turns
($H_g g/i = NE$). The Fourier component drops at high frequencies due to
the Wallace spacing loss and gap nulls.

There are three loss terms associated with the above expression:

1. *Spacing loss.* The term $e^{-k(d + a)}$ is originated from the 2D Laplace
   equation. The Fourier transform of the field with respect to x decays

(

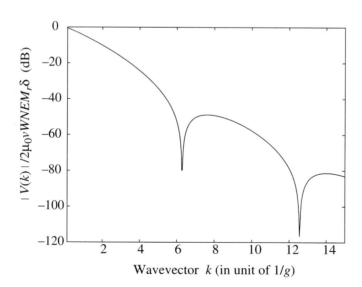

FIGURE 3.2.   Fourier transform of an isolated pulse from an arctangent transition
versus wavevector, where we have assumed $\delta << (a + d) = g/2$.

exponentially as a function of $y = d + a$ if all the field sources are in one side as shown in Chapter 2. At a constant y, the spacing loss drops with the wavevector exponentially, as clearly seen from Fig. 3.2.

2. *Gap loss.* The term $\sin(kg/2)/(kg/2)$ is originated from the Karlqvist head approximation. The head surface field is assumed to be a single square pulse, so its Fourier transform produces this gap loss term and gap nulls ($kg = 2\pi, 4\pi, 6\pi, \ldots$). Considering that the Karlqvist approximation is not precise, the first gap null actually occurs at a wavelength of $\sim 1.136g$ instead of g. In other words, the first gap null occurs at $kg \approx 1.8\pi$ instead of 2π.

3. *Thickness loss.* The term $(1 - e^{-k\delta})/k\delta$ is also originated from the Laplace equation. It simply says that the thinner the medium is, the closer the bottom of the medium is from the head, so the less the spacing loss is.

These loss mechanisms are universal in all types of magnetic recording, and have to be taken into careful consideration when designing or testing a magnetic recording system.

For thin film heads with finite pole tips, the outer corners of the finite pole tips produce two undershoots in readback voltage pulse. The undershoots cause ripples in the Fourier transform. This subject will be treated in Chapter 5.

3.3 READBACK FROM MULTIPLE TRANSITIONS

Magnetic information contains numerous magnetic transitions, therefore we need to consider the readback process from multiple transitions. The interaction among neighboring transitions is an extremely important aspect of magnetic recording. The magnetic read and write processes are not strictly linear. The write process is actually very nonlinear as will be shown in Chapter 4. However, a restricted linearity applies to the read process as long as transitions are not written too closely. The *linear superposition principle* states that the readback voltage from a sequence of magnetic transitions is given by the sum of the single pulses corresponding to the individual magnetic transitions:

$$V(x) = \sum_n (-1)^n V_{sp}(x - x_n), \qquad (3.12)$$

where x_n is the center of the nth individual magnetic transition, $(-1)^n$

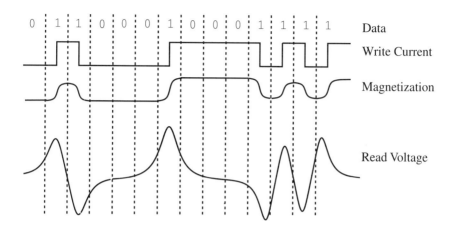

FIGURE 3.3. NRZI data '01100010001111' and the corresponding write current, magnetization pattern and voltage waveform. The dashed lines represent clock windows.

represents that the polarity of the magnetic transitions must alternate, and the subscript "sp" stands for single pulse, which has been derived earlier. The Fourier transform of the readback voltage is

$$V(k) = V_{sp}(k)\sum_{n}(-1)^{n}e^{-ikx_{n}}.\qquad(3.13)$$

The voltage waveform of a nonreturn-to-zero modified (NRZI) current pattern is shown in Fig. 3.3. NRZI coding (see Chapter 8) simply means that a bit "1" corresponds to a write current change, while a bit "0" corresponds to no change. The current is either positive or negative, but never zero. As shown in Fig. 3.3, the magnetization transitions and voltage pulses are no longer as idealized as before. Linear bit shift and amplitude roll-off due to linear superposition will be discussed next.

3.3.1 Linear bit shift

For a *dibit* consisting of two transitions written at a spacing of B, the readback voltage and Fourier transform are:

$$V_{dibit}(x) = V_{sp}(x - B/2) - V_{sp}(x + B/2),\qquad(3.14)$$
$$V_{dibit}(k) = 2iV_{sp}(k)\sin(kB/2).$$

The Fourier transform of the dibit is essentially the Fourier transform of the single pulse *modulated* by sin($kB/2$).

When $B \to 0$, the dibit amplitude will approach zero too because the two opposite pulses will cancel each other. This is called the amplitude roll-off. Closely spaced multiple transitions also cause bit shift (peak shift), or intersymbol interference: the peaks of alternating polarity occur at locations slightly shifted away from the intended locations. Since these bit shifts occur due to the linear superposition of multiple pulses, they are termed linear bit shifts, which can affect the detection accuracy of a digital recording channel. Consider two bits of opposite polarity (i.e., dibit) written at $x = -B/2$ and $x = B/2$, and assume that the single pulse is a Lorentzian pulse:

$$V_{sp}(x) = \frac{V_{sp}^{peak}(PW_{50}/2)^2}{x^2 + (PW_{50}/2)^2}. \tag{3.8}$$

Then the net voltage based on linear superposition is

$$V_{dibit}(x) = V_{sp}^{peak}\left[\frac{(PW_{50}/2)^2}{(x - B/2)^2 + (PW_{50}/2)^2} - \frac{(PW_{50}/2)^2}{(x + B/2)^2 + (PW_{50}/2)^2}\right].$$

The resulting peaks will be shifted away from $x = +B/2$ and $x = -B/2$ due to the linear superposition of the two original pulses. Generally speaking, *if considering linear bit shift alone, dibits in longitudinal recording tend to repel each other.*

3.3.2 Square-wave recording and roll-off

Square-wave recording refers to an alternating series of step write current changes at fixed amplitude and bit spacing. It is also called all "111s" data pattern in NRZI coding. If the bit spacing or bit length is B, then the recording density is $D = 1/B$, and the wavelength is $\lambda = 2B$. If the linear superposition applies, then the readback voltage can be expressed as a sum of single pulses:

$$V(x) = \sum_{n=-\infty}^{+\infty} (-1)^n V_{sp}(x - nB). \tag{3.15}$$

Due to the symmetry in the square-wave recording, all the pulses are uniformly spaced, and no linear bit shifts exist.

If the medium is very thin, and the head gap is small, then the single pulse response is approximately Lorentzian. The readback voltage from square-wave recording is then

$$V(x) = V_{sp}^{peak}(PW_{50}/2)^2 \sum_{n=-\infty}^{\infty} \left[\frac{(-1)^n}{(x - nB)^2 + (PW_{50}/2)^2} \right]. \qquad (3.16)$$

The amplitude of the readback voltage at $x = mB$, where m is any integer, is[4]

$$V^{peak} = V(x = mB) = V_{sp}^{peak}(PW_{50}/2)^2 \sum_{n=-\infty}^{\infty} \left[\frac{(-1)^n}{(nB)^2 + (PW_{50}/2)^2} \right]$$

$$= V_{sp}^{peak}(PW_{50}/2B) \sum_{n=-\infty}^{\infty} \left[\frac{(-1)^n(PW_{50}/2B)}{n^2 + (PW_{50}/2B)^2} \right]$$

$$= V_{sp}^{peak}(PW_{50}/2B) \frac{\pi}{\sinh(\pi PW_{50}/2B)}.$$

Note that we have used the following relation:

$$\sum_{n=-\infty}^{\infty} (-1)^n \frac{z}{n^2 + z^2} = \frac{\pi}{\sinh(\pi z)}.$$

The readback *resolution* for square-wave recording is defined as

$$R \equiv \frac{V^{peak}(B)}{V_{sp}^{peak}} = \frac{\pi PW_{50}/2B}{\sinh(\pi PW_{50}/2B)}. \qquad (3.17)$$

The roll-off curve, R vs. square-wave recording density ($D = 1/B$), is shown in Fig. 3.4. If $R = 50\%$, the half-amplitude density D_{50} is then given by

$$\frac{\pi PW_{50}D_{50}/2}{\sinh(\pi PW_{50}D_{50}/2)} = 50\%,$$

$$\pi PW_{50}D_{50}/2 \approx 2.1773, \qquad (3.18)$$

$$PW_{50}D_{50} \approx 1.39.$$

This means that *the 50% roll-off density for square wave recording is approximately inversely proportional to PW_{50} of the isolated pulse.* The linear recording density that can be reliably read back in magnetic recording using a peak detection channel is about D_{50}. Since PW_{50} and D_{50} are fundamental parameters that determine the possible recording density, they are widely quoted in the literature.

The above roll-off curve is strictly valid for a zero-gap Karlqvist head. If the read gap g is finite, it is approximately correct at low densities when

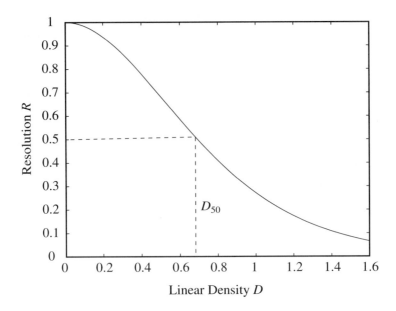

FIGURE 3.4.   Roll-off curve for square wave recording. The density is normalized by $2/PW_{50}$.

the gap loss term is not significant. At high densities ($D \geq D_{50}$), the roll-off curve can be modified as[5,6]

$$R \approx \frac{\pi P W_{50}/2B}{\sinh(\pi P W_{50}/2B)} \frac{\sin(k_0 g/2)}{k_0 g/2} = \frac{P W_{50}\sin(\pi g/2B)}{g\sinh(\pi P W_{50}/2B)}, \qquad (3.19)$$

where $k_0 = \pi/B$. The gap nulls will occur when $B = g/2n$, where n is a positive integer. In other words, the square wave recording patterns with these bit lengths will generate no signals, again originated from the Karl-qvist head approximation. For example, if $B = g/2$, the readback voltage of square-wave recording in a thin medium can be expressed as

$$V(x) = \sum_{n=-\infty}^{+\infty} (-1)^n V_{sp}(x - ng/2)$$

$$= \sum_{m=-\infty}^{+\infty} V_{sp}(x - mg) - \sum_{m=-\infty}^{+\infty} V_{sp}(x - mg - g/2)$$

$$\propto \sum_{m=-\infty}^{+\infty} \left[\tan^{-1}\left(\frac{x - mg + g/2}{d + a}\right) - \tan^{-1}\left(\frac{x - mg - g/2}{d + a}\right) \right]$$

$$- \sum_{m=-\infty}^{+\infty} \left[\tan^{-1}\left(\frac{x-mg}{d+a}\right) - \tan^{-1}\left(\frac{x-mg-g}{d+a}\right) \right]$$

$$= \sum_{m=-\infty}^{+\infty} \left[\tan^{-1}\left(\frac{x-mg+g/2}{d+a}\right) - \tan^{-1}\left(\frac{x-mg}{d+a}\right) \right]$$

$$- \sum_{m=-\infty}^{+\infty} \left[\tan^{-1}\left(\frac{x-mg-g/2}{d+a}\right) - \tan^{-1}\left(\frac{x-mg-g}{d+a}\right) \right]$$

$$= 0.$$

Indeed, the readback voltage disappears in this case.

Now we consider the frequency spectrum of the readback voltage of square-wave recording, which can be expressed as a convolution of the single pulse and the series of alternating δ-functions:

$$V(x) = \sum_{n=-\infty}^{\infty} (-1)^n V_{sp}(x-nB) = V_{sp}(x) * \sum_{n=-\infty}^{\infty} (-1)^n \delta(x-nB).$$

Since

$$FT\left[\sum_{m=-\infty}^{\infty} \delta(x-mB) \right] = \frac{2\pi}{B} \sum_{m=-\infty}^{\infty} \delta\left(k-m\frac{2\pi}{B}\right), \qquad (3.20)$$

by summing the even and odd terms separately, we obtain that the Fourier transform of the readback voltage is

$$V(k) = V_{sp}(k)\frac{2\pi}{2B} \sum_{m=-\infty}^{\infty} \delta\left(k-m\frac{2\pi}{2B}\right) - V_{sp}(k)e^{-ikB}\frac{2\pi}{2B} \sum_{m=-\infty}^{\infty} \delta\left(k-m\frac{2\pi}{2B}\right)$$

$$= (1-e^{-ikB})V_{sp}(k)\frac{\pi}{B} \sum_{m=-\infty}^{\infty} \delta\left(k-m\frac{\pi}{B}\right),$$

which leads to

$$V(k) = \begin{cases} 0 & \text{for } k = 2n\pi/B \quad \text{or} \quad \neq m\pi/B, \\ V_{sp}(k)\frac{2\pi}{B} \sum_{n=-\infty}^{\infty} \delta\left[k-(2n+1)\frac{\pi}{B}\right] & \text{for } k = (2n+1)\pi/B. \end{cases} \qquad (3.21)$$

where $\pi/B = 2\pi/2B \equiv k_0$ is the fundamental wavevector. This equation means that *the spectrum of square-wave recording is composed of odd harmonics,* a well-known experimental fact.

Spectrum analyzers are widely used in magnetic recording and electronic measurements. A spectrum analyzer can measure the root-mean-square (rms) power in a bandwidth Δf centered about frequency f, plus the corresponding quantity at $-f$. In other words, a spectrum analyzer can display $2V(f)\Delta f/\sqrt{2}$ as a function of f. Since

$$V(f) \equiv \int_{-\infty}^{\infty} dt V(t) e^{-i2\pi ft} = V(k)/v, \quad V(t) \equiv V(x), f = vk/2\pi, \quad (2.5)$$

then the rms spectrum peak intensity is

$$V_{\text{rms}}(f) = \frac{2V(f)\Delta f}{\sqrt{2}} = \frac{V(k)\Delta k}{\sqrt{2}\pi} = \frac{2k_0|V_{\text{sp}}(k)|}{\sqrt{2}\pi}, \quad k = (2n+1)k_0. \quad (3.22)$$

The spectrum is proportional to the Fourier transform of the single pulse by a factor of $2k_0/\sqrt{2}\pi$. As an example, the spectrum in frequency domain and the corresponding square wave recording pattern, captured by a Guzik spin stand and a LeCroy oscilloscope, are shown in Fig. 3.5, where the signal peaks (odd harmonics) are modulated by the spectrum of the single pulse, $|V_{\text{sp}}(k)|$. In addition, there are noises in both the time domain (oscilloscope) readback waveform and the frequency domain spectrum. How to detect signal peaks with extremely low errors in the presence of noises, is a central theme in magnetic storage research and development. Both signal and noise spectra will be further discussed in Section 9.1.

Note that the spectrum shown is accurate at low- and medium-frequency range. It is not accurate around the first gap null and beyond because the spectrum shown in Fig. 3.5 is obtained by fast Fourier transform (FFT) with a somewhat limited precision. More accurate spectrum beyond the first gap null can be captured by a more powerful spectrum analyzer like HP 4195A.

The total signal power in square-wave recording is distributed in the odd harmonics in the frequency spectrum, among which the fundamental harmonic is most important. The fundamental harmonic of square wave recording as a function of the fundamental wavevector is

$$V_{\text{rms}}(f_0) = \frac{2k_0|V_{\text{sp}}(k_0)|}{\sqrt{2}\pi}, \quad f_0 = vk_0/2\pi = v/2B. \quad (3.23)$$

The Fourier transforms of a single pulse (isolated pulse) $V_{\text{sp}}(k_0)$ and the fundamental harmonic of square wave recording $V_{\text{rms}}(f_0)$ are shown together in Fig. 3.6. The latter spectrum is similar to the Fourier transform of the isolated pulse except that it approaches zero at low frequencies because of the $2k_0/\sqrt{2}\pi$ factor. In other words, a direct current (DC) magnetic pattern will not generate any readback signal.

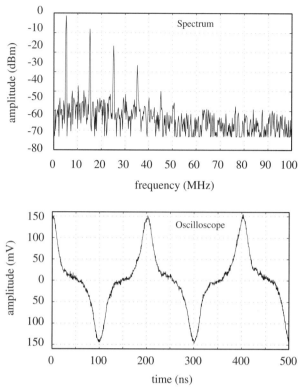

FIGURE 3.5. Experimental spectrum in frequency domain (top) of a square wave recording at 5 MHz (bottom).

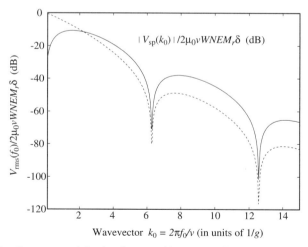

FIGURE 3.6. Spectrum of the fundamental harmonic for square wave (solid line) and Fourier transform of its corresponding single pulse (dashed line).

References

1. H. N. Bertram, *Theory of Magnetic Recording,* (Cambridge, UK: Cambridge University Press, 1994, pp. 113–119).

2. N. Smith, "Reciprocity principles for magnetic recording theory," *IEEE Trans. Magn.,* **23,** 1995, 1987.

3. J. C. Mallinson, "Some comments on the reciprocity integral used in magnetic recording," *IEEE Trans. Magn.,* **33,** 2403, 1997. N. Smith, "Further comments on the reciprocity integral for magnetic recording," *IEEE Trans. Magn.,* **34,** 772, 1998.

4. R. L. Comstock and M. L. Williams, "Frequency response in digital magnetic recording," *IEEE Trans. Magn.,* **9,** 342, 1973.

5. H. N. Bertram, *Theory of Magnetic Recording,* Cambridge University Press, 1994, p. 157.

6. G. J. Tarnopolsky and P. R. Pitts, "Media noise and SNR estimates for high areal density recording," *J. Appl. Phys.,* **81,** 4837, 1997.

CHAPTER 4

Write Process in Magnetic Recording

In order to reproduce well-defined voltage pulses from a magnetic medium, we must first write sharp magnetic transitions. In this chapter an analytical model on digital magnetic write process, the William-Comstock model, is described. Then, the concept of nonlinear distortions in the write process including partial erasure, nonlinear transition shift, and hard transition shift are introduced.

Digital magnetic write process is based on so-called *saturation recording*. The magnetization of the medium is more or less saturated by the write field, i.e., any increase of input current (write current) will not result in a proportional increase in magnetization (or readback voltage). This is of course one of the basic characteristics of a nonlinear process, and thus the write process in digital magnetic recording is generally nonlinear as opposed to the readback process. Furthermore, interactions between magnetic grains and the existence of a demagnetizing field lead to nonlinear distortions. These phenomena must be well understood and controlled to achieve high-density magnetic recording.

4.1 MODELS OF WRITE PROCESS

A simplified scheme of writing a magnetic transition (without considering the demagnetizing field in the medium) is shown schematically in Fig. 4.1. The DC erased magnetic medium is moving to the right relative to the recording head. The current initially magnetizes the medium to the right. At the moment the current switches the direction, the head field direction switches to the left. The head field rise time (see Chapter 5) is actually finite, but assumed to be zero here. A section of magnetic medium enclosed by the $H_x = -H_c$ contour, often called the "write bubble," is now magnetized to the left, and a magnetic transition is formed at the

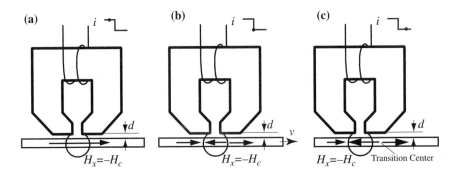

FIGURE 4.1. Formation of a magnetic transition. (a) Magnetic medium is DC erased (magnetized to the right), and current is positive. (b) Current and write bubble switch directions. (c) Written transition moves out of write bubble.

trailing edge of the write bubble. As the medium moves to the right, the write bubble stays with the head gap, and the transition at the trailing edge moves out of the influence of the write bubble. In other words, a magnetic transition is written. In contrast, the transition at the leading edge of the write bubble gets erased by the write bubble, so it does not get written.

A magnetic transition is often approximated by an arctangent function. Experimental investigations by Lorentz microscopy and electron holography vividly reveal that magnetic transitions in thin-film magnetic disk are typically in the form of "zigzag" domain boundaries[1] as shown in Fig. 4.2, while a "saturated" thin-film medium is marked with ripple patterns and vortex formation.[2] The zigzag domain boundary can be understood in several ways. First, the magnetostatic energy of the zigzag domain wall is much less than the straight domain wall represented by

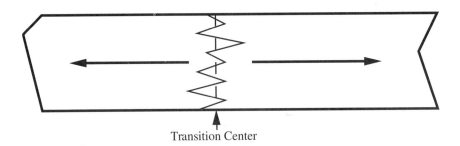

FIGURE 4.2. Zigzag magnetic transition.

the dashed line because the latter has a much higher concentration of magnetic charges. Second, the magnetic medium in thin film form is made of small grains. The exchange interaction among magnetic grains tends to produce magnetic clusters of grains that lead to the formation of zigzag domain boundaries. These experimental observations and the nature of magnetic transitions are elucidated with micromagnetic modeling, notably by Zhu and Bertram.[3]

4.1.1 Williams-Comstock model

An analytical model first proposed by Mason Williams and Larry Comstock provides an insightful and relatively simple analysis of write process.[4] It has been surprisingly successful in predicting magnetic transition width. In this model, the "zigzag" nature of the magnetic transition is ignored, and the shape of magnetic transition is presumed to be described by a function like arctangent function (assumption #1). It is experimentally found that the transition parameter correlates well with the average width of the zigzag domain wall, namely, the width of the zigzag wall is approximately the transition width πa of an arctangent transition.

Consider the write process in longitudinal recording as illustrated in Fig. 4.3. The head field $H_x(x) \equiv H_x(x, y = d + \delta/2)$ is the x-component of the head field along the center plane of the magnetic medium, and the Karlqvist head equation is often used for the simplest analysis. The written magnetization pattern is represented by $M_x(x)$, which produces the demagnetizing field $H_d(x)$. At a given point x in magnetic medium (the open circle in Fig. 4.3), the demagnetizing field and the head field together determine the magnetization $M_x(x)$ through the intrinsic hysteresis loop $M_x(H)$ of the magnetic recording medium (assumption #2):

$$H(x) = H_{tot}(x) = H_x(x) + H_d(x),$$ (4.1)
$$M_x(x) = M_x[H_{tot}(x)] = M_x[H_x(x) + H_d(x)]$$

The second assumption rests on the first assumption that the shape of magnetization is already known. Otherwise, there is no way to express the demagnetizing field. It follows that the longitudinal gradient of the magnetization is

$$\frac{dM_x}{dx} = \frac{dM_x}{dH_{tot}}\frac{dH_{tot}}{dx} = \frac{dM_x(H)}{dH}\left[\frac{dH_x(x)}{dx} + \frac{dH_d(x)}{dx}\right].$$ (4.2)

The derivative $dM_x(H)/dH$ can be obtained from the intrinsic hysteresis loop of the magnetic medium, so we dropped the subscript "tot" here. It

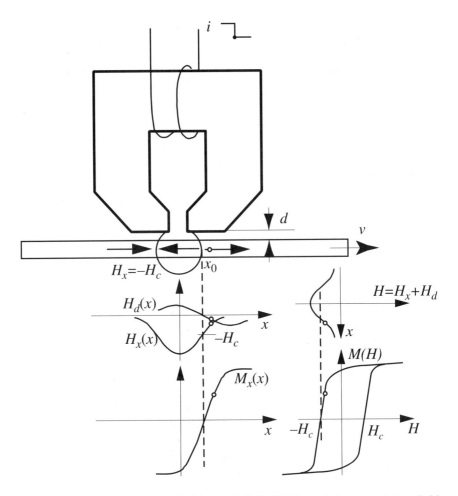

FIGURE 4.3. Magnetization $M_x(x)$, head field $H_x(x)$, and demagnetizing field $H_d(x)$ along the center plane of magnetic medium, where x_0 (dashed line) is the transition center. Total head field $H(x)$ and hysteresis loop are also shown.

is apparent that a sharp magnetization transition means a larger magnetization gradient. Therefore, *a near-square M-H loop and a large head field gradient are desirable.* Since the demagnetization field gradient opposes the head field gradient, *the demagnetizing field tends to broaden the magnetic transition in the write process* and should be sufficiently small.

The transition written at the instant of switching current is still under the influence of the head field; it *relaxes* as the head write bubble moves away. In other words, the final transition shape must be determined by taking the magnetization relaxation into consideration.[5] As shown in Fig. 4.4, the magnetic transition at the instant of current switching is described by the left side of the hysteresis loop. When the write bubble is moved away, the head field diminishes, and only the demagnetizing field remains. If the initial and final magnetic transition is presumed to be an analytical function such as arctangent function, the demagnetizing field is then a function of magnetization as shown in Fig. 4.4, opposing magnetization as expected. In the relaxation process, every point in the magnetic transition follows its own "minor" hysteresis loop to reach its final magnetic state. A minor loop is the hysteresis loop formed when the magnetic material is not fully saturated, while the major loop is formed when it is fully saturated.

At the transition center $x = x_0$, the magnetization and demagnetizing field should be both zero. It was implied earlier that the transition center was given by the trailing edge of the write bubble $H_x = -H_c$, which is not exactly true. A very subtle aspect of the Williams-Comstock model emerges after carefully examining the hysteresis loops of the magnetic medium shown in Fig. 4.4. If $x = x_0$ position of the medium is to have zero remanence, it must be subjected to a field $-H_{cr}$ ($H_{cr} > 0$) (just slightly below the coercivity point) on the hysteresis loop, so that the magnetiza-

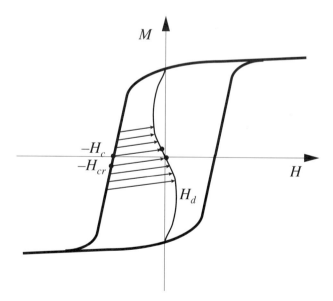

FIGURE 4.4. Magnetization relaxation. The arrows indicate minor loops.

tion will return to zero along a minor loop as the head field is removed. Therefore, the true location of the transition center is slightly shifted to the left from what is originally determined by the trailing edge of the write bubble. That is, the true transition center x is given by

$$H_x(x, d + \delta/2)|_{x=x_0} = -H_{cr}. \tag{4.3}$$

The slope of the major hysteresis loop near the coercivity point is given by

$$\frac{dM}{dH} = \frac{M_r}{H_c(1 - S^*)}, \tag{1.26}$$

where S^* is the coercive squareness. The slope of the minor loop near the zero remanence, χ, is much smaller than the major loop slope:

$$\chi << \frac{M_r}{H_c(1 - S^*)}, \quad \chi \frac{H_c(1 - S^*)}{M_r} << 1.$$

According to Fig. 4.4, we can write that

$$\chi(H_{cr} - 0) = \frac{M_r}{H_c(1 - S^*)}(H_{cr} - H_c), \quad H_{cr} = \frac{H_c}{r}, r \equiv 1 - \chi\frac{H_c(1 - S^*)}{M_r}. \tag{4.4}$$

Since $r \approx 1$, $H_{cr} \approx H_c$, we will assume $H_{cr} = H_c$ hereafter unless specially noted!

Generally speaking, the write field follows the write current almost instantaneously. (This assumption breaks down only at very high frequencies, which will be treated in Section 5.4 and Chapter 16.) As long as the write bubble penetrates the magnetic medium, a magnetic transition will be written. However, the location of the transition center depends on the size of the write bubble, which is in turn determined by the write current. The write current is a variable in magnetic recording. In principle, the write current i can be adjusted such that the maximum head field gradient also occurs at the transition center $x = x_0$. In other words, the optimum write current (or optimum deep gap field) is given by the maximum field gradient requirement:

$$\left.\frac{d^2H_x(x, d + \delta/2)}{dx^2}\right|_{x=x_0} = 0. \tag{4.5}$$

Now we use the William-Comstock model to derive the transition parameter of a transition written with a Karlqvist head. The longitudinal head field is

$$H_x = \frac{H_g}{\pi}\left[\tan^{-1}\left(\frac{x + g/2}{y}\right) - \tan^{-1}\left(\frac{x - g/2}{y}\right)\right], \tag{2.17}$$

where $y = d + \delta/2$. At the transition center, the head field is

$$H(x, y)|_{x=x_0} = -H_{cr} \approx -H_c. \tag{4.6}$$

The head field gradient is

$$\frac{dH_x}{dx}\bigg|_{x_0} = \frac{H_g}{\pi y}\left[\frac{1}{1 + \left(\dfrac{x_0 + g/2}{y}\right)^2} - \frac{1}{1 + \left(\dfrac{x_0 - g/2}{y}\right)^2}\right] = \frac{H_c Q}{y}, \tag{4.7}$$

where Q is a function of y and x_0, with typical values of 0.65-0.85:

$$
\begin{aligned}
Q(y) &= \frac{H_g}{\pi H_c}\left[\frac{y^2}{y^2 + (x_0 + g/2)^2} - \frac{y^2}{y^2 + (x_0 - g/2)^2}\right] \\
&= \left[\frac{y^2}{y^2 + (x_0 - g/2)^2} - \frac{y^2}{y^2 + (x_0 + g/2)^2}\right] \div \\
&\quad \left[\tan^{-1}\left(\frac{x_0 + g/2}{y}\right) - \tan^{-1}\left(\frac{x_0 - g/2}{y}\right)\right].
\end{aligned}
\tag{4.8}
$$

It is assumed that an arctangent transition with a transition parameter of a is written at the instant of current switching, and the transition parameter relaxes to a_f after moving away from the head. The magnetization gradient at the transition center at the instant of transition formation is

$$\frac{dM_x}{dx}\bigg|_{x_0} = \frac{2M_r}{\pi a}. \tag{4.9}$$

The demagnetizing field along the center of the medium is (see Chapter 2)

$$
\begin{aligned}
H_d(x) &= -\frac{2M_r}{\pi}\left[\tan^{-1}\left(\frac{a + \delta/2}{x - x_0}\right) - \tan^{-1}\left(\frac{a}{x - x_0}\right)\right] \\
&= -\frac{2M_r}{\pi}\left[\tan^{-1}\left(\frac{x - x_0}{a}\right) - \tan^{-1}\left(\frac{x - x_0}{a + \delta/2}\right)\right].
\end{aligned}
\tag{2.30}
$$

Therefore, the demagnetizing field gradient at the transition center is

$$\frac{dH_d}{dx}\bigg|_{x_0} = -\frac{2M_r}{\pi}\left[\frac{1}{a} - \frac{1}{a + \delta/2}\right] = -\frac{M_r\delta}{\pi}\frac{1}{a(a + \delta/2)}. \tag{4.10}$$

Note that at the transition center, the demagnetizing field is zero, but its gradient is nonzero. Substituting Equations (4.9), (1.26), (4.7), and (4.10) into (4.2), we get

$$\frac{2M_r}{\pi a} = \frac{M_r}{H_c(1 - S^*)}\left[\frac{H_cQ}{y} - \frac{M_r\delta}{\pi a^2}\right] \quad \text{for thin medium } (\delta << a). \quad (4.11)$$

This results from the William-Comstock model, with the additional approximation of arctangent transition and Karlqvist head. Multiplying both sides of Equation (4.11) by a^2 and rearranging the equation, we obtain a quadratic equation of a:

$$a^2 - \frac{2(1 - S^*)y}{\pi Q}a - \frac{M_r\delta}{\pi H_c}\frac{y}{Q} = 0. \quad (4.12)$$

The positive solution to Equation (4.12) is

$$a = \frac{(1 - S^*)y}{\pi Q} + \sqrt{\left[\frac{(1 - S^*)y}{\pi Q}\right]^2 + \frac{M_r\delta}{\pi H_c}\frac{y}{Q}}. \quad (4.13)$$

where y is taken as the distance from the head pole tip to the center of the medium, $y = d + \delta/2$, and d is the magnetic spacing.

To achieve a high linear density, we need to reduce the transition parameter a as much as possible. This leads to the following requirements:

1. Large medium coercivity H_c (but still recordable with an inductive head).
2. Small $M_r\delta$ product (but large enough to maintain readback signal level).
3. Small magnetic spacing d (but large enough so that there is no excessive wear or head crash).
4. Large coercive squareness S^*. It turns out that thin film disks with $S^* > 0.9$ tend to have excessive media noise, so S^* values ranging from 0.7 to 0.9 are typical in digital magnetic recording.

For a square M-H loop, $S^* = 1$, Equation (4.13) becomes

$$a = \sqrt{\frac{M_r\delta}{\pi H_c}\frac{y}{Q}}. \quad (4.14)$$

Even for $S^* < 1$, the other two terms tend to be relatively small. Therefore,

$$a \cong \sqrt{\frac{M_r\delta}{\pi H_c}\frac{y}{Q}} \quad \text{(SI)} \quad \text{or} \quad a \cong \sqrt{\frac{4M_r\delta}{H_c}\frac{y}{Q}} \quad \text{(Gaussian)}.$$

This is an equation widely cited in the literature. However, attentions must be paid to all the approximations and assumptions we made to arrive at this greatly simplified expression.

4.1.2 Head imaging and relaxation effects

When a transition is under a recording head, the transition shape could be modified due to the imaging effect of the high permeability material of the recording head, as shown in Fig. 4.5. A negative magnetic charge at $y = 0$ will have an image of positive magnetic charge at $y = 2d + \delta$. The field from the image charge will modify the transition shape. The demagnetizing field in the medium is now reduced to

$$H_d^{\text{net}} = H_d(x)\big|_{y=d+\delta/2} - H_d(x)\big|_{y=-d-\delta/2}. \tag{4.15}$$

The second term comes from the imaged transition. Using the results from Chapter 2, we can derive that the modified demagnetizing field gradient at the transition center is[6]

$$\frac{dH_d^{\text{net}}}{dx}\bigg|_{x=0} = -\frac{M_r\delta}{\pi}\frac{1}{a_{im}(a_{im}+\delta/2)} + \frac{M_r\delta}{\pi}\frac{1}{(a_{im}+2d+\delta)^2 - (\delta/2)^2} \tag{4.16}$$

$$\cong -\frac{M_r\delta}{\pi}\left[\frac{1}{a_{im}^2} - \frac{1}{(a_{im}+2d+\delta)^2}\right] \quad \text{(for thin medium)},$$

where a_{im} represents the transition parameter after taking the imaging effect into consideration.

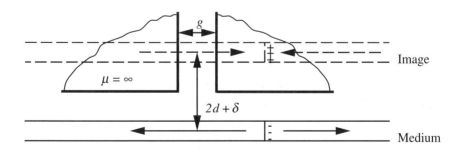

FIGURE 4.5. Imaged transition of high permeability recording head. Negative charges are imaged into positive charges.

If the magnetic medium has a nearly square loop, then the following relation must hold:

$$\frac{dH_x}{dx} + \frac{dH_d}{dx} = 0. \tag{4.17}$$

Namely,

$$-\frac{H_c Q}{y} = \begin{cases} -\dfrac{M_r \delta}{\pi a^2} & \text{without imaging effect,} \\[2ex] -\dfrac{M_r \delta}{\pi}\left[\dfrac{1}{a_{im}^2} - \dfrac{1}{(a_{im} + 2d + \delta)^2}\right] & \text{with imaging effect.} \end{cases}$$

Therefore, the transition parameter ignoring imaging effect is related to that considering imaging effect by the following equation:

$$\frac{1}{a^2} = \frac{1}{a_{im}^2} - \frac{1}{(a_{im} + 2d + \delta)^2}. \tag{4.18}$$

The net imaging effect reduces the transition parameter written by a recording head because $a_{im} < a$. The imaging effect is quite significant when d is small ($d \sim \delta$), but negligible if d is not so small ($d > 3\delta$)! In the past, digital magnetic recording technology mostly belonged to the latter case, but now it is approaching the former case as magnetic spacing is being reduced.

What is the final transition parameter a_f after the relaxation? The relaxed magnetization M_f is related to the initial magnetization M through a minor loop susceptibility of χ, as shown in Fig. 4.4:

$$M_f - M = \chi(H_{df} - H), \tag{4.19}$$

where H_{df} and H are the final demagnetizing field after relaxation and the total field at the instant of transition formation, respectively. Taking derivatives of Equation (4.19) with respect to x, we have

$$\frac{dM_f}{dx} - \frac{dM}{dx} = \chi\left(\frac{dH_{df}}{dx} - \frac{dH}{dx}\right). \tag{4.20}$$

Assume that the initial and final transition centers are very close (χ is small), then the above equation can be simplified as

$$\frac{2M_r}{\pi a_f} - \frac{2M_r}{\pi a} = \chi\left(-\frac{M_r \delta}{\pi a_f^2} - \frac{H_c(1 - S^*)}{M_r}\frac{2M_r}{\pi a}\right), \tag{4.21}$$

$$a_f^2 - \frac{a}{r}a_f - \frac{\chi \delta a}{2\,r} = 0.$$

The positive solution to the quadratic equation (4.21) is

$$a_f = \frac{a}{2r} + \sqrt{\left(\frac{a}{2r}\right)^2 + \frac{\chi \delta a}{2r}} \quad \text{(SI)}, \quad \text{or} \quad a_f = \frac{a}{2r} + \sqrt{\left(\frac{a}{2r}\right)^2 + \frac{2\pi\chi\delta a}{r}} \quad \text{(Gaussian)}.$$

(4.22)

where $r < 1$ as defined previously. It is easily seen that the final relaxed transition parameter is greater than the initial value written. However, $\chi \ll 1$, $r \approx 1$, so $a_f \approx a$.

4.1.3 Demagnetization limit

After the transition moves away from the recording head, if the demagnetizing field is larger than the medium coercivity, the medium will demagnetize and lose the transition. The demagnetizing field at the center plane of the medium is

$$H_d(x) = -\frac{2M_r}{\pi}\left[\tan^{-1}\left(\frac{a + \delta/2}{x - x_0}\right) - \tan^{-1}\left(\frac{a}{x - x_0}\right)\right]$$

$$= -\frac{2M_r}{\pi}\tan^{-1}\left[\frac{(x - x_0)\delta/2}{(x - x_0)^2 + a(a + \delta/2)}\right].$$

(2.30)

where x_0 is the transition center. The maximum amplitude of the demagnetizing field is

$$H_{d,\max} = \frac{2M_r}{\pi}\tan^{-1}\left[\frac{\delta/2}{2\sqrt{a(a + \delta/2)}}\right] \cong \frac{2M_r}{\pi}\tan^{-1}\frac{\delta}{4a} \quad \text{for thin medium}.$$

(4.23)

Let a_d represent the transition parameter limited by the demagnetization, and $H_{d,\max} = H_c$, then

$$\frac{2M_r}{\pi}\tan^{-1}\frac{\delta}{4a_d} = H_c, \quad a_d = \frac{\delta/4}{\tan(\pi H_c/2M_r)} \approx \frac{M_r\delta}{2\pi H_c}.$$

(4.24)

The demagnetization limit of transition parameter is typically smaller than the transition parameter given by the Williams-Comstock model. This becomes untrue only at very small magnetic spacing when

$$a_d \approx \frac{M_r\delta}{2\pi H_c} > a \approx \sqrt{\frac{M_r\delta(d + \delta/2)}{\pi Q H_c}},$$

i.e.,

$$\frac{M_r}{H_c} > \frac{2\pi}{Q}\left(\frac{2d}{\delta} + 1\right). \tag{4.25}$$

The demagnetization limit should be kept in mind in the case of low-coercivity disks or thick magnetic medium. *For ultrahigh-density recording towards near-contact recording, the demagnetization limit is typically not reached.* For example, consider $d \sim \delta \sim 200$ Å, $Q \sim 1$, $H_c \sim 3000$ Oe $\sim$ 2.4×10^5 A/m, $M_r \sim 800$ emu/cc $\sim 8 \times 10^5$ A/m. The right side of the preceding inequality (4.25) is ~ 19, while the left side is $M_r/H_c \sim 3$.

The effect of demagnetizing field is far more complicated than just imposing a demagnetization limit to the transition parameter. It really makes the Williams-Comstock model an oversimplification because the shape of transition, demagnetizing field, and head field are interdependent, thus the shape of transition can not be known *a priori* as assumed in the Williams-Comstock model. More rigorous modeling of write process resorts to *self-consistent* calculations in which only head field is assumed to be known, while the magnetization transition and demagnetization field are calculated by an iteration process until the solutions converge.[7]

4.2 NONLINEARITIES IN WRITE PROCESS

Nonlinearity refers to a phenomenon that causes linear superposition to be invalid. In the magnetic write process, commonly observed nonlinearities include partial erasure, nonlinear transition shift, and hard transition shift. These concepts are introduced here, but much more in-depth discussions will be given in Chapter 10.

Nonlinear Amplitude Loss (Partial Erasure).   As shown in Fig. 4.6, "percolation" occurs at high densities, i.e., regions across the track become partially erased and the amplitude loss results from an effective trackwidth narrowing.

Numerical micromagnetic modeling and experimental observations confirmed that large exchange coupling produces large, but few, percolation regions while low exchange coupling produces many narrow percolation regions.[8]

Nonlinear Transition Shift (NLTS).   Nonlinear transition shift or bit shift refers to transition shift which occurs due to the demagnetizing field

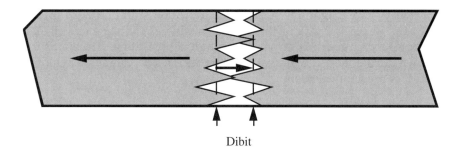

Dibit

FIGURE 4.6. Partial erasure of two closely spaced transitions (dibit).

from previously written transitions. Since the demagnetizing field drops quite fast with the distance, the location of a transition is shifted mostly by the perturbing demagnetizing field from the nearest-neighbor previous transition.

The neighboring magnetic transitions must have opposite magnetic charge signs, and their mutual magnetostatic interaction is manifested by demagnetizing field. Since opposite charges attract each other, the two magnetic transitions (dibit) must be shifted closer! In other words, *NLTS always causes the transition to be written earlier than desired.* One need not worry about this effect if the dibit is far apart, but it is very serious at high recording densities. One way to solve the problem of NLTS is to employ *write precompensation* (precom), i.e., to intentionally delay switching the write current so that the resulting transition center is in the desired location.

Hard Transition Shift. In digital magnetic recording the erasure of old information is accomplished by directly writing new data pattern over old data pattern. In this process, a *hard transition* is written if the head field is opposing the incoming magnetization, while an *easy transition* is written when the head field is in the direction of the incoming magnetization. The hard transition is more "difficult" to write because it requires more head field to saturate the magnetization under the head gap. The additional difficulty is due to the demagnetizing field from the transition formed at the leading edge of the write bubble, which does not occur in the writing of easy transition.

Similar to nonlinear bit shift due to previous transitions, the demagnetizing field from the leading edge of the write bubble leads to another type of nonlinear bit shift, called the hard transition shift, and *the hard transition always gets shifted later than desired* in the absence of other nonlinear effects. The magnitude of the hard transition shift decreases greatly with the increasing size of the writing bubble as the magnetic charges become farther apart. Since the incoming magnetization direction is unknown during the write process, hard transition shift can not be removed by write precompensation.

Overwrite.   In digital magnetic recording the erase of old information is accomplished by direct overwrite of new data pattern on the old data pattern. The basic requirement here is that the write field must be sufficient to reduce any residual original information to levels low enough not to cause errors while reading the new data.

The most commonly cited *overwrite ratio* is defined as follows. First write a square-wave pattern at frequency f_1, then overwrite with a square-wave pattern at frequency f_2 (most often, $f_2 = 2f_1$). The overwrite ratio ("f_1/f_2" ratio) is the level of the residual f_1 signal $V_{f_2}(f_1)$ divided by the original f_1 signal level $V_{f_1}(f_1)$. Normally the overwrite ratio is given in decibels (dB):

$$OW = 20 \log \frac{V_{f_2}}{V_{f_1}}.$$

The absolute value of the overwrite ratio is

$$|OW| = 20 \log \frac{V_{f_1}}{V_{f_2}}. \tag{4.26}$$

Note that both signals are measured at old frequency f_1. For a typical magnetic recording system, the required OW is < -30 dB, or $V_{f_2} < 0.032 V_{f_1}$. The common sources of overwrite residual signal are listed in the following.

1. Residual recorded signal due to incomplete erasure of the f_1 signal when the write current is not sufficient
2. Hard transition shifts of f_2 signal at f_1 rate due to the demagnetizing fields from the incoming f_1 pattern
3. Residual track edge effect due to incomplete erasure of track-edge magnetization

The first two sources can be minimized by using a "good" write head with a sufficient write field, and the second source can be reduced by a "write wide, read narrow" scheme, to be discussed in Chapter 14.

References

1. H. C. Tong *et al.,* "The micromagnetics of thin-film disk recording tracks," *IEEE Trans. Magn.,* **20,** 1831, 1984.

2. T. Chen, "The micromagnetic properties of high coercivity metallic thin films and their effects on the limit of packing density in digital recording," *IEEE Trans. Magn.,* **17,** 1831, 1981.

3. J. G. Zhu and H. N. Bertram, "Recording and transition noise simulations in thin-film media," *IEEE Trans. Magn.,* **24,** 2706, 1988.

4. M. L. Williams and R. L. Comstock, "An analytical model of the write process in digital magnetic recording," 17th *Annu. AIP Conf. Proc.,* Part I, No. 5, 738, 1971.

5. B. K. Middleton, "Recording and reproducing processes," in *Magnetic Recording Technology,* edited by C. D. Mee and E. D. Daniel, (New York, NY: McGraw-Hill, 1996, Chapter 2).

6. H. N. Bertram, *Theory of Magnetic Recording,* (Cambridge, UK: Cambridge University Press, 1994, p. 218).

7. R. M. White, *Introduction to Magnetic Recording,* (New York, NY: IEEE Press, 1985, pp. 37 and 198).

8. J. G. Zhu, "Noise of interacting transitions in thin film recording media," *IEEE Trans. Magn.,* **27,** 5040, 1992.

CHAPTER 5

Inductive Magnetic Heads

The inductive read/write head is an essential component in all magnetic recording systems, including digital (hard, floppy, and tape drives), audio, and video recorders. Although inductive recording heads were introduced in Chapter 2, they are actually such complicated devices that their in-depth discussion requires a separate chapter. Inductive head models, fabrication process, read/write head designs, and new materials for inductive heads will be examined here. The chapter will be mainly concerned with thin-film inductive heads.

As mentioned previously, there are mainly two categories of inductive heads. One is the ring heads such as ferrite and metal-in-gap (MIG) heads, in which the magnetic poles are usually machined, the coils are mechanically wound around a magnetic pole, and the recording head gaps are formed by glass bonding. The other category is thin-film heads, in which the magnetic poles and coils are batch fabricated in a fashion similar to that in integrated circuit (IC) technology. No machining is involved except for head slider fabrication. Ring heads have been extensively used in hard disk drives in the beginning. By the early 1990s, however, thin-film heads have largely replaced ring heads in hard disk drives, although the latter are still used in audio/video recording and some niche digital recording applications. Thin-film heads offer several advantages over ring heads[1]:

1. Thin-film heads can achieve higher data rates because they have lower inductance due to miniaturized head coils, and because their magnetization response rolls off at higher frequencies.
2. The linear recording density can be higher due to narrower head field gradients and higher readback sensitivity.
3. The track density can be higher due to the miniaturization of the poles.

4. The batch fabrication of magnetic thin film heads and the fabrication of multielement thin-film heads are realized by using processes and equipment similar to semiconductor IC technology.

5.1 MODELING OF INDUCTIVE HEADS

Theoretical studies of thin-film inductive heads and ring heads have been published extensively in the literature. The approaches adopted include analytical methods, transmission line modeling, finite element modeling, boundary element modeling, and micromagnetic modeling, which are very useful in the design and testing of inductive heads. These modeling approaches will be discussed in this section, with an emphasis in analytical and transmission line modeling.

5.1.1 Analytical models

Very useful analytical expressions of head fields can be obtained for thin-film heads and ring heads by using the Green's function method[2] and conformal mapping method.[3–5] Semiempirical analytical expressions for head fields of thin-film heads[6] and ring heads[7] have been obtained by experimentally measuring the fields of large-scale model heads. To minimize the mathematics involved, here we will describe an analytical expression of thin-film head field based on an approach by Bertero, Bertram and Barnett.[8]

The Fourier transforms of the fringe field components generated by a 2D inductive head suffer the Wallace exponential spacing loss (Chapter 2):

$$H_x(k, y) = H_x(k,0)e^{-|k|y} , \tag{5.1}$$
$$H_y(k, y) = H_y(k,0)e^{-|k|y} = i\mathrm{sgn}(k)H_x(k,0)e^{|k|y} ,$$

where y (>0) is the distance from the head air-bearing surface (ABS), and k is the wave vector. Running the Mathematica® software gives:

$$FT\left[\frac{y}{\pi(x^2 + y^2)}\right] = e^{-|k|y}, \quad (y > 0)$$

$$FT\left[-\frac{x}{\pi(x^2 + y^2)}\right] = i\mathrm{sgn}(k)e^{-|k|y}. \quad (y > 0)$$

Performing the inverse Fourier transform to Equation (5.1) and use the convolution theorem, we obtain that

$$H_x(x, y) = H_x(x,0) * \frac{y}{\pi(x^2 + y^2)} = \frac{y}{\pi} \int_{-\infty}^{\infty} dx' \frac{H_s(x')}{(x - x')^2 + y^2}, \quad (5.2)$$

$$H_y(x, y) = H_x(x,0) * \frac{-x}{\pi(x^2 + y^2)} = -\frac{1}{\pi} \int_{-\infty}^{\infty} dx' \frac{H_s(x')(x - x')}{(x - x')^2 + y^2},$$

where $H_s(x) = H_x(x,0)$ is the x-component of the magnetic field at $y = 0$ surface, i.e., the ABS of the inductive head. For the thin-film head with asymmetrical poles shown in Fig. 5.1, Bertero et al.[8] obtained the surface magnetic field from the conformal mapping method:

$$H_s(x) \approx \begin{cases} \dfrac{H_g}{2}\left[1 + \dfrac{2}{\pi\sqrt{1 - (2x/g)^2}}\right] & \text{if } |x| \leq g/2, \text{ (gap region)} \\[3ex] -\dfrac{3^{1/6}H_g g}{2^{5/3}\pi^{1/3}L_2^{2/3}(x - L_2/2)^{1/3}} & \text{if } x \geq L_2/2, \text{ (right pole edge)} \\[3ex] +\dfrac{3^{1/6}H_g g}{2^{5/3}\pi^{1/3}L_1^{2/3}(x + L_1/2)^{1/3}} & \text{if } x \leq -L_1/2, \text{ (left pole edge)} \\[3ex] 0 & \text{otherwise,} \quad (5.3) \end{cases}$$

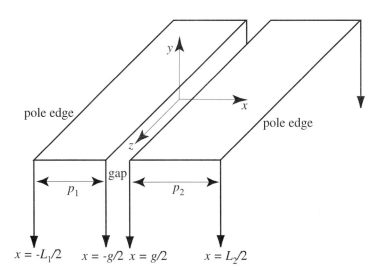

FIGURE 5.1. Schematic cross section of thin-film head with asymmetrical poles.

where H_g is the deep gap field. Substituting this into Equation (5.2), we obtain the head field:

$$\frac{H_x(x,y)}{H_g} = \frac{1}{2\pi}\left[\tan^{-1}\left(\frac{x+g/2}{y}\right) - \tan^{-1}\left(\frac{x-g/2}{y}\right)\right]$$

$$+ \frac{g}{2\sqrt{2}\pi}\left\{\frac{\sqrt{\sqrt{[x^2-y^2-(g/2)^2]^2+(2xy)^2}-x^2+y^2+(g/2)^2}}{\sqrt{[x^2+y^2-(g/2)^2]^2+(gy)^2}}\right.$$

$$-\frac{g\sin\left\{\frac{1}{3}\left[\frac{\pi}{2}-\tan^{-1}\left(\frac{x+L_1/2}{y}\right)\right]\right\}}{2^{2/3}3^{1/3}\pi^{1/3}L_1^{2/3}[(x+L_1/2)^2+y^2]^{1/6}}$$

$$\left.-\frac{g\sin\left\{\frac{1}{3}\left[\frac{\pi}{2}+\tan^{-1}\left(\frac{x-L_2/2}{y}\right)\right]\right\}}{2^{2/3}3^{1/3}\pi^{1/3}L_2^{2/3}[(x-L_2/2)^2+y^2]^{1/6}}\right\}$$

(5.4)

$$\frac{H_y(x,y)}{H_g} = -\frac{1}{4\pi}\ln\left[\frac{(x+g/2)^2+y^2}{(x-g/2)^2+y^2}\right]$$

$$-\text{sgn}(x)\frac{g}{2\sqrt{2}\pi}\left\{\frac{\sqrt{\sqrt{[x^2-y^2-(g/2)^2]^2+(2xy)^2}+x^2-y^2-(g/2)^2}}{\sqrt{[x^2+y^2-(g/2)^2]^2+(gy)^2}}\right.$$

$$+\frac{g\cos\left\{\frac{1}{3}\left[\frac{\pi}{2}-\tan^{-1}\left(\frac{x+L_1/2}{y}\right)\right]\right\}}{2^{2/3}3^{1/3}\pi^{1/3}L_1^{2/3}[(x+L_1/2)^2+y^2]^{1/6}}$$

(5.5)

$$\left.-\frac{g\cos\left\{\frac{1}{3}\left[\frac{\pi}{2}+\tan^{-1}\left(\frac{x-L_2/2}{y}\right)\right]\right\}}{2^{2/3}3^{1/3}\pi^{1/3}L_2^{2/3}[(x-L_2/2)^2+y^2]^{1/6}}\right\}$$

The first two terms are due to the gap region, which were approximated by the Karlqvist equations in Chapter 2. The last two terms are due to the left and right pole edges, respectively, resulting in the so-called undershoots in the longitudinal head sensitivity function at $x = -L_1/2$ and $L_2/2$ (Section 5.3.1). The analytical expressions are somewhat more complicated than the Karlqvist head equations shown in Chapter 2, but more accurate near the ABS.

5.1.2 Transmission line models

The effects of head geometry, magnetic properties of head materials, and operating frequency on head performance can be analyzed by a *transmission line model* in which magnetic flux is propagated by the bottom and top magnetic poles similar to the transmission lines carrying alternating current (AC) and voltage.[9] The magnetic flux and field in a transmission line model are solved from the Maxwell's equations.

If the current into the head coil is *quasi-static*, a 2D quasi-static transmission line model can be readily applied to a multiturn thin film head of a fixed width to calculate its gap field, head efficiency, and head inductance. The head width is assumed to be sufficiently large so that edge effects can be ignored. Consider the cross section of a thin-film inductive head with spiral coils as shown in Fig. 5.2a. Assume that (1) the head structure is infinite in the direction perpendicular to the cross section, (2) the permeability of magnetic poles is μ, and (3) the current is quasi-static, then the differential equation governing the flow of flux (valid at low frequencies) can be derived from Ampere's law. The transmission line model properly accounts for the magnetic flux leakage between the magnetic poles, which reduces head efficiency. The predicted efficiency for a thin-film head with N-turn spiral coils is[10]

$$E = \frac{1}{N} \sum_{i=1}^{N} \frac{2\lambda_1}{Td} \sinh\frac{d}{2\lambda_1} \cosh\frac{s_i}{\delta_1}, \tag{5.6}$$

where

$$T = \cosh\frac{l_1}{\lambda_1} \cosh\frac{l_2}{\lambda_2} + \frac{g_1\lambda_2}{g_2\lambda_1} \sinh\frac{l_1}{\lambda_1} \sinh\frac{l_2}{\lambda_2},$$

$$\lambda_1 = \sqrt{\mu t g_1/2}, \quad \lambda_2 = \sqrt{\mu t g_2/2}.$$

Based on the transmission line model, an efficient thin-film head design requires the following:

1. A high-permeability head material, and thick (but short) magnetic yokes, both lead to small magnetic yoke reluctance
2. A short throat height and a large gap length, both lead to large gap reluctance
3. A large yoke separation, which reduces the magnetic flux leakage between the yokes

In particular, the head efficiency is very sensitive to the throat height, as shown in Fig. 5.2. The sensitivity is even more pronounced when the gap length is small, or the yoke permeability is small. Therefore, the throat

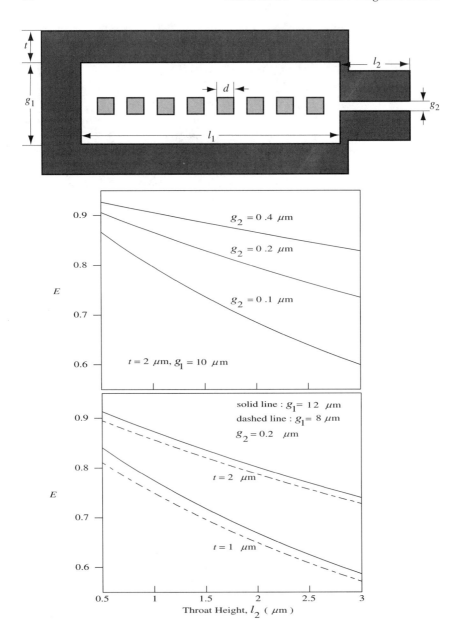

FIGURE 5.2. (a) Top: multiturn spiral head cross section, where t is the magnetic pole thickness, d the coil width, l_1 the yoke length, g_1 the yoke separation, l_2 the throat height (or pole tip depth), g_2 the gap length, and w the pole width. (b) Middle: head efficiency vs. throat height and gap length. (c) Bottom: head efficiency vs. throat height, pole thickness, and yoke separation.

height of thin film heads must be carefully controlled during head fabrication. High head efficiency requires a throat height of ~1 μm or smaller.
Since the pole tip and head gap of the thin film head end at the ABS,
which is formed at a mechanical lapping step, elaborate optical and electronic lapping guides are devised to achieve a precise control of throat
height.[11,12]

The 2D transmission line model is not appropriate for a real 3D thin
film head with narrow pole tips but wide yokes. In Fig. 5.2 the neck (or
apex) region connecting pole tips and magnetic yoke is assumed to be
very short, which is not true in a real thin-film head. Such a thin-film
head can be more realistically simulated by a 3D *modular transmission line
model*,[13,14] as shown in Fig. 5.3. The head is divided into transmission line

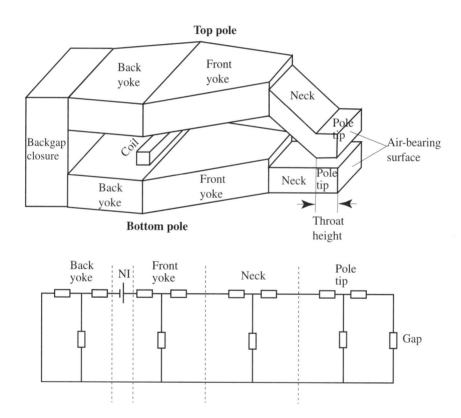

FIGURE 5.3. Modular transmission line model with four segments. The back gap
reluctance is neglected. (a) Top: schematic 3D thin-film head. (b) Bottom: equivalent magnetic circuit. (N = # of coil turns, I = write current.)

segments with various gaps and widths. Each segment (including both the top pole and bottom pole portions) can be represented by a T-network of three reluctances, i.e., a two-port Thevenin equivalent circuit. The coils are modeled by a magnetomotive force inserted between the front yoke and back yoke. Through the analysis of the equivalent magnetic circuit, the magnetic flux and head field can be obtained. It is found that saturation of head material often occurs at the neck region first.

At high frequencies magnetic yoke permeability tends to be reduced, and eddy currents are generated in the thin film heads. The quasi-static transmission line model is generally no longer valid at high frequencies. Several models accounting for both geometry and high-frequency effects have been described in the literature.[15,16]

5.1.3 Finite element and boundary element modeling

Both 2D and 3D *finite element methods* have been used extensively to simulate thin-film heads.[17] These methods generate finite element meshes and use variational techniques in which the potential is approximated by a sequence of functions defined over the entire domain of the problem. Alternatively, *boundary element methods* can also be used in which the potential is not solved directly, but an equivalent source located on the boundaries and interfaces is found by forcing it to satisfy given boundary conditions.[18]

For example, thin-film heads with NiFe ($4\pi M_s = 10$ KG for $Ni_{80}Fe_{20}$) or high-saturation FeAlN ($4\pi M_s \approx 20$ KG) pole pieces can be readily simulated by the Maxwell Field Simulator, a finite element software package from Ansoft Corp. The hard axis (HA) magnetic hysteresis loops of typical NiFe and FeAlN sheet films were used to define the permeability of the head materials in the nonlinear simulations. For an eight-turn NiFe head with a magnetic pole thickness of 2 μm, a gap length of 0.2 μm, and a throat height of 1 μm, modeling indicates that the onset of local saturation occurs at a write current of ~20 mA, resulting in a sharp drop in the head efficiency and a leveling-off of the deep gap field, as shown in Fig. 5.4. The deep gap field is about 6 KOe at a write current of 30 mA. For an FeAlN head with the same geometry, the onset of local saturation occurs at a write current around 30 mA. The deep gap field of the FeAlN head is about 10 KOe at a write current of 30 mA, significantly larger than that of the NiFe head. In either case, the onset of saturation usually occurs at the sloping neck region connecting pole tip and the magnetic yoke.

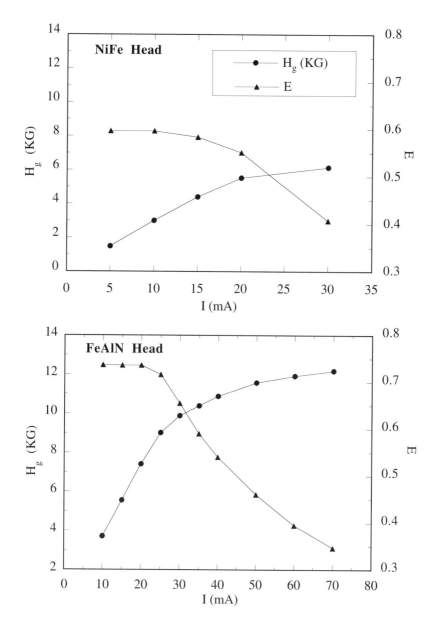

FIGURE 5.4.   Simulated deep gap field and head efficiency vs. write current of the $Ni_{80}Fe_{20}$ head (top) and the FeAlN head (bottom). In this example, gap length $g = 0.2$ μm, throat height = 1 μm, pole thickness = 2 μm, and track width is assumed to be >10 μm.

5.1.4 Micromagnetic models

All of the preceding theoretical models have ignored the magnetic domains and domain walls, which however become more and more important as the head dimension shrinks. In a micromagnetic model, the magnetization vector is assumed to be a function of the space variables with a constant magnitude. The domains and domain walls are treated identically, the former being identified as regions of uniform magnetization while the latter as regions of rapid changes in the magnetization directions. Micromagnetic modeling is frequently coupled with finite difference methods, and dynamic studies are often based upon the Landau-Lifshitz equation. Due to the complexity of treating thick soft magnetic poles, micromagnetic modeling tends to be less successful for thin-film inductive heads than for magnetoresistive read heads. The latter contains much thinner magnetic layers.[19]

5.2 FABRICATION OF THIN-FILM INDUCTIVE HEADS

Ferrite and MIG heads are manufactured with a combination of thin film processing, machining, and bonding processes, while thin film inductive heads are manufactured with thin-film processes similar to semiconductor IC technology. However, thin-film head fabrication process is very unique, and involves both very thin films (<0.2 μm) and very thick films ($>2\mu$m).

5.2.1 Wafer fabrication process

The substrates commonly used for thin film heads are ceramic alumina-TiC wafers with serial numbers for each head laser written on one side while the other side is polished mechanically. The main process flow for a thin-film head with a single-layer spiral coil can be described as follows:

1. Coat thick alumina ($\sim$15 μm) on polished ceramic Al_2O_3-TiC substrate, and then lap back to $\sim$10 μm, as shown in Fig. 5.5, where *the cross-sectional view of a typical head being processed is shown at the left, and the top view is at the right*. The same convention is used throughout Figs. 5.5–5.16.

2. Deposit and pattern the bottom magnetic pole that is 2–4 μm thick, as shown in Fig. 5.6. The magnetic pole is in a pancake shape to increase the head efficiency. Only the narrow pole tip dimension is important in determining the written trackwidth. The commonly used NiFe magnetic poles are electroplated,[20] while amorphous Co-based alloy and nanocrys-

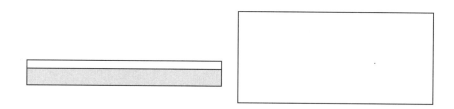

FIGURE 5.5.   Alumina-TiC substrate coated with alumina.

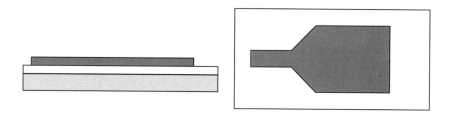

FIGURE 5.6.   Bottom magnetic pole, electroplated or sputter deposited.

talline FeN-based alloy are sputter deposited in a vacuum chamber. The former is often referred as a *wet* process because it involves placing wafers in an acidic plating bath, and the latter as a *dry* process. These materials will be discussed further in Section 5.6.

3. Deposit the gap layer that is typically 0.1–0.5 μm of sputtered alumina as shown in Fig. 5.7. The gap layer thickness greatly affects the linear resolution and side write/read effects of the recording heads.

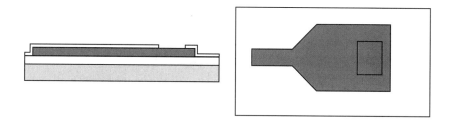

FIGURE 5.7.   Gap layer deposited, and via hole etched.

4. Pattern hard-cured photoresist insulator which is ~5 μm thick to provide electrical insulation and a relatively planar surface to plate copper coils, as shown in Fig. 5.8.

5. Electroplate spiral copper coils that are ~3 μm thick, 3–4 μm wide, and ~2 μm apart in the critical region above the magnetic pole, as shown in Fig. 5.9. Before plating a seed layer must be sputtered, and afterward it must be selectively etched. The coils are made wider outside the magnetic poles to reduce overall coil resistance. It is interesting that around 1997 the semiconductor IC industry just started to electroplate copper (Cu) interconnects instead of sputtering aluminum (Al) interconnects, but electroplated Cu has been used in magnetic recording heads for a long time.

6. Pattern hard-cured photoresist insulator ~5 μm thick to provide electrical insulation for copper coils and planarized surface for top magnetic pole deposition, as shown in Fig. 5.10.

7. Deposit and pattern the top magnetic pole that is 2–4 μm thick, and overpass lead that provides electrical connection to the central tap of the coils, as shown in Fig. 5.11. The back regions of the magnetic poles are often made thicker than the pole tips to achieve high head efficiency, and to avoid magnetic saturation.

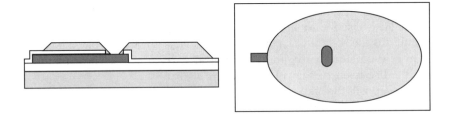

FIGURE 5.8.   Hard-cured photoresist pattern added.

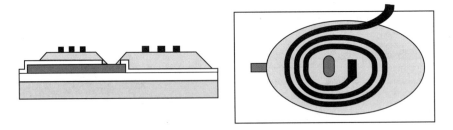

FIGURE 5.9.   Copper coils electroplated.

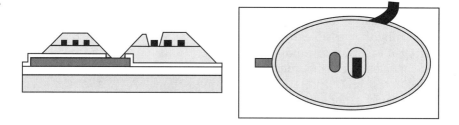

FIGURE 5.10. Hard-cured photoresist insulator added for insulation and planari-zation.

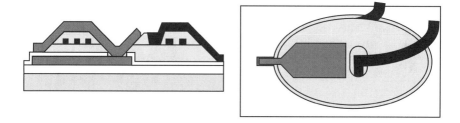

FIGURE 5.11. Top magnetic pole and overpass lead added.

8. Electroplate 20–40 μm thick copper studs, deposit 10–15 μm thick alumina overcoat, and then lap open copper studs, as shown in Fig. 5.12. The thick alumina is necessary to protect the magnetic and coil structures during machining and lapping processes, and to prevent corrosion.

9. Electroplate soft gold bonding pads with seed layer. The gold pads are necessary to protect the underlying devices and to facilitate wire

FIGURE 5.12. Copper studs and alumina overcoat deposited. Note that the copper studs are placed away from the magnetic poles (cf. Fig. 5.14.).

bonding, as shown in Fig. 5.13. Note that *wafer level testing* of thin film heads can be performed at this point or earlier. This is very useful for diagnosing the possible problems during the head fabrication. For example, by measuring self-inductance of each head, pole permeability and head efficiency can be inferred.[21] Excessive values of head inductance and coil resistance indicate possible nonfunctional devices. A 4" square wafer would contain more than ten thousands thin film heads in "nanoslider" form factor (to be explained later), thus automatic wafer probing must be adopted.

10. Scribe and slice the wafer into row bars, which are then bonded to tooling bars that hold them during mechanical processing. As shown in Fig. 5.14, the row bars are first diced into individual sliders (but without debonding), then the rails at the air bearing surface are defined by machining, ion milling, or reactive ion etching according to fly height requirements, finally the row bars are lapped to the desired throat height (gap depth). The throat height should be controlled to 1–2 μm, with a precision of ± 0.25 μm. It must be continuously monitored during lapping by elec-

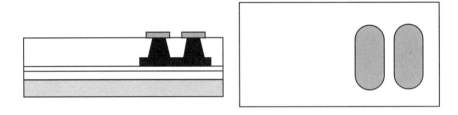

FIGURE 5.13.   Gold bonding pads electroplated.

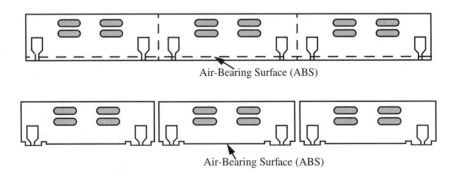

FIGURE 5.14.   (a) Top: a part of row bar before dicing and lapping. (b) Bottom: individual sliders and air bearing surface defined.

tronic lapping guides, and lapping process is stopped automatically when an optimum throat height is reached.

11. Randomly sample heads for pole-tip geometry measurements after lapping and then sort out good individual sliders. For sorting sliders, they are first debonded from the tooling bars. Then they are selected by robots based on wafer probing results. Sliders with no good heads are discarded.

12. Bond good sliders to flexures, as shown in Fig. 5.15. The length of each slider is approximately equal to the original substrate thickness. In this case, each slider has two symmetrical heads, but only one of them is wire bonded. Most head sliders manufactured today only contain one write/read transducer. A twisted pair of wires are bonded to the exposed electroplated gold pads, and the bonding joints are covered with a conformal coating of epoxy cement for physical strength and protection against corrosion. Each slider is then epoxy-bonded to a stainless-steel *gimbal*, or *flexure*. This suspension flexure provides the required load force (*gramload*) and damping to interact with the air bearing, which is formed between the slider contour and disk surface, resulting in the desirable flying characteristics. The products at this point are often called *head gimbal assemblies* (HGAs). In the end, a gentle lapping process (kisslap) is used to achieve a desired curvature (the crown) of the air bearing surface to fine-tune the flying and start-stop performance of sliders.

Before shipping, HGAs are usually subjected to a 100% write/read performance test which measures overwrite, signal amplitude, pulse

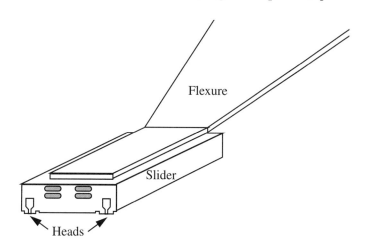

FIGURE 5.15. Slider bonded to flexure (head gimbal assembly).

shape, Barkhausen noise, error rate, etc. A random sample of HGAs from each lot is also tested for flying height and flying characteristics. The entire lot is either passed or rejected based on the mean and variance of the test results.

Most thin-film inductive heads actually employ multilayer spiral coils to achieve large readback signals, as shown in Fig. 2.1. This means that the Cu plating and hard-cured photoresist steps must be reiterated. Electrical connections between the neighboring coil layers must also be implemented. These connections greatly add to the complexity of thin-film head fabrication.

5.2.2 Head-slider form factor

Over the years, thin-film head track width and gap length are scaled downward progressively to write/read smaller and smaller data bits. At the same time, the dimension of head sliders has also gone through miniaturization. However, the lateral size of substrates has been scaled up while their thickness has been scaled down. The larger substrates and smaller slider form factors allow more head sliders to be produced per wafer, a major boost to productivity. Smaller and lighter sliders are also beneficial to their tribological performance and servo control.

The first generation of flying sliders were designed for the ferrite ring heads used in the original Winchester hard disk drives, so they have been referred as the "full-size" Winchester ferrite sliders. Many generations of slider form factors have been introduced since, as listed in Table 5.1. The percentage of each form factor is the linear scaling factor from the minislider. For example, the commonly used 50% nanoslider today has linear dimensions that are approximately half of those in 100% minislider. The volume of the nanoslider is about 1/8 that of the minislider. For the same wafer size, one can produce nanosliders four times as much as minisliders. The air-bearing designs of sliders have also gone through many revolutions. The geometry of typical sliders is shown in Fig. 5.16, where the representative slider form factors are illustrated.[22]

The rails of the air-bearing surface shown in Fig. 5.16, along with disk rotating speed and gram load, determines the fly height of the slider above the disk. Some innovations did not adhere to Fig. 5.16. For example, supersmall and light Micro Flexhead™ sliders were proposed to achieve "contact recording" in 1991.[23] In addition, "tripad" nanosliders with ABS similar to that of picoslider in Fig. 5.16 were designed for "proximity recording."[24] Proximity recording allows the slider to fly near the glide height of disk (Chapter 7), and the low fly height maximizes the recording

TABLE 5.1. Form Factors of Flying Sliders

Form factor	Full-size slider	Minislider 100%	Microslider 70%	Nanoslider 50%	Picoslider 30%	Future
Dimension (mm)	$5.6 \times 4 \times 1.93$	$4 \times 3 \times 0.8$	$2.84 \times 2.23 \times 0.61$	$2.05 \times 1.6 \times 0.43$	$1.25 \times 1 \times 0.3$	
Slider materials	Ferrite	Ferrite	Al_2O_3-TiC	Al_2O_3-TiC	Al_2O_3-TiC	Thin-film/composite
Air-bearing features	Machined rails	Machined rails	Machined rails	Etched rails	Etched rails	Head near or at contact

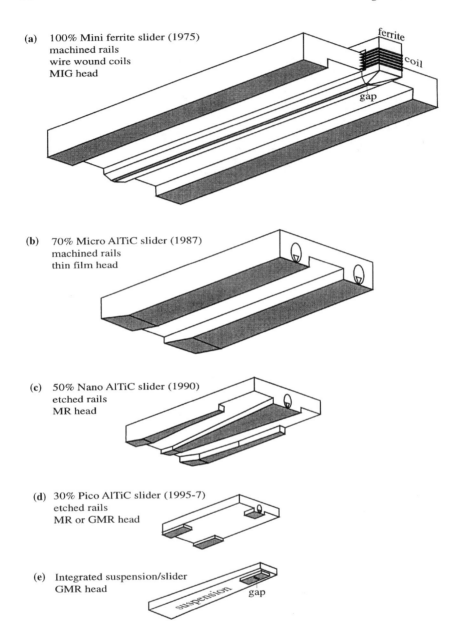

(a) 100% Mini ferrite slider (1975)
    machined rails
    wire wound coils
    MIG head

(b) 70% Micro AlTiC slider (1987)
    machined rails
    thin film head

(c) 50% Nano AlTiC slider (1990)
    etched rails
    MR head

(d) 30% Pico AlTiC slider (1995-7)
    etched rails
    MR or GMR head

(e) Integrated suspension/slider
    GMR head

FIGURE 5.16.   Evolution of slider/air-bearing surface.

densities achievable by thin-film inductive read/write heads. In comparison, inductive write/magnetoresistive (MR) read heads (Chapter 6) are usually designed to fly above the glide height because of the thermal asperities of MR heads.

5.3 INDUCTIVE HEAD AS READER

A thin-film inductive head can be used either as a write head (writer) or as a read head (reader). The current in the head coil in the read mode is typically ~10 mA, much smaller than that in the write mode (~10 mA). The physics in the write and read processes are subtly distinct. For example, the magnetic permeability of head materials is basically independent of applied field in the read mode, but it is field-dependent in the write mode because the large write current tends to drive part of the head poles into saturation. In other words, the reader is more or less a linear device, but the writer is a nonlinear device. In this sense, the reciprocity principle in Chapter 3 is valid only when both the read and write currents are similar. Not surprisingly, the distinct functionality leads to the diverging requirements and different practical issues for inductive reader and writer. Here we first examine the important issues relevant to thin-film inductive readers.

5.3.1 Undershoot

From Section 5.1 and the reciprocity principle, we can easily derive the typical isolated readback pulse of a thin-film inductive head as shown in Fig. 5.17. The two small negative peaks beside the main positive peak, often called the undershoots, are due to the two outer edges of the magnetic poles, while the main peak is due to the head gap. When a magnetic transition passes an outer edge, the magnetic flux picked up by the head coil changes suddenly, resulting in a small negative undershoot. The opposite polarity between the undershoots and the main peak is caused by the opposite changes in the signal flux into head coils. The distance between an undershoot and the main peak is equal to the corresponding pole thickness.

The Fourier transform of the above isolated pulse is that of Fig. 3.2 modulated by the ripples associated with the undershoots. If the thin film head in question has an equal pole thickness of p, then the ripple period is $2\pi/(p + g/2)$, as shown in Fig. 5.18. The undershoots and ripples of

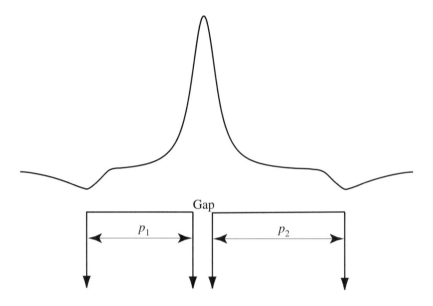

FIGURE 5.17.   Isolated readback pulse (top) of a thin-film inductive head (bottom).

thin-film inductive readers are largely harmless in the peak detection channels which detect main pulse peaks (Chapter 11). The undershoots do not affect the position of the main pulses which are very close to one another at high recording densities. However, when the thin-film inductive readers are used in the partial-response maximum likelihood (PRML) channel (Chapter 12), the undershoots and ripples make it much more difficult to equalize the isolated pulse into the desired pulse shape of the partial response channel. To solve this problem, the outer edges of magnetic pole tips can be elaborately shaped to reduce or even eliminate the undershoots. For example, the undershoots have been successfully suppressed by pole tip trimming from the ABS to cant the leading and trailing pole tip outer edges.[25]

5.3.2 Readback noises of inductive readers

Ideally a thin-film inductive reader would be a linear and noise-free device, but in reality, it does generate noises during the readback process. The noises of an inductive reader can generally be classified into two categories:

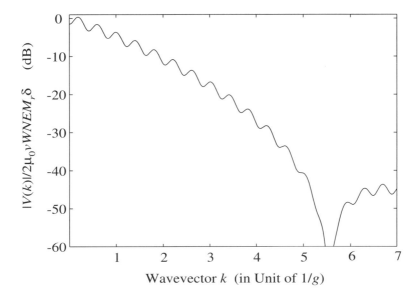

FIGURE 5.18. Fourier transform of the thin-film head readback pulse from an arctangent transition. We assumed $\delta \ll (a + d) = g/2$, $p = 15g$, where δ is the medium thickness, a the transition parameter, d the magnetic spacing, and g the gap length. In this case, the ripple period is $2\pi/15.5g \approx 0.41/g$, the first gap null occurs at $2\pi/1.136g \approx 5.5/g$.

1. *Barkhausen noise* due to the domain wall motion in magnetic poles. Thin-film inductive readers are designed to operate under magnetization rotation only. However, domain walls may move under circumstances, resulting in spurious changes in the flux through head coil and noises in readback waveform.
2. *Johnson thermal noise* which is a random noise present in all linear and lossy physical systems. The sources of thermal noise in thin film inductive readers have been attributed to head coil DC resistance and energy losses associated with domain wall motion, magnetization rotation, and air coil.[26]

Thermal noise is usually smaller than Barkhausen noise and recording medium noise, so they are allowed in magnetic disk drives. In comparison, Barkhausen noise is a major concern in the design and fabrication of thin film inductive readers. Many efforts are made to reduce or eliminate Barkhausen noises, which can be traced back to unwanted domain wall motion in magnetic poles. Typical domain configurations in magnetic yoke and pole tip (including neck) are shown in Fig. 5.19. Since the thin-film heads are designed to conduct flux along the vertical hard axis (HA), the magnetic domains transverse to the HA are desirable.

The Barkhausen noises in thin-film inductive heads often occur after large signal excitations, such as a write or erase operation. Barkhausen noise can also occur at any time during a read operation, particularly for the heads with undesirable domain configurations. In general, the Barkhausen noises observed in thin-film inductive heads can be classified into the following forms:

1. *Read instability.* Spurious transients riding on the readback pulses (*wiggles*) and read amplitude variations can occur long after write, on the order of ms or longer, and in the presence of external signal excitation. Wiggles can occur at or near the top of an isolated pulse, or at its leading edge, or at its trailing edge. Wiggles can also occur after the isolated pulse has decayed. The largest isolated pulse variations seem to occur after magnetic yoke saturation. This type of Barkhausen noise is generally attributed to signal-induced irreversible domain wall jumps due to shallow domain wall pinning sites.[27] Wiggles can also be caused by a delayed signal flux contribution from 90° edge closure domains.[28] Readback pulse amplitude and shape can also vary in a "bad" thin-film inductive head even if the head has not gone through a previous write-excitation. This

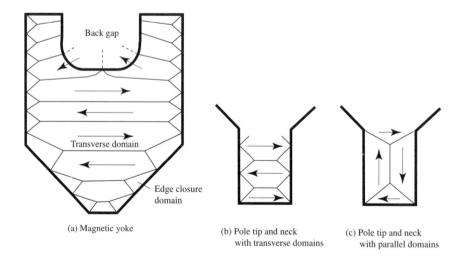

(a) Magnetic yoke

(b) Pole tip and neck with transverse domains

(c) Pole tip and neck with parallel domains

FIGURE 5.19. Schematic magnetic domains in the magnetic poles of thin-film inductive heads: (a) magnetic yoke, (b) pole tip and neck with walls *transverse* to flux conduction, (c) pole tip and neck with walls *parallel* to flux conduction.

type of read instability is generally attributed to undesirable domain configurations in magnetic poles, such as the irreversible jumps of 180° walls in magnetic pole tips,[29] as shown in Fig. 5.19c.

2. *Write instability.* Spike noises (*popcorn* noises) can occur shortly after write, on the order of μs, and in the absence of external signal excitation. This type of noise is also referred as *noise-after-write.* It is a major concern in disk drives with sector servo in which read-after-write within the same data track is necessary to generate position error signals (Chapter 15). Popcorn noise is generally attributed to stress-induced delayed relaxation of the edge closure domains, spikelike domains, or unstable domain configurations in thin-film head magnetic yokes.[30] It occurs after write because write current pulses cause a *thermal transient* in thin-film heads even after the write, resulting in additional stress variations in magnetic yokes. Near-zero magnetostriction or thermal stress reduction of magnetic pole material is beneficial for reducing stress-induced anisotropy energy ($\propto \lambda_s \sigma$) and popcorn noise, where λ_s is magnetostriction, σ is stress.

5.4 INDUCTIVE HEAD AS WRITER

Thin-film inductive heads can be used either as write heads or as readback heads. However, thin-film MR readback heads (to be discussed in Chapter 6) have been proven to offer much larger signal amplitude per unit trackwidth, thus supporting much greater areal recording densities. Although it is possible to extend thin-film inductive write/read heads to higher recording densities by using three or more layers of coils to enhance signal amplitude, the approach is ultimately limited by the complexity of fabricating multilayer coils and the large inductance associated with the coils. *Inductive-write and MR-read* dual-element heads[31] have become the dominant form of recording transducers for rigid disk drives by the late 1990s for the following reasons:

1. The read and write functions of the dual-element heads can be optimized separately. The read gap can be made smaller than the write gap to achieve high linear density without the expense of overwrite performance.
2. Magnetoresistive read heads can provide larger signal per unit trackwidth than inductive read heads. Thus very high track density becomes possible. Furthermore, MR head technology can be easily extended to giant magnetoresistive (GMR) read head technology,[32] offering even higher recording densities.

3. The read trackwidth can be made slightly narrower than the write trackwidth to offer better off-track capability (Chapter 14).
4. The readback signal amplitude is velocity independent because the MR head is flux-sensitive and does not depend on the time rate of flux change. This is increasingly important as the disk drives are adopting ever smaller form factors (i.e., smaller diameters).

An *inductive-write, GMR-read* head with recording densities of >10 Gb/in.^{2} was announced by an IBM research laboratory in December 1997, which underscores the advantage of the dual-element head technology over the inductive read/write head technology.

Generally speaking, a thin-film inductive head which can deliver a large write field with a sharp field gradient at a desired fly height is considered a good write head. In write mode, poor head efficiency can be compensated by a large write current, and Barkhausen noises are of no consequences. Therefore, many requirements for read heads can be somewhat relaxed here.

5.4.1 Write head design considerations

To analytically derive some simple but useful relations for write heads, we approximate the head field of a thin-film inductive head with the Karlqvist equations. Strictly speaking, this is valid only when the magnetic spacing d is much smaller than the pole thickness, and the point of interest is not too close to the gap corners. More precise numerical values can be obtained by using the complete analytical models or numerical models in Section 5.1.

The maximum longitudinal head field H_x at the bottom of the medium must be large enough to overcome medium coercivity H_c and demagnetizing field to write effectively. It can be shown that a typical requirement is[33]

$$H_x(x, y)|_{\max} = H_x(0, d + \delta) \geq [1.3 + 4(1 - S^*)]H_c, \qquad (5.7)$$

where d is the magnetic spacing, δ is the medium thickness, and S^* is the medium coercive squareness. The required write field amplitude depends on squareness of the medium, as shown in Table 5.2. Not surprisingly, it is easier to switch magnetic media with more square M-H loops.

TABLE 5.2.   Write Field Requirements for Selected Squareness

Coercive squareness S^*	0.7	0.8	0.85	0.9
$H_x(0, d + \delta)/H_c$	≥ 2.5	≥ 2.1	≥ 1.9	≥ 1.7

The deep gap field cannot exceed the saturation magnetization M_s of magnetic pole materials. As a rule of thumb, it is can be chosen to be less than $0.8M_s$ to avoid severe head saturation:

$$H_g \leq 0.8M_s. \quad \text{(SI)} \tag{5.8}$$

Consider a medium with a coercive squareness of 0.85 (IBM 5 Gb/in.[2] demonstration[34]), we can derive the maximum overwritable medium coercivity based on the Karlqvist equations:

$$H_{c,\text{max}} \cong 0.268M_s \tan^{-1}\left(\frac{g/2}{d + \delta}\right), \quad \text{(SI)}$$

$$\tag{5.9}$$

$$H_{c,\text{max}} \cong 0.268(4\pi M_s) \tan^{-1}\left(\frac{g/2}{d + \delta}\right). \quad \text{(Gaussian)}$$

For high saturation head materials like FeXN ($4\pi M_s$ = 20 KG) and Ni$_{45}$Fe$_{55}$ ($4\pi M_s$ = 16 KG), the maximum medium coercivity allowed as a function of $d + \delta$ is shown in Fig. 5.20. For a given medium coercivity, appropriate

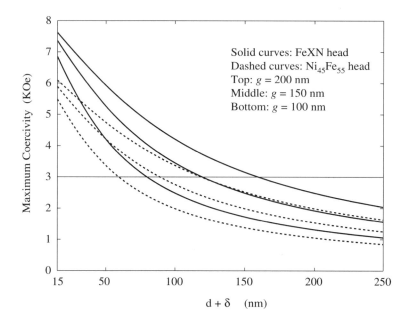

FIGURE 5.20. Estimated maximum medium coercivity overwritable by FeXN and Ni$_{45}$Fe$_{55}$ heads in three cases: (a) g = 200 nm, (b) g = 150 nm, (c) g = 100 nm.

write head material and magnetic spacing must be carefully chosen. For example, the maximum medium coercivity allowed by FeXN write head is about 5.27 KOe if $d + \delta$ is 50 nm (2 μin) and the gap length is 150 nm.

When heads are designed, the head coils should be able to sustain the write current required to provide a sufficient write field. A write current density of $\sim 5 \times 10^5$ A/cm^2 is generally considered to be acceptable in the plated copper coils of thin-film heads.[35] If the copper coil is 3 μm wide and 2.5 μm thick, then a reasonable write current should be $\leq \sim 63$ mA (0-peak). A larger aspect ratio of each coil turn, more coil turns, higher head efficiency, or better heat sink capability should be designed into the thin-film heads if larger write currents are desired. In practice, write current rarely exceeds 50 mA (0-peak).

One of the limits to linear recording density is the read-back pulse width, which is determined by the read head gap, transition parameter, magnetic spacing, etc. The transition parameter a is determined by the write process, and can be calculated from the head field gradient based on the Williams-Comstock model (Chapter 4). The minimum written bit length is approximately the transition width (πa), regardless of readback methods. Therefore, it is very important to raise head field gradient and to reduce transition parameter. In practice, the head write current can be chosen according to one of the following criterions:

1. Write current is adjusted until the maximum field gradient occurs at the write bubble trailing edge (Section 4.1).
2. Write current is adjusted until the field at the bottom of the medium is sufficiently large: 1.9 times medium coercivity if $S^* = 0.85$.
3. Write current is adjusted until the overwrite ratio at specified recording densities exceed 30 dB. The last criterion is the most convenient to use in experiments because overwrite ratios are easy to measure with a magnetic spin stand. In contrast, head fields are more difficult to measure directly.

In the latter two cases, the write field gradient at the trailing edge may be slightly different from that in the first criterion, but the order of magnitude should be the same. Using the first criterion, the maximum head field gradient as a function of head gap length and magnetic spacing can be calculated, as shown in Fig. 5.21. It is seen that reducing magnetic spacing is a much more effective way than reducing gap length to raise field gradient. In this sense, inductive write heads could have a somewhat wider head gap than read heads, for a wider read gap length would immediately lead to a larger pulse width.

From the field gradients we can calculate transition parameter as a function of magnetic spacing (neglecting relaxation and imaging effect),

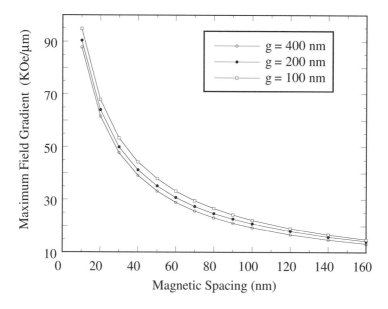

FIGURE 5.21. Maximum head field gradient versus magnetic spacing at selected head gap lengths. Assume $\delta = 25$ nm, $H_c = 3000$ Oe.

as shown in Fig. 5.22. The requirements on transition parameter and overwrite ratio together set an important bound in write head design.

5.4.2 Head field rise time

Neither write current nor write field waveforms are perfect square wave functions implied in previous chapters. A step voltage pulse applied to a thin-film inductive head results in a finite *current rise time* τ_c as determined by the available write driver voltage and the electrical characteristics of write head and interconnects. Even if the write current is perfectly square, it still results in a finite *intrinsic head field rise time* τ_h due to the eddy current damping and Landau-Lifshitz damping of domain wall motion and magnetization rotation.[36] The rise time is usually defined as the time span for the current or field to pass from 10% to 90% of the steady-state value. The actual *field rise time* τ_f can be typically expressed as[37]

$$\tau_f = \sqrt{\tau_h^2 + \beta^2 \tau_c^2} \quad (\beta \le 1), \tag{5.10}$$

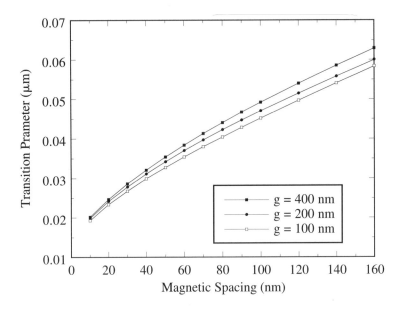

FIGURE 5.22.   Transition parameter versus magnetic spacing at selected head gap lengths. Assume $M_r \delta = 0.7$ memu/cm^2, $S^* = 0.8$.

where both β and τ_h depend on write current amplitude. For small current excitation (below neck saturation), $\beta \approx 1$; for larger currents (above neck saturation), $\beta < 1$. As write current increases, the field rise time depends less and less on current rise time because the yoke magnetization swings faster and faster past the neck saturation "window."

Field rise time is an important parameter for write heads, particularly for high data rate applications. To write transitions T apart, the field rise time has to be $\tau_f \leq 0.8T$. A finite field rise time can contribute to nonlinear transition shifts (Chapter 10) and ultimately limit data rates and linear densities achievable in magnetic disk drives (Chapter 16).[38,39] The experimentally observed[40–42] field rise time typically ranges from ~2 ns to ~10 ns, depending on experimental conditions and head materials. To achieve a small field rise time, one needs a large write current with a short current rise time, short and thin magnetic yokes, and magnetic pole materials with high saturation magnetization and high resistance (to reduce eddy current effect), etc. The lamination of magnetic pole materials should also help.

A simple time-dependent write model taking the head rise time into consideration can be constructed. As the write field rises with time, the write bubble (the $H_x = H_c$ contour, see Fig. 4.1) expands. Assume that the write field in the medium is in phase with the deep gap field $H_g(t)$, The time-dependent Karlqvist field in the medium can be expressed as

$$H_x(x, y, t) = \frac{H_g(t)}{\pi}\left(\tan^{-1}\frac{x' + vt + g/2}{y} - \tan^{-1}\frac{x' + vt - g/2}{y}\right), \quad (5.11)$$

where x' is a given point on the medium relative to the gap center at the instant $t = 0$ (when the write current is switched), $x = x' + vt$ is its new position at the instant t, and v is the disk velocity.

In the absence of field rise time, the deep gap field switches instantaneously at $t = 0$:

$$H_g(t = 0_-) = H_g(-\infty) = -H_g,$$
$$H_g(t = 0_+) = H_g(\infty) = H_g,$$

and the trailing edge of the write bubble x' is given by $H_x(x,y,t = 0_+) = H_c$, so Equation (5.11) gives

$$x' = \sqrt{\left(\frac{g}{2}\right)^2 - y^2 + \frac{gy}{\tan[\pi H_c/H_g]}}. \quad (5.12)$$

In the presence of field rise time, the trailing edge of the write bubble occurs at $|H_x(x,y,t)| = H_c$. Therefore, Equation (5.11) leads to

$$x' = \sqrt{\left(\frac{g}{2}\right)^2 - y^2 + \frac{gy}{\tan[\pi H_c/|H_g(t)|]}} - vt. \quad (5.13)$$

The field rise behavior can be expressed as

$$H_g(t) = H_g(0) + [H_g(\infty) - H_g(0)]\tanh(t/\tau) \quad (5.14)$$
$$= H_g\tanh(t/\tau),$$

where τ is the rise time constant, which is related to the field rise time by the following relation:

$$\tau_f = \tau(\tanh^{-1}0.9 - \tanh^{-1}0.1) \approx 1.37\,\tau. \quad (5.15)$$

When $t > 0$, the gap field first rapidly increases such that the write bubble trailing edge is moving to the right. When the field rise slows down, the trailing edge of the write bubble eventually starts to shift to the left, opposite to the disk speed. Equation (5.13) is plotted in Fig. 5.23. Neglecting the medium switching time, the transition center is formed at the

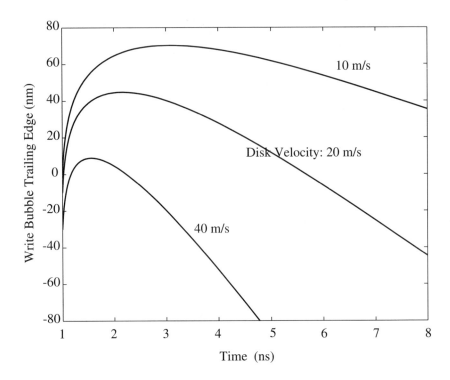

FIGURE 5.23. Write bubble trailing edge in the medium versus time and disk velocity. Assume the field rise time is 3 ns, $g = 150$ nm, $y = 50$ nm, $H_c/H_g = 0.2$.

maximum x' value. Observing from the write head, the transition center is formed where the write bubble expansion speed dx/dt is equal to the medium velocity v.

As the disk velocity increases, the transition center is shifted from the gap corner ($x' = g/2$) toward the gap center ($x' = 0$). It can be shown that along with the shift in transition center, the transition parameter will also increase with field rise time and disk velocity.[39]

5.5 SIDE READ, SIDE WRITE, AND SIDE ERASE

Thin-film recording heads have been largely treated as 2D devices by assuming that it is uniform along the z-axis (cross-track direction). In reality, the thin-film heads are of course finite in the z-direction, and submicron trackwidth heads designed for very high track densities have begun to appear in the literature. The 2D model is often sufficient for analyzing on-track performance of the device when the head and data track are aligned. However, the heads do often go off-track in disk drives due to mechanical disturbances (Chapter 15), which dictates that we must treat recording heads as 3D devices. Even if a head is on-track, its side read, side write, or side erase widths affects how close we can write data tracks.

5.5.1 Side fringing field

The air-bearing surface (ABS) view of a thin-film inductive head is shown in Fig. 5.24. Finite element or boundary element modeling is very suitable for calculating 3D field components. Analytical expressions of head side field can also be derived for certain head geometry.[43,44] When the head track width W is much greater than the gap length g so that the effect of one head side on the other can be neglected, an analytical expression of side fringing field was derived by Hughes and Bloomberg:[43]

$$H_x = \frac{H_g}{\pi}\left[\tan^{-1}\left(\frac{R_-^{2/3} - R_-^{-2/3}}{2\sin(2\theta/3)}\right) - \tan^{-1}\left(\frac{R_+^{2/3} - R_+^{-2/3}}{2\sin(2\theta/3)}\right)\right], \quad (5.16a)$$

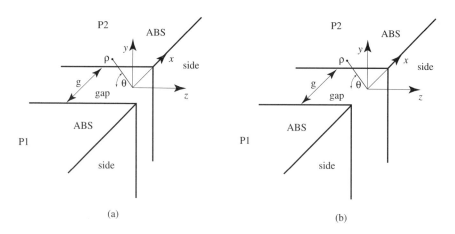

FIGURE 5.24. Thin-film heads (a) without trimming, and (b) with trimming.

$$H_z = \frac{2H_g}{\pi}\left\{\frac{1}{4}\ln\right.$$

$$\left[\frac{(1 + R_+^{1/3}\sin\theta/3)^2 + R_+^{2/3}\cos^2\theta/3}{(1 - R_+^{1/3}\sin\theta/3)^2 + R_+^{2/3}\cos^2\theta/3} \cdot \frac{(1 - R_-^{1/3}\sin\theta/3)^2 + R_-^{2/3}\cos^2\theta/3}{(1 + R_-^{1/3}\sin\theta/3)^2 + R_-^{2/3}\cos^2\theta/3}\right]$$

$$\left. - (R_+^{1/3} + R_+^{-1/3} - R_-^{1/3} - R_-^{-1/3})\frac{\sin\theta}{3}\right\},\qquad (5.16b)$$

$$H_y(x, \rho, \theta) = H_z\left(x, \rho, \frac{3\pi}{2} - \theta\right),\qquad (5.16c)$$

where

$$\rho = \sqrt{y^2 + z^2},\quad \theta = \pi/2 + \tan^{-1}(z/y),$$

$$R_\pm = \rho/[x \pm g/2 + \sqrt{\rho^2 + (x \pm g/2)^2}].$$

The above equations should give the side fringing field of the trimmed head (Fig. 5.24b) when $z > 0$. Generally speaking, the side fringing field does not drop to a negligible level until the point of interest has past the head edge by a few times of gap length.

5.5.2 Side read/write/erase

Based on the reciprocity principle, the presence of side fringing field indicates that the head will not only read the magnetic charges located at $z \leq 0$, but also $z \geq 0$. The latter is referred as the *side read*.

Extending the Wallace spacing loss (Chapter 2) to the z-direction, the Fourier transform with respect to x of the side read voltage signal from a magnetic charge at $z \gg y$ or g will decay exponentially:

$$V(k, z) \propto e^{-kz}.\qquad (5.17)$$

In other words, the side read voltage signal at a wavelength of λ will decay $-54.6z/\lambda$ dB. Therefore, side read is more serious for low-frequency signals than for high-frequency ones!

Since the write field on track is typically several times that of medium coercivity, the side fringing field can exceed medium coercivity, causing *side write* and *side erase* effects.[45] For thin-film media, the head field can generally write at the side to a distance defined by the $H_x = H_c$ contour. Beyond this point, the head field can generally erase the old data to a distance defined by the $H_x = S^*H_c$ contour, where S^* and H_c are medium coercive squareness and coercivity, respectively. The area between the two contours is often called the *side erase band*, which is beneficial to the off-track capability of the head (Chapter 14). The area from the head physical edge to the first contour is called the *side write band.* The written data track width on thin-film media is generally considered to be the head physical track width plus the two side write bands. Consequently, side write bands add to the track pitch required in disk drives. The maximum side write band (one side) can be obtained by letting

$$H_x(x = 0, y = 0, z) = H_c, \tag{5.18}$$

which gives the side write distance at the ABS:

$$z_0 \approx 0.38g/\tan(\pi H_c/2H_g) \propto g. \tag{5.19}$$

As shown in Fig. 5.24a, the bottom pole is often much wider than the top pole in thin film heads. In this case, the side write distance is nearly double of that in trimmed head as shown in Fig. 5.24b. However, trimming the bottom pole P1 (using P2 as a mask) down to a depth of ~$2g$ would reduce the side write distance like that of a completely trimmed head.[46] Alternatively, focused ion beam etching at the ABS can trim both P1 and P2 pole tips to a desired narrow trackwidth, with an etch depth of ~1 μm to reduce side write/erase.

5.5.3 Track profile

The written track width, read track width, and erase band can be obtained by measuring track profiles with a magnetic spin stand. This involves taking track-averaged amplitude (TAA), i.e., the isolated pulse amplitude averaged over a whole data track, as a function of the head center offset from the written data track.[47] If a test track is written by the head on an AC-erased background, we obtain a *full track profile.* As shown schematically in Fig. 5.25a, the written width w_w and the read width w_r can be easily calculated from DE and CF. Typically $w_w > w_r$ for a full track profile, so the written width is equal to the full track profile width at the half-amplitude:

$$w_w = (DE + CF)/2. \tag{5.20}$$

If the test track is side-erased to a width much narrower than the head physical track width, the track profile obtained is called the *microtrack profile,* which is essentially the read head sensitivity function across the track. A schematic microtrack profile is shown in Fig. 5.25b. The microtrack width at half-amplitude is generally equal to the *read track width:*

$$w_r = (GJ + HI)/2. \tag{5.21}$$

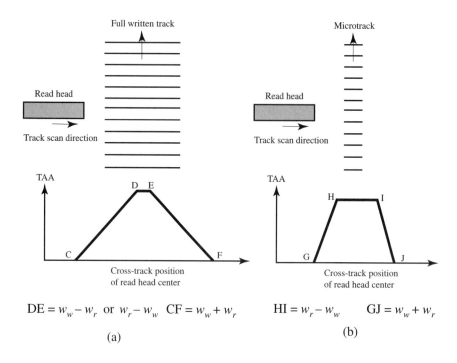

$$DE = w_w - w_r \text{ or } w_r - w_w \quad CF = w_w + w_r$$

$$HI = w_r - w_w \qquad GJ = w_w + w_r$$

(a)   (b)

FIGURE 5.25. (a) Full-track profile. (b) Microtrack profile.

To measure the erase band of a thin film head, we can use it to overwrite a wider data track. The track profile across the resulted data tracks is shown in Fig 5.26. The central track and side tracks have different recording frequencies, so their readback signals can be captured simultaneously by a spectrum analyzer. If the erase band does not exist, the head should start to read the maximum signal from the left side track (B) and at the same time start to read the signal from the central track (C). The delay between C and B is equal to the erase band e. The erase band can also be directly observed by magnetic imaging tools such as magnetic force microscope (MFM).

5.6 HIGH-SATURATION INDUCTIVE HEAD MATERIALS

High-saturation soft magnetic materials are necessary for thin-film write heads to write high-coercivity media, thus achieving high linear densities. In addition, such materials are very beneficial to attaining high data rates.

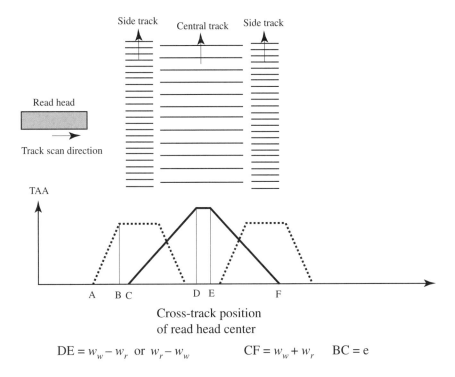

$$DE = w_w - w_r \ \text{or} \ w_r - w_w \qquad CF = w_w + w_r \qquad BC = e$$

FIGURE 5.26. Schematic diagrams of side and central tracks written by a thin-film head and predicted track profile. Note that the original wide track can be written by the same head writing two tracks side by side, as long as their recording densities are the same.

Therefore, there has been a long history of searching and researching high-saturation magnetic materials, or high-moment magnetic materials, with a saturation magnetization well above that of $Ni_{80}Fe_{20}$ permalloy (1 T or 10 KG).

5.6.1 Requirements of inductive head materials

Unfortunately, soft magnetic materials must satisfy many requirements in addition to high saturation, to be useful for thin-film inductive read/write heads. These requirements include the following:

   1. The head materials for magnetic poles must be magnetically very soft, typically with coercivities less than ~1 Oe. The low coercivity is

necessary not only to minimize the heat generation (hysteresis loss) and the corresponding thermal noise but also to readily induce a uniaxial anisotropy in the film plane. Such uniaxial anisotropy is very important to the head operation, but often difficult to induce in the films with larger coercivities.

2. The soft magnetic materials must have a large permeability (>1000 at small drive fields before saturation) to achieve sufficiently large head efficiency. For high data rate heads, the large permeability must be maintained over a wide frequency range (hundreds of MHz). To reduce the permeability roll-off due to eddy current losses at high frequencies, highly resistive magnetic films or laminated magnetic materials with insulating spacers are preferred.

3. The thin-film heads are designed to conduct magnetic flux along the hard axis of the magnetic poles and orthogonal to the magnetization. Thus, the magnetic flux propagates mainly by magnetization rotation that is preferable to domain wall motion, for domain wall motion is noisy and tends to be severely damped at high frequencies. When magnetic films are patterned into pole pieces, the effective uniaxial anisotropy constant (excluding the demagnetizing field) is

$$K_{u,\text{eff}} = K_u + \frac{3}{2}\lambda_s(\sigma_e - \sigma_h), \tag{5.22}$$

where K_u is the as-deposited anisotropy constant, λ_s is the saturation magnetostriction constant, and $\sigma_e - \sigma_h$ is the difference in the stresses along the easy axis (EA) and hard axis (HA) (assuming that the stress is biaxial). To ensure that the magnetization of the main magnetic domains will be parallel to the as-deposited easy-axis, the effective anisotropy must be positive, i.e.,

$$K_u + \frac{3}{2}\lambda_s(\sigma_e - \sigma_h) > 0,$$

For example, if $K_u \approx 4 \times 10^3$ erg/cm^3, $\sigma_e - \sigma_h \approx 3 \times 10^9$ dynes/cm^2, then the requirement leads to

$$\lambda_s > -0.9 \times 10^{-6}.$$

However, the effective permeability of magnetic poles along the HA is equal to

$$\mu_{\text{eff}} = \frac{4\pi M_s}{H_{k,\text{eff}}} = \frac{2\pi M_s^2}{K_{u,\text{eff}}} \quad \text{(Gaussian)} \tag{5.23}$$

so K_u or λ_s should not be a large positive number in order to maintain a sufficiently large head efficiency. In addition, the net stress in head poles could be of opposite sign under certain circumstances, then the requirement for magnetostriction becomes:

$$\lambda_s < +0.9 \times 10^{-6}.$$

The rule of thumb is that the magnetostriction constant of a good read/write head material should be controlled within $\pm 1 \times 10^{-6}.$

4. Head materials with critical damping and fast switching speed are desirable for high data rates (cf. Chapter 16).

5. During inductive head fabrication processes, head materials are usually subjected to high-temperature procedures at ~250°C or higher. Furthermore, the thin-film heads are expected to operate for a long life at elevated temperatures due to self-heating. Therefore, it is necessary that the head materials have sufficient thermal stability in their properties.

6. The head materials in a finished slider are exposed to the ambient atmosphere at the ABS, and any degradation due to corrosion will shorten the life of the device. The head materials may also be exposed to corrosive species during head fabrication. A good head material is required to withstand certain accelerated corrosion tests.

7. During the mechanical polishing of ABS to define the throat height, the head pole tips are polished along with alumina undercoat, overcoat, and gap layer. Therefore, a good magnetic pole material must be mechanical hard enough to avoid head gap "smearing" and to reduce "pole tip recession" relative to the alumina ABS. The latter will add to the magnetic spacing between the head pole tips and the magnetic recording layer in the medium.

If inductive heads are used for both write and read, the properties of permalloy are often used as the reference to those of new high-saturation materials. However, *the requirements on coercivity, permeability, and magnetostriction for write-only head materials are less strict than for read/write head materials* because the head noises and signal amplitudes in read heads are of no concern in write-only heads. Ultimately the suitability of a high saturation material for write head applications should be established by performing electrical and magnetic testing of the fabricated heads at the wafer level, as well as write/read testing of the head-sliders.

5.6.2 Survey of new inductive head materials

The search and characterization of inductive head materials is among the ongoing research and development activities in academia and data storage industries. The properties of selected head pole materials are listed in Table 5.3. Electroplating is more cost effective, and electroplated materials

TABLE 5.3. Properties of Magnetic Pole Materials[a]

Materials	Deposition method	$4\pi M_s$ (KG)	H_c (Oe)	λ_s (Oe)	H_k (Oe)	ρ ($\mu\Omega \cdot$ cm)	Corrosion resistance[b]
$Ni_{80}Fe_{20}$	Plating	10	0.3	−1	2.5–4	20–25	
$Ni_{45}Fe_{55}$	Plating	16	0.4	20	9.5	48	Good
$Co_{88}Fe_{12}$	Plating	20	3		10–13		Poor
$Co_{65}Fe_{23}Ni_{12}$	Plating	20–21	1.2	1.8	30	21	Good
CoFeVB	Plating	18	8		12.5	34	Poor
CoZr-based amorphous	Sputtering	14	<1	<1	6–20	~120	Poor
FeXN[c]	Sputtering	18–20	<1	<2	4–10	~40	Good to poor
FeMO[c]	Sputtering	13	2		9	~410	Good

[a]Refs. 48–55.
[b]Good roughly means comparable with NiFe; poor means more corrosive.
[c]X = Al, Ta, Rh, Mo, Zr, Si, etc.; M = Hf, Zr, rare-earth metals.

are relatively easier to process and pattern, so electroplating is widely used in the data storage industry. In comparison, sputtering is more flexible with target material choices and easier to handle amorphous or laminated materials. Over the years, the industry has grown to be more receptive to sputtering head pole materials. It has recently been found that FeXN materials switch faster than $Ni_{80}Fe_{20}$, which has reinvigorated the research on high-saturation head materials.

References

1. IBM Publication: *Disk Storage Technology,* (IBM General Products Division, San Jose, CA 1980).

2. R. I. Potter, "Analytic expression for the fringe field of finite pole-tip length recording heads," *IEEE Trans. Magn., 11,* **80,** 1975.

3. W. K. Westmiji, "Studies of magnetic recording," *Philips Res. Rep.,* **8**(3), 148 & 161, 1953.

4. L. Elabd, "A study of the field around magnetic heads of finite length," *IEEE Trans. Audio,* **11,** 21, 1963.

5. D. A. Lindholm "Image fields for two-dimensional recording heads," *IEEE Trans. Magn.,* **13,** 1463, 1977.

6. T. J. Szczech, "Analytic expressions for field components of nonsymmetrical finite pole length magnetic head based on measurements on large-scale model," *IEEE Trans. Magn.,* **15,** 1317, 1979.

7. T. J. Szczech, D. M. Perry, and K. E. Palmquist, "Improved field equations for ring heads," *IEEE Trans. Magn.,* **19,** 1740, 1983.

8. G. A. Bertero, H. N. Bertram, and D. M. Barnett, "Fields and transforms for thin film heads," *IEEE Trans. Magn.,* **29,** 67, 1993.

9. A. Paton, "Analysis of the efficiency of the thin-film magnetic recording heads," *J. Appl. Phys.,* **42,** 5868, 1971. G. F. Hughes, "Thin film recording head efficiency and noise," *J. Appl. Phys.,* **54,** 4168, 1983.

10. R. E. Jones, Jr., "Analysis of the efficiency and inductance of multi-turn thin film magnetic recording heads," *IEEE Trans. Magn.,* **14,** 509, 1978.

11. C. D. Abbott, G. W. Brock, N. L. Robinson, Shelledy, and S. H. Smith, "Apparatus for batch-fabricating magnetic film heads and method therefor," US Patent 3,821,815, 1974.

12. K. Kawakami, M., M. Aihara, and H. Fukuoka, "Electrical detection of end point in polishing process of thin-film heads," *J. Appl. Phys.,* **61,** 4163, 1987.

13. T. C. Arnoldussen, "A modular transmission line/reluctance head model," *IEEE Trans. Magn.,* **24,** 2482, 1988.

14. W. P. Jayasekara, Inductive Write Heads Using High Moment Pole Materials, Ph.D. Dissertation, Dept. of Electrical and Computer Engineering, Carnegie Mellon University, 1998.

15. B. Corb, "High frequency head and playback model for thin-film recording heads," *IEEE Trans. Magn.*, **30,** 394, 1994.

16. E. P. Valstyn and H. Huang, "An extended dynamic transmission line model for thin film heads," *IEEE Trans. Magn.*, **29,** 3870, 1993.

17. E. R. Katz, "Finite element analysis of the vertical multi-turn thin film head," *IEEE Trans. Magn.*, **14,** 506, 1978; D. Shenton, *Adaptive Finite Elements for Magnetic Field Computation,* Ph.D. Dissertation, Dept. of Electrical and Computer Engineering, Carnegie Mellon University, 1987.

18. M. H. Lean and A. Wexler, "Accurate field computation with the boundary element method," *IEEE Trans. Magn.*, **18,** 331, 1982; M. H. Kryder and W.-Y. Lai, "Modeling of narrow track thin film write head fields," *IEEE Trans. Magn.*, **30,** 3873, 1994.

19. For example, S. W. Yuan, *Micromagnetics of Domains and Walls in Soft Ferromagnetic Materials,* Ph.D. dissertation, Department of Physics, University of California, San Diego, 1992.

20. L. T. Romankiw and P. Simon, "Batch fabrication of thin film magnetic recording heads: a literature review and process description for vertical single turn heads," *IEEE Trans. Magn.*, 11, **50,** 1975.

21. N. C. Anderson and R. E. Jones, Jr., "Substrate testing of film heads," *IEEE Trans. Magn.*, **17,** 2896, 1981.

22. R. F. Hoyt, "Head-disk Interface," in *Magnetic Disk Drive Technology,* edited by K. G. Ashar, (Piscataway, NJ: IEEE Press, 1997).

23. H. Hamilton, R. Anderson, and K. Goodson, "Contact perpendicular recording on rigid media," *IEEE Trans. Magn.*, **27,** 4921, 1991.

24. T. Q. Doan, "Proximity recording design considerations," *IEEE Trans. Magn.*, **33,** 903, 1997.

25. M. Yoshida, M. Sakai, K. Fukuda, N. Yamanaka, T. Koyanagi, and M. Matsuzaki, "Edge eliminated head," *IEEE Trans. Magn.*, **29,** 3837, 1993.

26. K. B. Klaassen and J. C. L van Peppen, "Noise in thin film inductive heads," *IEEE Trans. Magn.*, **28,** 2097, 1992.

27. K. B. Klaassen and J. C. L van Peppen, "Irreversible wall motion in inductive recording heads," *IEEE Trans. Magn.*, **25,** 3209, 1989; K. B. Klaassen and J. C. L van Peppen, "Magnetic instability of thin-film recording heads," *IEEE Trans. Magn.*, **29,** 375, 1994.

28. R. E. Jones, Jr., "Domain effects in the thin film head," *IEEE Trans. Magn.*, **15,** 1619, 1979.

29. P. Kasiraj and R. D. Holmes, "Effect of magnetic domain configuration on read-back amplitude variations in inductive thin film heads," *J. Appl. Phys.*, **69**(8), 5423, 1991.

30. K. B. Klaassen and J. C. L van Peppen, "Delayed relaxation in thin-film heads," *IEEE Trans. Magn.*, **25,** 3212, 1989; F. H. Liu, *High Frequency Magnetization Dynamics and Popcorn Noise in Thin Film Heads,* Ph.D. Dissertation, Department of Electrical and Computer Engineering, Carnegie Mellon University, 1994.

31. C. A. Bajorek, S. Krongelb, L. T. Romankiw, and D. A. Thompson, "An integrated magnetoresistive read, inductive write high density recording head," *AIP Conf. Proc., Magn. Magn. Mater.,* **24,** 548, 1974.

32. For example, D. E. Heim, K. E. Fontana, Jr., C. Tsang, V. S. Speriosu, B. A. Gurney, and M. L. Williams, "Design and operation of spin valve sensors," *IEEE Trans. Magn.,* **30,** 316, 1994.

33. C. D. Mee and E. D. Daniel, *Magnetic Recording, Vol II: Computer Data Storage,* (New York, NY: McGraw-Hill, 1988) p. 119.

34. C. Tsang, T. Lin, S. MacDonald, M. Pinarbasi, N. Robertson, H. Santini, M. Doerner, T. Reith, L. Vo, T. Diola, and P. Arnett, "5 Gb/in^2 recording demonstration with conventional AMR dual element heads and thin film disks," *IEEE Trans. Magn.,* **33,** 2866, 1997.

35. P. Ciureanu and H. Gavrila, *Magnetic Heads for Digital Recording,* (Amsterdam, Holland: Elsevier, 1990).

36. S. Chikazumi and S. H. Charap, *Physics of Magnetism,* (Malabar, FL: Krieger, 1964, pp. 321, 333).

37. K. B. Klaasen, R. G. Hirko, and J. T. Contreras, "High speed magnetic recording," *IEEE Trans. Magn.,* **34,** 1822, 1998.

38. R. Wood, M. Williams, and J. Hong, "Considerations for high data rate recording with thin-film heads," *IEEE Trans. Magn.,* **26,** 2954, 1990.

39. P. Thayamballi, "Modeling the effects of write field rise time on the recording properties in thin film media," *IEEE Trans. Magn.,* **32,** 61, 1996.

40. T. C. Arnoldsussen, C. Vo, M. Burleson, and J. G. Zhu, "A simple recording head write field rise time measurement," *IEEE Trans. Magn.,* **32,** 3521, 1996.

41. M. R. Freeman, A. Y. Elezzabi, and J. A. H. Stotz, "Current dependence of the magnetization rise time in thin film heads," *J. Appl. Phys.,* **81**(8), 4516, 1997.

42. M. L. Mallary, "Effective field gradient dependence on write field rise time," *IEEE Trans. Magn.,* **32,** 3527, 1996.

43. G. F. Hughes and D. S. Bloomberg, "Recording head side read/write effects," *IEEE Trans. Magn.,* **13,** 1457, 1977.

44. I. D. Mayergoyz and D. S. Bloomberg, "Analytical solutions for the three-dimensional side-fringing field of magnetic recording heads," *J. Appl. Phys.,* **63**(8), 3381, 1988.

45. T. C. Arnoldussen, L. L. Nunnelley, F. J. Martin, and R. P. Ferrier, "Side write/reading in magnetic recording," *J. Appl. Phys.,* **69**(8), 4718, 1991.

46. S. X. Wang and P. R. Webb, "Modeling of submicron trackwidth inductive write head designs," *IEEE Trans. Magn.,* **31,** 2687, 1995.

47. T. Lin, J. A. Christner, T. B. Mitchell, J. S. Gas, and F. K. George "Effects of current and frequency on write, read, and erase widths for thin-film inductive and magnetoresistive heads," *IEEE Trans. Magn.,* **25,** 710, 1989.

48. M. Jurisch, K. Rook, M. Nenderson, and D. Liu, "Alternative pole materials for MR tape heads," *IEEE Trans. Magn.,* **32,** 156, 1996.

49. N. Robertson, B. Hu, C. Tsang, "High performance write head using NiFe 45/55," *IEEE Trans. Magn.*, **33,** 2818, 1997.

50. J. W. Chang, P. C. Andricacos, B. and L. T. Romankiw, "Electrodeposition of soft CoFe alloys," *Proc. of the Symp. on Mangetic Materials, Process & Devices,* the Electrochemical Society, **90**(8), 1990, p. 361.

51. K. Ohashi, Y. Yosue, M. Saito, K. Yamada, M. Takai, K. Hayashi, and T. Osaka, "Newly developed inductive write head with electroplated CoNiFe film," *IEEE Trans. Magn.*, **34,** 1462, 1998.

52. H. Fujimori, N. S. Kazama, K. Hirose, J. Zhng, M. Morita, I. Sato, and H. Sugawara, "Magnetostriction of Co-base amorphous alloys and high frequency permeability in their sputtered thin films," *J. Appl. Phys.*, **55**(6), 1769, 1984.

53. M. H. Kryder, S. X. Wang, and K. Rook, "FeAlN/SiO_2 and FeAlN/Al_2O_3 multilayers for thin-film recording heads," *J. Appl. Phys.*, **73**(10), 6212, 1993; W. P. Jayasekara, S. X. Wang, and M. H. Kryder, "4 Gbit/in^2 inductive write heads using high moment FeAlN poles," *J. Appl. Phys.*, **79**(8), 1996, p. 5880.

54. L. Nguyentran, K. Sin, J. Hong, P. P. Pizzo, and S. X. Wang, "Corrosion resistance of low coercivity, high moment FeXN thin film head materials," *IEEE Trans. Magn.*, **33,** 2848, 1997.

55. Y. Hayakawa, A., H. Fujimori, and A. Inoue, "High resistive nanocrystalline Fe-M-O (M=Hf, Zr, rare-earth metals) soft magnetic films for high-frequency applications," *J. Appl. Phys.*, **81**(8), 3748, 1997.

CHAPTER 6

Magnetoresistive Heads

In the beginning of the 1990s the annual compound growth rate of areal densities in hard disk drives changed from ~30% to ~60%, largely thanks to the introduction of magnetoresistive (MR) read heads in drives. This revolution was first lead by IBM, and gradually followed by the entire data storage industry. Although inductive read head technology was extended to recording densities that were never thought possible, MR heads showed a clear advantage in read signal amplitude, and a demonstrated extensibility to much higher densities because of the discovery of giant magnetoresistance (GMR). It is now widely accepted that future growth in recording densities will be met by advanced MR heads and GMR heads.

This chapter examines the basic principle and designs of MR heads, with an emphasis on anisotropic MR (AMR) heads. The widely practiced biasing schemes to linearize the MR sensor and to eliminate Barkhausen noise are discussed. Small signal analysis of MR sensors will also be given. Reciprocity principle and transmission line model will be applied to MR heads. The practical engineering issues of MR heads such as side read asymmetry, readback nonlinearity, baseline shift, thermal asperity, and electrostatic discharge (ESD) sensitivity, and MR head fabrication process will be introduced. Since the GMR materials and head technology is still rapidly evolving, only a brief discussion is given. The chapter ends with an introduction to spin-dependent tunneling junction which may be used in the proposed tunneling MR (TMR) heads.

6.1 ANISOTROPIC MAGNETORESISTIVE (AMR) EFFECT

Due to spin-orbit coupling, some ferromagnetic materials display *anisotropic* resistivity, i.e., the resistivity is a function of the orientation between current and the magnetization.[1] Consider the configuration as shown in Fig. 6.1. According to Ohm's law, the electrical fields parallel and perpendicular to the magnetization are as follows:

$$E_\| = \rho_\| j_\|, \quad E_\perp = \rho_\perp j_\perp,$$

where

$$j_\| = j\cos\theta, \; j_\perp = j\sin\theta.$$

The electrical field $\vec{E}$ is *not* parallel to the current density $\vec{j}$. Its component along the current direction is

$$E_j = E_\| \cos\theta + E_\perp \sin\theta = \rho_\| j \cos^2\theta + \rho_\perp j \sin^2\theta.$$

The resistivity along the current direction is

$$\begin{aligned}
\rho_j \equiv \frac{E_j}{j} &= \rho_\| \cos^2\theta + \rho_\perp \sin^2\theta \\
&= (\rho_\| - \rho_\perp)\cos^2\theta + \rho_\perp \qquad\qquad (6.1) \\
&= \rho_0 + \Delta\rho_{max}\cos^2\theta,
\end{aligned}$$

where

$$\rho_0 \equiv \rho_\perp, \; \Delta\rho_{max} \equiv \rho_\| - \rho_\perp.$$

The value of $\Delta\rho_{max}/\rho_0$ is often called the magnetoresistance (MR) ratio. In general, the resistivity of an anisotropic MR material will vary according to a cosine square function if we rotate the magnetization with respect to the current direction or rotate the current with respect to the magnetization direction.

The MR ratio of a ferromagnetic film depends on film thickness and deposition method. The most commonly used MR sensor material today is the permalloy at a composition near 81% Ni and 19% Fe. It has near-zero magnetostriction, low crystalline anisotropy, and high-saturation magnetization, thus, it has a very high intrinsic permeability of >2000.

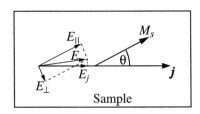

FIGURE 6.1.   Electrical resistance anisotropy between the parallel and normal directions of magnetization.

The electrical and magnetic properties of permalloy, like many other magnetic materials, are dependent upon the film thickness and deposition method. For ion beam sputtered $Ni_{81}Fe_{19}$ alloy,[2] the typical MR ratios at selected thicknesses are listed in Table 6.1. The typical resistivity is 20–25 $\mu\Omega\cdot cm$, and it increases at very small thicknesses when surface scattering and grain size become important factors in electrical conduction. It is reported that ion beam deposited permalloy films have somewhat higher MR ratios than conventionally sputtered ones, particularly when the films are thinner than 200 Å.

6.2 MAGNETORESISTIVE READ HEAD AND BIAS SCHEMES

The concept of the MR head was first proposed by Hunt in 1971.[3] A schematic of magnetoresistive (MR) head and its response to applied field (transfer curve) are shown in Fig. 6.2. The MR head is a magnetic flux-sensing device that responds to the vertical field above the magnetic medium averaged over the MR element height, D. If the sense current, I_s, is kept constant, then the change in resistance due to the vertical applied field results in a change in voltage across the element.

Note that the slope of the transfer curve at $H = 0$ is zero, namely, the MR element is not sensitive to an applied field. Therefore, a bias field of H_b is required for the MR element to operate at the maximum slope of the transfer curve. Since

$$\rho = \rho_0 + \Delta\rho_{max}\cos^2\theta,$$

$$\frac{\partial\rho}{\partial\theta} = -\Delta\rho_{max}\sin(2\theta),$$

the bias angle should be $\theta_0 = 45°$ for the slope to be maximum. The required bias field should be perpendicular to the easy axis and sense current, so this is called *transverse biasing*. The skirts in the transfer curve

TABLE 6.1. Typical MR Ratios, Sheet Resistance, and Resistivity of Permalloy Deposited on a Ta Seed Layer By Ion Beam Sputtering[a]

Thicknness (Å)	250	150	100	50
MR ratio $\Delta\rho_{max}/\rho_0$ (%)	2.6	2.3	2.0	1.2
Sheet resistance R_s (Ω)	8.45	15.2	23.5	62.5
Resistivity $\rho_0 = R_s t$ ($\mu\Omega\cdot cm$)	21	23	24	31

[a]Ta sheet resistance is ~450 Ω.

are due to nonuniform demagnetization field in the MR element and easy axis (EA) dispersion.

Compared with inductive read heads, one of the major characteristics of MR read heads is that it reproduces recording signals which are fundamentally independent of head-medium relative speed. Furthermore, an MR element in Fig. 6.2 will sense not only the transition underneath the element, but also the neighboring transitions. Therefore, the MR element need to be shielded by high-permeability shields. Here we will consider

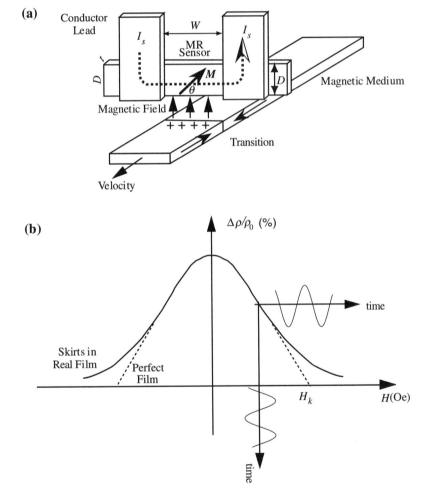

FIGURE 6.2.   Schematic of MR head (a) and its transfer curve (b).

unshielded MR heads as shown in Fig. 6.2. The *shielded* MR heads will be discussed later in Section 6.3.

The readback voltage of an MR head depends on the magnetic field in the MR element from a recording medium, which rotates the magnetization of the MR element. If the magnetic element is biased at an angle, θ_0, then the readback voltage is

$$V(\bar{x}) = I \, \Delta R = JW \, \Delta\rho_{\max}(\cos^2\theta - \cos^2\theta_0), \tag{6.2}$$

where W is the length of the MR element (trackwidth) defined by the conductor leads, $\bar{x} = vt$ is the relative head-medium position, and J is the current density, assumed to be uniform across the MR element. Note that the physical width of the MR element must be equal to or greater than its trackwidth. Since θ and θ_0 are normally not uniform across the MR element, the readback voltage can be approximately represented by an average $\langle \, \rangle$ over the film plane (y, z):

$$V(\bar{x}) = I \, \Delta R = JW \, \Delta\rho_{\max} \left(\langle\cos^2\theta\rangle - \langle\cos^2\theta_0\rangle\right) \tag{6.3}$$

$$= JW \, \Delta\rho_{\max} \cdot \Delta\langle\cos^2\theta\rangle.$$

For example, consider an MR element with a track width of 1 μm and a thickness of 150 Å. The sense current density is $\sim 5 \times 10^7$ A/cm^2 (mainly limited by the Joule heating and electromigration of the MR element). If $\Delta\langle\cos^2\theta\rangle = 1/4$, (which is a half of the maximum value to maintain linearity), then a zero-to-peak readback voltage of ~ 575 μV is obtained.

The idea of the MR head is simple and elegant, yet the design and mass production of MR heads require sophisticated engineering work and micromagnetic (or finite element) modeling. For pedagogical reasons, we will focus on analytical analyses of MR heads in the following sections. *Quiescent analysis* in the absence of medium field signal is carried out to assess if the MR element is biased properly. *Dynamic analysis* in the presence of the medium field signal is necessary to evaluate the MR head response and performance. In this regard, *small-field signal analysis* near the bias point (as determined by the quiescent analysis) is particularly useful to obtain the field sensitivity of MR heads. Neal Bertram has used linear signal analysis to calculate MR head readback signal based on a proper extension to the small field signal analysis.[4]

The sensitivity of the MR head readback voltage to a small field signal H_{sig} is

$$\frac{\partial V(\bar{x})}{\partial H_{sig}} = -2JW \, \Delta\rho_{\max} \cos\theta_0 \, \sin\theta_0 \frac{\partial\theta}{\partial H_{sig}} \tag{6.4}$$

$$= -2JW \, \Delta\rho_{\max} \sin\theta_0 \frac{\partial(\sin\theta)}{\partial H_{sig}}.$$

Then, the average small-signal sensitivity over the MR element is

$$\left\langle \frac{\partial V(\overline{x})}{\partial H_{sig}} \right\rangle = -2JW\,\Delta\rho_{max} \left\langle \sin\theta_0 \frac{\partial(\sin\theta)}{\partial H_{sig}} \right\rangle . \tag{6.5}$$

6.2.1 Transverse bias of magnetoresistive heads

In this section we will analyze the transverse biasing schemes and read-back sensitivity of unshielded MR heads.

Magnetoresistive heads can be transversely biased by a permanent magnet or a soft adjacent layer (SAL). Either biasing scheme generally produces nonuniform bias fields, but in simple analytical models we can approximate that a fixed average bias field acts on an MR element. If the external transverse field is H_t and the exchange coupling energy can be neglected, then the magnetic energy per unit volume for a very thin MR element $(t \ll D \ll W)$ is

$$E = -\mu_0 M_s H_t \sin\theta + K_u \sin^2\theta + \frac{1}{2}\mu_0 \vec{M}_s \cdot \vec{H}_d \tag{6.6}$$

$$= \mu_0 M_s \left(-H_t \sin\theta + \frac{1}{2} H_k \sin^2\theta + \frac{1}{2} N_{dy} M_s \sin^2\theta \right),$$

where the first term is the magnetic energy due to the external field, the second term is the intrinsic anisotropy energy, and the last term is the demagnetizing energy. The orientation of the magnetization is determined by

$$\partial E/\partial\theta = 0\,,$$

which yields

$$\sin\theta = \frac{H_t}{H_k + N_{dy} M_s}\,. \tag{6.7}$$

If the external transverse field can be represented as a uniform signal field in addition to a bias field over the MR element:

$$H_t = H_{sig} + H_{bias}\,,$$

then the magnetization orientation is $(H_{dy} \equiv N_{dy} M_s)$

$$\sin\theta = \frac{H_{sig} + H_{bias}}{H_k + N_{dy} M_s} = \frac{H_{sig} + H_{bias}}{H_k + H_{dy}}\,. \tag{6.8}$$

Transverse Demagnetizing Factor. Assuming that the MR element is very long in the z-direction (across track), the demagnetizing factor along the x-direction (down track, perpendicular to the thickness) is[5]

$$N_{dx} = \frac{1}{\pi}\left[\tan^{-1}\left(\frac{D-2y}{t}\right) + \tan^{-1}\left(\frac{D+2y}{t}\right)\right], \tag{6.9}$$

where D is the MR element height. Then the demagnetizing factor in the y-direction (height) is

$$\begin{aligned} N_{dy} &= 1 - N_{dx} \quad (\because N_{dz} \approx 0) \\ &= \frac{1}{\pi}\left[\tan^{-1}\left(\frac{t}{D-2y}\right) + \tan^{-1}\left(\frac{t}{D+2y}\right)\right]. \end{aligned} \tag{6.10}$$

At the center of MR element with a stripe height of $D = 1\ \mu$m and a thickness of 200 Å, $N_{dy} = \frac{2}{\pi}\tan^{-1}\left(\frac{t}{D}\right) \approx 0.013$, $H_{dy} \approx 130$ Oe.

The average demagnetizing factor along the MR element height (y-direction) can be calculated based on a micromagnetic model of the MR element.[6] The MR element under bias reveals a nonuniform magnetization distribution. Namely, the magnetization near the edge tends to align with the film surface as if a portion of the MR element with a distance of Δ (~75 nm for 200 Å NiFe) from each edge were "dead" with little response to an applied signal. Therefore, the average of N_{dy} is performed over $-D/2 + \Delta$ to $D/2 - \Delta$. Typically $t \ll D/2$, then

$$\begin{aligned} \langle N_{dy} \rangle &= \frac{1}{D}\int_{-D/2+\Delta}^{D/2-\Delta} N_{dy}\,dy \approx \frac{1}{\pi D}\int_{-D/2+\Delta}^{D/2-\Delta}\left(\frac{t}{D-2y}+\frac{t}{D+2y}\right)dy \\ &= \frac{t}{2\pi D}\left(\ln|y+D/2|-\ln|y-D/2|\right)\big|_{-D/2+\Delta}^{D/2-\Delta} = \frac{t}{\pi D}\ln\left(\frac{D-\Delta}{\Delta}\right). \end{aligned} \tag{6.11}$$

For example, if $t = 200$ Å, $D = 1\ \mu$m, $\Delta = 750$ Å, $M_s = 800$ kA/m for permalloy, then $\langle N_{dy} \rangle \approx 0.016$, and $\langle H_{dy} \rangle = \langle N_{dy} \rangle M_s \approx 12.8$ kA/m $= 160$ Oe. The average demagnetizing field is more than 20% greater than the demagnetizing field at the center of the MR element.

The demagnetizing field is much larger than the intrinsic anisotropy field of the permalloy film, so H_k can be neglected and Equation (6.8) becomes

$$\sin\theta \cong \frac{H_{sig} + H_{bias}}{H_{dy}}. \tag{6.12}$$

The resistivity of the MR element is

$$\rho = \rho_0 + \Delta\rho_{max}\left[1 - \left(\frac{H_{sig} + H_{bias}}{H_{dy}}\right)^2\right], \tag{6.13}$$

where

$$H_{bias} \approx H_{dy}/\sqrt{2} \text{ for } \theta \approx 45°.$$

The small field signal sensitivity of the MR element is

$$\frac{\partial V}{\partial H_{sig}} = JW\frac{\partial\rho}{\partial H_{sig}} = JW\frac{\sqrt{2}\,\Delta\rho_{max}}{H_{dy}}. \tag{6.14}$$

The average small-signal sensitivity over the MR element is

$$\left\langle\frac{\partial V}{\partial H_{sig}}\right\rangle \approx JW\frac{\sqrt{2}\,\Delta\rho_{max}}{\langle H_{dy}\rangle} = \frac{\sqrt{2}\,JW\,\Delta\rho_{max}}{\langle N_{dy}\rangle M_s}. \tag{6.15}$$

It is very important to recognize that the sensitivity of MR element is greatly influenced by the demagnetizing field, thus by the shape of the MR element.

SAL Transverse Bias. Another approach for transverse biasing is to use a SAL instead of a permanent magnet. This turns out to be the best method for single MR element read heads in engineering practices. As shown in Fig 6.3, the MR element and SAL are assumed to be uniformly magnetized, and they are also assumed to be electrically isolated from each other. The SAL experiences four sources of fields: a demagnetizing field (from its own magnetization), a signal field of H_{sig}, a field H_J due to the sense current, and a magnetostatic coupling field from the MR element. The last one is equal to

$$(1 - \alpha)H_{dy}^{MR} = (1 - \alpha)N_{dy}^{MR}M_s^{MR}\sin\theta_{MR}, \tag{6.16}$$

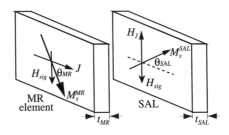

FIGURE 6.3. Schematic of MR element coupled with SAL.

where $(1 - \alpha)$ is the magnetostatic coupling coefficient, and α is the flux leakage factor between the MR element and SAL. From Equation (6.12), we obtain

$$\sin\theta_{SAL} = \frac{-H_{sig} + H_J + N_{dy}^{MR} M_s^{MR} (1 - \alpha)\sin\theta_{MR}}{N_{dy}^{SAL} M_s^{SAL}}, \tag{6.17}$$

where the minus sign of H_{sig} arises from H_{sig} reference direction being opposite to that of H_J. Likewise, we obtain a relation for the MR element (it is not subjected to a field due to current)

$$\sin\theta_{MR} = \frac{H_{sig} + N_{dy}^{SAL} M_s^{SAL} (1 - \alpha)\sin\theta_{SAL}}{N_{dy}^{MR} M_s^{MR}}. \tag{6.18}$$

In the absence of signal field, the MR element is supposed to be biased at $\theta_{MR} \approx 45°$, and the SAL should be saturated, i.e., $\theta_{SAL} = 90°$. If $\alpha = 0$, then Equation (6.18) leads to the following relation:

$$\frac{N_{dy}^{SAL} M_s^{SAL}}{N_{dy}^{MR} M_s^{MR}} \approx \frac{t_{SAL} M_s^{SAL}}{t_{MR} M_s^{MR}} \approx \sin 45° = \frac{1}{\sqrt{2}}. \tag{6.19}$$

This equation can be understood as *the magnetic flux along the y-direction is conserved between the MR element and SAL* since zero flux leakage is assumed. If permalloy is used for both MR element and SAL, then the SAL thickness should be ~0.707 times the MR element thickness.

If the SAL does not stay in saturation, then we can solve θ_{MR} from the coupled equations of θ_{MR} and θ_{SAL}, (6.17) and (6.18). Cancel θ_{SAL}, we obtain an equation of θ_{MR}:

$$\sin\theta_{MR} = \frac{H_{sig} + (1 - \alpha)\left[-H_{sig} + H_J + N_{dy}^{MR} M_s^{MR} (1 - \alpha)\sin\theta_{MR}\right]}{N_{dy}^{MR} M_s^{MR}}$$

$$= \frac{\alpha H_{sig} + (1 - \alpha)H_J + N_{dy}^{MR} M_s^{MR} (1 - \alpha)^2 \sin\theta_{MR}}{N_{dy}^{MR} M_s^{MR}}.$$

Note that $1 - (1 - \alpha)^2 \approx 2\alpha$ for $\alpha \ll 1$, then

$$\sin\theta_{MR} \approx \frac{H_{sig}}{2N_{dy}^{MR} M_s^{MR}} + \frac{H_J}{2\alpha N_{dy}^{MR} M_s^{MR}}. \tag{6.20}$$

Compared with the saturated SAL, the field sensitivity of the MR head with unsaturated SAL is reduced by a factor of 2 since the energy from the signal field is spent on rotating the magnetization in both MR element and SAL. This situation must be absolutely avoided in the design of the single element SAL-biased MR heads. This means that the sense current must be large enough so that the SAL stays saturated. The required condition is

$$\frac{H_J + N_{dy}^{MR} M_s^{MR} (1 - \alpha) \sin\theta_{MR}}{N_{dy}^{SAL} M_s^{SAL}} \geq 1 , \tag{6.21}$$

which, when MR element is biased at 45° leads to

$$H_J \geq \alpha N_{dy}^{SAL} M_s^{SAL} \approx \frac{2\alpha t_{SAL}}{\pi D_{SAL}} M_s^{SAL} . \tag{6.22}$$

According to Ampere's law, the maximum field generated by the sense current is

$$H_J \approx J t_{MR}/2 , \tag{6.23}$$

then

$$J \geq \frac{4\alpha t_{SAL}}{\pi D_{SAL} t_{MR}} M_s^{SAL} . \tag{6.24}$$

For example, if $\alpha \approx 0.2$, $t_{SAL}/t_{MR} \approx 0.7$, $D_{SAL} \approx 1$ μm, $M_s^{SAL} \approx 8 \times 10^5$ A/m, the minimum required sense current density is 1.4×10^{11} A/m^2 (1.4×10^7 A/cm^2), which is lower than the typical sense current density used in MR heads ($\sim 5 \times 10^7$ A/cm^2).

Dual-Stripe Biasing.   In another transverse biasing scheme called the *dual-stripe* MR head, two *identical* MR elements are closely spaced but electrically isolated. Transverse biasing is achieved through magnetostatic coupling between the two MR elements and sense currents, as shown in Fig. 6.4. Similar to the SAL MR heads, we have the following relations:

$$\sin\theta_{MR1} = \frac{H_{sig} + H_J + N_{dy}^{MR} M_s^{MR} (1 - \alpha) \sin\theta_{MR2}}{N_{dy}^{MR} M_s^{MR}} , \tag{6.25}$$

$$\sin\theta_{MR2} = \frac{-H_{sig} + H_J + N_{dy}^{MR} M_s^{MR} (1 - \alpha) \sin\theta_{MR1}}{N_{dy}^{MR} M_s^{MR}} .$$

If both the MR elements are biased at $\theta_{MR1} \approx \theta_{MR2} \approx 45°$, then the required condition is that

$$\frac{1}{\sqrt{2}} = \frac{H_J + N_{dy}^{MR} M_s^{MR} (1 - \alpha)/\sqrt{2}}{N_{dy}^{MR} M_s^{MR}} ,$$

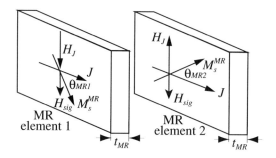

FIGURE 6.4.   Schematic of dual-stripe MR heads.

or

$$H_J = \alpha N_{dy}^{MR} M_s^{MR} / \sqrt{2}.$$ (6.26)

Similar to the derivation in SAL MR heads, the required sense current density is

$$J \approx \frac{4\alpha}{\sqrt{2}\,\pi D_{MR}} M_s^{MR}.$$ (6.27)

If $\alpha \approx 0.2$, $D_{MR} \approx 1\ \mu\text{m}$, $M_s^{MR} \approx 8 \times 10^5$ A/m, the required sense current density is about 1.4×10^{11} A/m² (1.4×10^7 A/cm²), which is the same as the minimum sense current density in the corresponding SAL MR heads shown earlier in Equation (6.24).

A major difference between the dual stripe head and the SAL head is that the former requires a certain level of sense current density. Excessive sense current density will "overbias" the two MR elements. In addition, two MR elements must be matched very well. Therefore, the design for dual stripe MR heads is more constrained than for SAL MR heads. On the other hand, the subtraction of the voltage drop of the two MR elements allows differential detection of magnetic field, which is advantageous in rejecting common mode signal and doubling field sensitivity.

6.2.2 Longitudinal bias of magnetoresistive heads

If the MR element is in a multidomain state, it will produce *Barkhausen noise* when subjected to an applied field. This is a major challenge in the manufacture of MR heads. To avoid Barkhausen noise, the MR head is required to be in a single domain state. Therefore, a longitudinal bias field along the EA (current direction) of the MR element is needed. There are two popular longitudinal bias schemes for MR heads: *exchange bias and hard bias.*

The *design criterion of longitudinal bias* includes: (1) The bias field is large enough to ensure a stable single-domain state; (2) The bias field is small enough not to "stiffen" magnetization response to the applied field; (3) The materials used must be corrosion resistant; (4) Ease of fabrication.

Exchange Bias.   Exchange bias utilizes the exchange coupling between a ferromagnet (MR element) and an antiferromagnet (or ferrimagnet) such as FeMn, NiMn, IrMn, NiO, and TbCo. When the antiferromagnet is properly deposited next to the ferromagnet, the hysteresis loop of the ferromagnet is shifted by H_{ex}, which is called the exchange bias field, as shown in Fig. 6.5. If the exchange bias field is large enough ($H_{ex} > H_c$), the ferromagnet will be in a saturated state, or single-domain state. For a MR element longitudinally biased by a *uniform exchange bias,* and transversely biased by a uniform field H_t, the magnetic energy per unit volume is

$$E = -\mu_0 M_s^{MR} H_t \sin\theta + K_u \sin^2\theta + \frac{1}{2}\mu_0 \vec{M}_s^{MR} \cdot \vec{H}_d - \mu_0 M_s^{MR} H_{ex} \cos\theta$$

$$(6.28)$$

$$= \mu_0 M_s^{MR} \left(-H_t \sin\theta - H_{ex} \cos\theta + \frac{1}{2}H_k \sin^2\theta + \frac{1}{2}N_{dy}^{MR} M_s^{MR} \sin^2\theta \right).$$

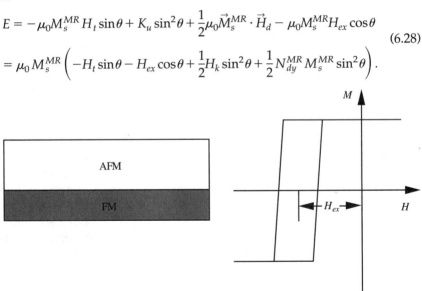

FIGURE 6.5.   Exchange-biased ferromagnet (left) and its hysteresis loop (right). The unidirectional anisotropy is called exchange-induced anisotropy. (AFM = antiferromagnet; FM = ferromagnet.)

where θ is the angle between the magnetization and the easy axis (longitudinal direction), and H_d is the demagnetizing field in the MR element. The angle θ can be found by minimizing E:

$$-H_t \cos\theta + H_{ex} \sin\theta + \frac{1}{2} N_{dy}^{MR} M_s^{MR} \sin(2\theta) = 0, \qquad (6.29)$$

where the intrinsic induced anisotropy H_k of MR element is neglected. Equation (6.29) will not yield a simple analytical expression for θ. To bias the MR element at ~45°, the required transverse bias field is

$$H_{t,bias} = H_{ex} + N_{dy}^{MR} M_s^{MR}/\sqrt{2}. \qquad (6.30)$$

The corresponding small-signal sensitivity near $\theta = \theta_0$ of the exchange-biased MR element can be found by taking the derivative of Equation (6.29) with respect to H_{sig}:

$$-\cos\theta_0 + H_{t,bias} \sin\theta_0 \frac{\partial\theta}{\partial H_{sig}} + H_{ex} \cos\theta_0 \frac{\partial\theta}{\partial H_{sig}}$$

$$+ N_{dy}^{MR} M_s^{MR} \cos(2\theta_0) \frac{\partial\theta}{\partial H_{sig}} = 0,$$

$$\frac{\partial\theta}{\partial H_{sig}} = \frac{\cos\theta_0}{H_{t,bias} \sin\theta_0 + H_{ex} \cos\theta_0 + N_{dy}^{MR} M_s^{MR} \cos(2\theta_0)}.$$

If $\theta_0 = 45°$, then the small-field sensitivity is

$$\frac{\partial(\sin\theta)}{\partial H_{sig}} = \cos\theta_0 \frac{\partial\theta}{\partial H_{sig}} = \frac{1/\sqrt{2}}{H_{t,bias} + H_{ex}} = \frac{1}{2\sqrt{2}H_{ex} + N_{dy}^{MR} M_s^{MR}}. \qquad (6.31)$$

In contrast, in the absence of the exchange bias, the small-field sensitivity of the magnetization orientation is

$$\frac{\partial(\sin\theta)}{\partial H_{sig}} = \frac{1}{N_{dy}^{MR} M_s^{MR}}. \qquad (6.32)$$

It is apparent that the excessive H_{ex} significantly reduces the sensitivity of MR element. The exchange bias H_{ex} aligns the magnetization along the positive EA ($\theta = 0$), while the demagnetizing field $N_{dy}^{MR} M_s^{MR}$ aligns the magnetization along the either direction of EA ($\theta = 0$ or π). This seems not drastically different, but uniform exchange bias "stiffens" MR magnetization response more than shape anisotropy by a factor of $2\sqrt{2}$. A clever scheme has been devised to bias MR element by *exchange tabs* at the two ends of the MR element, as shown in Fig. 6.6, which minimizes the loss of sensitivity.[7] The tail regions of the MR element under the exchange tabs are pinned to the right, and the middle region (the active read region) defined by the leads is forced into a single domain state due to magnetostatic and exchange interactions. In the absence of the transverse bias, the magnetization of the whole MR element is along the longitudinal direction in a single-domain state. When the magnetization in the middle region is biased at ~45°, magnetic charges are created in the middle/tail region boundaries, producing a demagnetizing field pointing to the right. This demagnetizing field minus the demagnetizing field due to the magnetic charges at the outer boundaries of the tail regions is the longitudinal bias field H_l, which affects the small field sensitivity of the MR element as the exchange field H_{ex} does in the case of uniform exchange bias. However, a proper design of the exchange tabs can make H_l much smaller than H_{ex}, thus enhancing the sensitivity of the MR element.

Hard Bias.   A popular longitudinal bias scheme is the *contiguous or abutted junction hard bias*, as shown in Fig. 6.7, which utilizes two hard magnets abutting the MR element along the longitudinal direction.[8] The hard magnets consist of a hard magnetic layer such as CoPtCr and appropriate

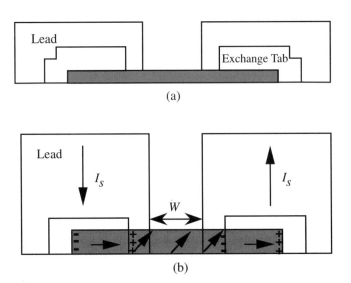

(a)

(b)

FIGURE 6.6.   Cross section (a) and side view (b) of longitudinal bias with exchange tabs, including leads. The lead and exchange tab could be made flush to the middle active read region of the MR element.

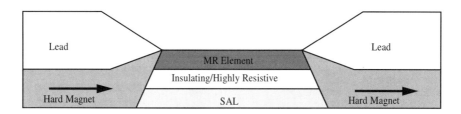

FIGURE 6.7. Cross-sectional view of abutted junction hard bias.

under- and/or overlayer for desirable magnetic and electrical properties. The fabrication of the MR head with hard bias will be described later in Section 6.5. The hard magnets are electrically connected to the MR element. The magnetostatic field generated by the hard magnets in the MR element serves as the longitudinal bias field H_l. Similar to exchange-biased AMR head, the small field sensitivity of the abutted junction AMR head is

$$\frac{\partial(\sin\theta)}{\partial H_{sig}} = \frac{1}{2\sqrt{2}H_l + N_{dy}^{MR} M_s^{MR}}. \tag{6.33}$$

The thickness and the magnetization of the hard magnetic layer are chosen such that H_l is sufficient to maintain the single-domain state of the MR element, but not excessive to sacrifice its sensitivity. This design is also advantageous in that the hard magnets will not respond to medium field, so side reading of MR element is greatly reduced (Section 6.4.1).

An alternative longitudinal bias scheme is the *overlay hard bias*,[9] as shown in Fig. 6.8, where the trackwidth is again defined by the leads, similar to the case of exchange tab bias. The tail regions of the MR element are exchange coupled to the hard magnet such that they are pinned to the right. The longitudinal bias field in the middle active read region can be adjusted by controlling the thickness and the magnetization of the hard magnetic overlayer. A variation of this bias scheme is to have a thin

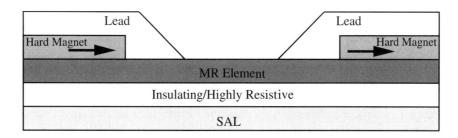

FIGURE 6.8. Cross-sectional view of overlay hard bias.

nonmagnetic spacer between the hard magnetic over-layer and the MR element so that the hard magnets and the MR element are magnetostatically coupled rather than exchange-coupled.[10] The spacer layer helps to attain the large coercivity in the hard magnet, and allows better control of the longitudinal bias field.

6.3 MAGNETORESISTIVE HEAD READBACK ANALYSIS

In addition to proper longitudinal/transverse biasing schemes and a high sensitivity, MR heads must have high-permeability shields to realize a good linear resolution of closely spaced magnetic transitions. This can be simply illustrated in Fig. 6.9 where transition #2 is being detected by the MR element. An unshielded MR will pick up magnetic flux from all three transitions. Since the flux from transition #1 & #3 oppose that of transition #2, the MR signal will be greatly weakened. On the other hand, the shielded MR element will pick up only the flux of transition #2 unless the spacing between the transitions is very close compared with the shield-to-shield gap. Therefore, the shielded MR head has a much higher linear resolution.

As mentioned before, MR heads are *flux-sensing* devices while inductive heads are *flux-derivative-sensing* devices. The former would yield a readback voltage even if there is no relative motion between MR head and magnetic transition. To analyze the spatial variation of the readback voltage, which is completely neglected in *quiescent analysis,* we need to find out how a *shielded* MR head responds to a spatially varying signal

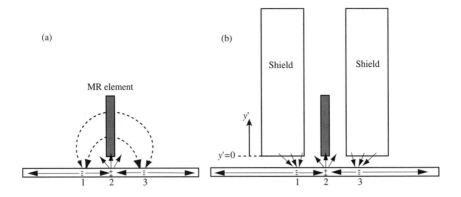

FIGURE 6.9.   Unshielded (a) and shielded (b) MR heads.

flux from a magnetic medium. As shown in Section 6.2, the voltage of an MR element under external field is

$$V_{MR}(\bar{x}) = I\Delta R = JW\Delta\rho_{max}\langle\cos^2\theta - \cos^2\theta_0\rangle$$
$$= -JW\Delta\rho_{max}\langle\sin^2\theta - \sin^2\theta_0\rangle.$$
(6.3)

where $\bar{x}$ is the relative position of magnetic medium vs. the MR head, and the average is over the whole active read region of the MR element. This is the formula one would use to calculate the readback voltage directly from the magnetization pattern in the recording medium by micromagnetic modeling or finite element modeling. However, here we do this analytically by using the *reciprocity principle* as we did in analyzing the inductive read process. Note that the $-$ sign in Equation (6.3) depends on whether the bias angle is $+45°$ or $-45°$. It is not critical in the context of single-pulse analysis.

6.3.1 Reciprocity principle for magnetoresistive heads

The instantaneous magnetization, the quiescent magnetization, and the magnetization change (i.e., signal magnetization) due to signal flux of the MR element are, respectively,

$$M_y = M_s\sin\theta, \quad M_{y0} = M_s\sin\theta_0, \quad M_{y,sig} = M_y - M_{y0}, \quad (6.34)$$

where M_s is the MR element saturation magnetization. Substituting Equation (6.34) into (6.3),

$$V_{MR}(\bar{x}) = -\frac{JW\Delta\rho_{max}}{M_s^2}\langle(M_y + M_{y0})(M_y - M_{y0})\rangle$$
$$\approx -\frac{2JW\Delta\rho_{max}}{M_s^2}\langle M_{y0}M_{y,sig}\rangle \quad (\text{if } M_{y,sig} << M_{y0}) \quad (6.35)$$
$$= -\frac{2JW\Delta\rho_{max}}{M_s^2}\frac{1}{D}\int_0^D M_{y0}(y')M_{y,sig}(\bar{x},y')\,dy',$$

where y' is the coordinate on the MR element. Because the MR element is very thin, we can assume that the signal flux is propagating along the y-direction into the MR element and is uniform in the x- and z-direction. Then it can be related to the signal magnetization as follows:

$$\Phi_{sig}(\bar{x},y') = \mu_0 M_{y,sig}(\bar{x},y')tW. \quad (6.36)$$

The readback voltage of the MR head becomes

$$V_{MR}(\bar{x}) = - \frac{2J \, \Delta\rho_{max}}{\mu_0 D t M_s^2} \int_0^D M_{y0}(y') \Phi_{sig}(\bar{x}, y') dy' . \qquad (6.37)$$

If the MR element is almost uniformly biased at $\sim 45°$, $M_{y0}(y') \approx M_s/\sqrt{2}$, then

$$V_{MR}(\bar{x}) \approx - \frac{\sqrt{2} J \, \Delta\rho_{max}}{\mu_0 D t M_s} \int_0^D \Phi_{sig}(\bar{x}, y') \, dy' . \qquad (6.38)$$

The signal flux $\Phi_{sig}(\bar{x}, y')$ into the MR element will decay from the air-bearing surface (ABS) ($y' = 0$) to the top of MR element ($y' = D$) due to flux leakage to the shields. The efficiency of the MR head can be defined as

$$E_{MR} \equiv \frac{\langle \Phi_{sig}(\bar{x}, y') \rangle}{\Phi_{sig}(\bar{x}, y' = 0)} = \frac{\int_0^D \Phi_{sig}(\bar{x}, y') \, dy'/D}{\Phi_{sig}(\bar{x})} . \qquad (6.39)$$

where $\Phi_{sig}(\bar{x}) \equiv \Phi_{sig}(\bar{x}, y' = 0)$ is the signal flux injected into the MR element *at the ABS*. The MR head efficiency can be evaluated by a transmission line model which will be described in Section 6.3.5. The expression of the MR head readback voltage can be simplified as

$$V_{MR}(\bar{x}) = - \frac{\sqrt{2} J \, \Delta\rho_{max}}{\mu_0 t M_s} E_{MR} \Phi_{sig}(\bar{x}) . \qquad (6.40)$$

Now the evaluation of the readback voltage boils down to that of the signal flux into the MR element at the ABS, which can be accomplished by the reciprocity principle.

The reciprocity principle as applied to a shielded MR head[11] can be demonstrated in Fig. 6.10a. The validity of the reciprocity principle does not depend on whether there is a real current in the coil or not. Imagine that a coil is wrapped around the MR element, then the MR head will generate a magnetic field if an *imaginary* current i is passed through the coil. Assume that the permeability of MR element and the shields is infinity and adopt the Karlqvist approximation similar to that of the inductive heads, then we can derive that the magnetic potential *above* the air-bearing surface (ABS) is given by Fig. 6.10b. Note that we have assumed that the MR element is located in the center of the shield-to-shield gap, so the deep gap fields in the two MR-element-to-shield gaps are the same in magnitude (H_g) but opposite in direction. The magnetic field, $\vec{H}^{MR}(x,y)$, generated by the MR head at the recording medium, can then be calculated from the surface magnetic potential as done in Chapter 2.

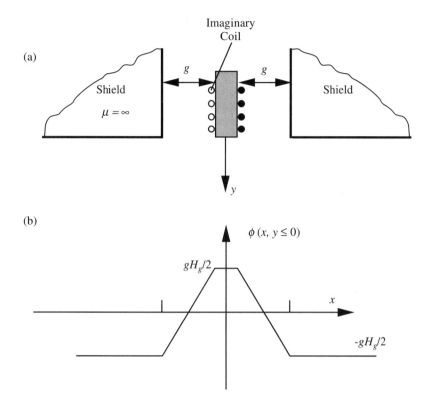

FIGURE 6.10. (a) Shielded MR head with imaginary coil. (b) Surface and deep gap magnetic potential under Karlqvist approximation.

Therefore, the signal flux into the MR element at the ABS can be expressed as

$$\Phi_{sig}(\bar{x}) = \mu_0 W \int_{-\infty}^{\infty} dx \int_{d}^{d+\delta} dy \, \frac{H_x^{MR}(x, y)}{i} M_x(x - \bar{x}, y), \qquad (6.41)$$

where i is the current in the imaginary coil, H_x^{MR} is the MR head longitudinal field generated by the imaginary coil, and $M_x(x,y)$ is the medium magnetization. This equation is called the *reciprocity formula for MR heads*. If the MR head magnetic field potential is $\phi_{MR}(x,y)$, then

$$\begin{aligned} \Phi_{sig}(\bar{x}) &= \frac{\mu_0 W}{i} \int_{-\infty}^{\infty} dx \int_{d}^{d+\delta} dy \left[-\frac{\partial \phi_{MR}(x, y)}{\partial x} \right] M_x(x - \bar{x}, y) \\ &= \frac{\mu_0 W}{i} \int_{-\infty}^{\infty} dx \int_{d}^{d+\delta} dy \, \phi_{MR}(x, y) \left[\frac{\partial M_x(x - \bar{x}, y)}{\partial x} \right], \end{aligned} \qquad (6.42)$$

where we did integration by parts, and used the fact that $M_x = 0$ at $x = \pm\infty$. In short, the reciprocity principle indicates that the signal flux picked by an MR head can be expressed either as the convolution of its head field and medium magnetization or as the convolution of its head magnetic potential and medium magnetic charge.

Before using the reciprocity formula to derive the MR head readback voltage, we first introduce Potter's "short-cut" approach by recognizing that the shielded MR head is essentially composed of two inductive Karlqvist heads with gap centers located at

$$x = \pm\,\frac{g+t}{2}.$$

It follows that the readback voltage of the imaginary coil is

$$V^*(\bar{x}) = -v\frac{d\Phi_{sig}}{dx} = V\left(\bar{x} - \frac{g+t}{2}\right) - V\left(\bar{x} + \frac{g+t}{2}\right), \qquad (6.43)$$

where $V(\bar{x})$ is the readback voltage of a single gap inductive Karlqvist head with a gap length of g. For an *arctangent* magnetization transition, the inductive readback voltage at thin-medium limit is

$$V(\bar{x}) = \frac{2\mu_0 v WH_g M_r \delta}{\pi i}\left(\tan^{-1}\frac{\bar{x}+g/2}{a+d} - \tan^{-1}\frac{\bar{x}-g/2}{a+d}\right)$$

$$\equiv C\left(\tan^{-1}\frac{\bar{x}+g/2}{a+d} - \tan^{-1}\frac{\bar{x}-g/2}{a+d}\right). \qquad (3.6)$$

Note that

$$\int\tan^{-1}(cx)\,dx = \frac{1}{c}\left[cx\tan^{-1}(cx) - \frac{1}{2}\ln(1+c^2x^2)\right],$$

and define a function

$$f_\Phi(x) \equiv x\tan^{-1}(x) - \frac{1}{2}\ln(1+x^2), \qquad (6.44)$$

then the signal flux injected into the MR element can be derived from Equation (6.43):

$$\Phi_{sig}(\bar{x}) = -\frac{1}{v}\int V^*(\bar{x})\,d\bar{x} = \frac{C}{v}(a+d)$$

$$\left[-f_\Phi\left(\frac{\bar{x}+t/2}{a+d}\right) - f_\Phi\left(\frac{\bar{x}-t/2}{a+d}\right) + f_\Phi\left(\frac{\bar{x}+g+t/2}{a+d}\right) + f_\Phi\left(\frac{\bar{x}-g-t/2}{a+d}\right)\right] \qquad (6.45)$$

$$\equiv \frac{2\mu_0 WM_r \delta}{\pi g}(a+d)f_V(\bar{x}),\ (\because H_g g/i = 1)$$

where $f_V(\bar{x})$ represents the expression in [ ]. Therefore, the MR head read-back voltage can be expressed as

$$V_{MR}(\bar{x}) = -\frac{2\sqrt{2}E_{MR}JW\,\Delta\rho_{max}M_r\delta(a + d)}{\pi t M_s}\,\frac{}{g}f_V(\bar{x}).\qquad(6.46)$$

Now we will apply the reciprocity formula Equation (6.42) to the MR head directly. For a thin medium ($\delta \ll d$) and a *step* transition at $x = 0$,

$$\frac{\partial M_x(x - \bar{x})}{\partial x} = 2M_r \cdot \delta(x - \bar{x}),\qquad(6.47)$$

$$\Phi_{sig}(\bar{x}) = \frac{2\mu_0 WM_r\delta}{i}\,\phi_{MR}(\bar{x}, d).\qquad(6.48)$$

An *arctangent* transition is equivalent to a step transition with an effective magnetic spacing of $d + a$, so the corresponding signal flux from an arctangent transition at $x = 0$ is

$$\Phi_{sig}(\bar{x}) = \frac{2\mu_0 WM_r\delta}{i}\,\phi_{MR}(\bar{x}, d + a).\qquad(6.49)$$

Under the Karlqvist approximation, the MR head field is the sum of the two inductive head fields:

$$H_x^{MR}(x, y) = H_x^{ind}\left(x - \frac{g + t}{2}, y\right) - H_x^{ind}\left(x + \frac{g + t}{2}, y\right),$$

where the negative sign arises from the fact that the head gap fields are opposite. Based on Equation (3.6) and the definition of f_V, we obtain

$$\begin{aligned}
\phi_{MR}(x, d + a) &= -\int H_x^{MR}(x, d + a)\,dx \\
&= -\int dx\,\frac{H_g}{\pi C}\left[V\left(x - \frac{g + t}{2}\right) - V\left(x + \frac{g + t}{2}\right)\right] \quad(6.50) \\
&= \frac{H_g}{\pi}(a + d)f_V(x).
\end{aligned}$$

This leads to identical expressions for the signal flux and MR head read-back voltage in Equations (6.45) and (6.46).

To see what the MR head readback pulse looks like, let us examine an example. When $a = 15$ nm, $g = 50$ nm, $d = 25$ nm, and $t = 10$ nm, the isolated pulse is as plotted in Fig. 6.11 (solid line). Note that its shape is not the same as a Lorentzian pulse (dashed line), although they do look similar. The dashed line is plotted using the small gap limit readback pulse expression derived in Section 6.3.2 [Equation (6.55)].

6.3.2 Small gap and thin medium limit

The readback voltage of the inductive head can be rewritten as

$$V(\bar{x}) = C \tan^{-1}\left[\frac{g(a+d)}{(a+d)^2 - (g/2)^2 + \bar{x}^2}\right]. \tag{3.6}$$

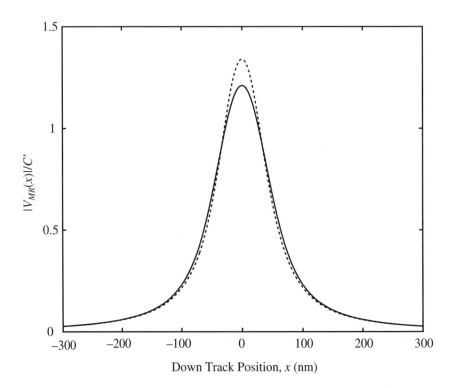

FIGURE 6.11. Isolated pulses of MR head, where the normalization constant is $C' = 2\sqrt{2}E_{MR}JW\Delta\rho_{max}M_r\delta/\pi tM_s$. The dashed line represents small gap read pulse.

Under the small gap limit ($g + t \ll a + d$), we can approximate the following Taylor expansions up to the first-order term of $g + t$:

$$V\left(\bar{x} - \frac{g+t}{2}\right) \approx V(\bar{x}) + \frac{dV(\bar{x})}{d\bar{x}}\left(-\frac{g+t}{2}\right),$$

$$V\left(\bar{x} + \frac{g+t}{2}\right) \approx V(\bar{x}) + \frac{dV(\bar{x})}{d\bar{x}}\left(+\frac{g+t}{2}\right).$$

Therefore,

$$V\left(\bar{x} - \frac{g+t}{2}\right) - V\left(\bar{x} + \frac{g+t}{2}\right) \approx -\frac{dV(\bar{x})}{d\bar{x}}(g+t).$$

This leads to

$$-\frac{dV(\bar{x})}{d\bar{x}}(g+t) = -v \cdot \frac{d\Phi_{sig}(\bar{x})}{d\bar{x}},$$

$$\Phi_{sig}(\bar{x}) = \frac{g+t}{v}V(\bar{x}). \tag{6.51}$$

For the MR head under the Karlqvist approximation, $gH_g/i = 1$, then

$$\Phi_{sig}(\bar{x}) = \frac{2\mu_0 WM_r\delta}{\pi}\frac{g+t}{g}\left(\tan^{-1}\frac{\bar{x}+g/2}{a+d} - \tan^{-1}\frac{\bar{x}-g/2}{a+d}\right). \tag{6.52}$$

The readback voltage of the MR head biased at ~45° becomes

$$V_{MR}(\bar{x}) = -\frac{2\sqrt{2}E_{MR}JW\Delta\rho_{max}M_r\delta}{\pi t M_s}\frac{g+t}{g}$$

$$\left(\tan^{-1}\frac{\bar{x}+g/2}{a+d} - \tan^{-1}\frac{\bar{x}-g/2}{a+d}\right). \tag{6.53}$$

This is a negative-going isolated pulse, and its shape is the same as that of an isolated pulse of an inductive head with a gap of g. The negative sign arises if the MR element transverse bias and the signal flux has the same direction. In this case, the signal flux into the MR element will rotate the magnetization further away from the current direction, so the resistance and voltage drops. On the contrary, if the signal flux opposes the MR element transverse bias, it will drive the resistance and voltage higher, resulting in a positive-going isolated pulse.

It follows that the half-amplitude pulse width of the MR readback voltage from an arctangent transition is

$$PW_{50} \approx \sqrt{g^2 + 4(a+d)^2}. \tag{6.54}$$

In other words, under the small gap and thin medium limit, the shielded MR head with a shield-to-shield spacing of $2g + t$ has a pulse width like an inductive Karlqvist head with a gap length equal to ~50% of $2g + t$.

In the extreme case of zero gap limit, $g \sim 0$, the MR head read pulse becomes

$$V_{MR}(\bar{x}) = -\frac{2\sqrt{2}E_{MR}JW\Delta\rho_{max}M_r\delta}{\pi t M_s}(g + t)\frac{(a + d)}{\bar{x}^2 + (a + d)^2}. \quad (6.55)$$

This is a Lorentzian pulse, and the pulse width is

$$PW_{50} = 2(a + d).$$

Under more general circumstances, it can be shown by a conformal mapping method[12] that the MR head has a pulse width like an inductive Karlqvist head with a gap length equal to ~60% of $2g + t$:

$$PW_{50} \approx \sqrt{[0.6(2g + t)]^2 + 4(a + d)^2}. \quad (6.56)$$

Similar to the case of inductive read heads, one must design small magnetic spacing, small transition parameter, and small gap length, *simultaneously*, to achieve a narrow readback voltage pulse.

6.3.3 Advantage of magnetoresistive over inductive read head

When a MR head reads an arctangent transition written in a thin magnetic medium, the peak of the corresponding isolated pulse under small gap limit is

$$V_{MR}^{peak}(\bar{x}) - \frac{4\sqrt{2}E_{MR}JW\Delta\rho_{max}M_r\delta}{\pi t M_s}\frac{g + t}{g}\tan^{-1}\frac{g/2}{a + d}. \quad (6.57)$$

As a comparison, assume that an inductive head has an efficiency of E_{ind}, a gap length of g, and N coil turns, then its peak readback voltage is

$$V_{ind}^{peak}(\bar{x}) = \frac{4\mu_0 v W N E_{ind}M_r\delta}{\pi g}\tan^{-1}\frac{g/2}{a + d}. \quad (3.7)$$

Thus

$$\frac{V_{MR}^{peak}}{V_{ind}^{peak}} = \frac{\sqrt{2}J\Delta\rho_{max}}{Nv\mu_0 M_s}\frac{E_{MR}}{E_{ind}}\frac{(M_r\delta)|_{MR}}{(M_r\delta)|_{ind}}\frac{g + t}{t}. \quad (6.58)$$

For a state-of-the-art inductive head, $g \sim 1000\text{Å}$, $v = 5$ m/s, $N \sim 50$. For a state-of-the-art MR head, $J \sim 5 \times 10^7$ A/cm^2, $\Delta\rho_{max} \sim 4 \times 10^{-7}$ Ω-cm, $t \sim 150\text{Å}$, $\mu_0 M_s = 1$ T. Since the MR element is much thinner than a regular pole in inductive heads, we may assume $E_{MR}/E_{ind} \sim 0.5$. To avoid saturation of the MR element, the $M_r\delta$ product of the magnetic medium must be limited, while the $M_r\delta$ product of the magnetic medium for inductive read head is not limited by the saturation consideration. For the purpose of this simple analysis, assume that

$$(M_r\delta)|_{ind} \approx 2(M_r\delta)|_{MR}.$$

Then

$$\frac{V_{MR}^{peak}}{V_{ind}^{peak}} = \frac{\sqrt{2} \times 5 \times 10^{11} \times 4 \times 10^{-9}}{50 \times 5 \times 1} \times 0.5 \times 0.5 \times \frac{1150}{150} \approx 19.$$

This example clearly demonstrates *a major advantage of the MR heads over the inductive heads* that the former can generate much greater signal amplitudes. This is why MR heads have superseded inductive heads as one scales down magnetic recording bit size!

If a square wave recording pattern (all 1s) is reproduced by an MR head, the voltage amplitude will drop with increasing recording density due to linear superposition of pulses of alternating polarity. Similar to inductive heads, if the isolated pulse can be approximated by a Lorentzian pulse, then the 50% roll-off linear density D_{50} of MR head is related to PW_{50} by

$$D_{50}PW_{50} \approx 1.39. \tag{3.18}$$

A more rigorous simulation without using Lorentzian pulse approximation gives that[13]

$$D_{50}PW_{50} \approx 1.46, \tag{6.59}$$

which is quite accurate if $d/g > 0.2$. Like inductive read heads, the linear density resolvable by MR heads is approximately inversely proportional to the isolated pulse width.

6.3.4 Fourier transform of magnetoresistive head read signal

Under the small gap limit, the Fourier transform of the MR head isolated pulse $V_{MR}(\overline{x})$ resembles that of $V_{ind}(\overline{x})$:

$$V_{MR}(k) \propto V_{ind}(k) \propto e^{-k(a+d)}\frac{1-e^{-k\delta}}{k\delta}\frac{\sin(kg/2)}{kg/2}. \tag{6.60}$$

Likewise, the fundamental harmonic of the MR head readback voltage waveform from square wave recording pattern at a bit length of B is

$$V_{MR,rms}\left(k_0 = \frac{\pi}{B}\right) \propto k_0 e^{-k_0(a+d)} \frac{1 - e^{-k_0\delta}}{k_0\delta} \frac{\sin (k_0 g/2)}{k_0 g/2}. \qquad (6.61)$$

The Fourier spectra are similar to those shown in Fig. 3.5, except that the normalization coefficients are different. In other words, MR heads also suffer from spacing loss, thickness loss, and gap nulls. Not surprisingly, the exact locations of the gap nulls experimentally observed are actually slightly shifted from $kg/2 = \pi, 2\pi, 3\pi, \ldots$ since the Karlqvist approximation needs to be modified near gap corners.

6.3.5 Transmission line model of magnetoresistive heads

The efficiency of the MR head was defined earlier in Section 6.3.1, but we did not elaborate how it could be calculated. Similar to inductive head, MR head efficiency can be calculated numerically by micromagnetic modeling and finite element modeling, or analytically by transmission line modeling. For pedagogical reasons, we choose to introduce a simple two-dimensional transmission line model for MR heads.[14]

Consider the MR head shown in Fig 6.12, where the MR element and the shields are assumed to be very long in the z-direction (across data track, i.e., normal to the page) and have a very large relative permeability. The

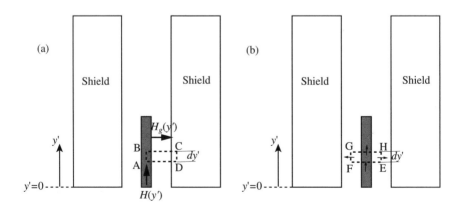

FIGURE 6.12. Transmission line model of shielded MR head: (a) applying Ampere's law, (b) applying Gauss's law, where the four arrows denote flux directions.

SAL layer is fully saturated, so it can be treated as air with a relative permeability of unity. Applying Ampere's law to loop ABCD in Fig. 6.12, we have

$$H(y') \cdot dy' + H_g(y' + dy') \cdot g - 0 \cdot dy' - H_g(y') \cdot g = 0, \qquad (6.62)$$

$$H(y') + \frac{dH_g(y')}{dy'} g = 0.$$

where $H(y')$ is the magnetic field in the MR element along the y'-direction,
$H_g(y')$ is the magnetic field in the gap along the x'-direction, and the
magnetic field in the shields are assumed to be negligible because of their
large cross-sectional area.

Applying Gauss's law of flux conservation to closed surface EFGH,
we obtain that

$$-\Phi(y') + \Phi(y' + dy') + 2\mu_0 H_g(y') \cdot W \, dy' = 0, \qquad (6.63)$$

$$\frac{d\Phi(y')}{dy'} = -2\mu_0 W H_g(y').$$

where $\Phi(y')$ is the magnetic flux at height y' in the MR element propagating along the y'-direction. If the relative permeability of the MR element
is μ, then

$$\Phi(y') = \mu\mu_0 H(y') \cdot tW. \qquad (6.64)$$

Combining Equations (6.62)–(6.64), we obtain the following wave equation which describes the transmission (or propagation) of magnetic flux
in MR element:

$$\frac{d^2\Phi(y')}{dy'^2} - \frac{\Phi(y')}{\lambda^2} = 0. \qquad (6.65)$$

where $\lambda \equiv \sqrt{\mu g t/2}$ is a characteristic length which describes how fast
magnetic flux leaks out of the MR element.

The magnetic flux signal injected into the MR element at the ABS is
$\Phi(\bar{x})$, and the magnetic flux signal at the top end of the MR element is
approximately zero because it is next to air with a large terminating
reluctance. Therefore, the boundary condition can be set as

$$\Phi(y' = 0) = \Phi(\bar{x}), \quad \Phi(y' = D) = 0. \qquad (6.66)$$

Equations (6.65) and (6.66) lead to a simple expression of magnetic flux
signal in MR element:

$$\Phi(\bar{x}, y') = \Phi(\bar{x}) \frac{\sinh[(D - y')/\lambda]}{\sinh(D/\lambda)}. \qquad (6.67)$$

where $0 \leq y' \leq D$. For example, if $t = 15$ nm, $g = 100$ nm, $\mu = 2000$, then $\lambda \approx 1.2$ μm. If $D \ll \lambda$, then the signal flux decays almost linearly along the y'-direction:

$$\Phi(\overline{x}, y') \approx \Phi(\overline{x}) \frac{D - y'}{D}. \tag{6.68}$$

The efficiency of the MR head can then be calculated as

$$E_{MR} \equiv \frac{\langle \Phi(\overline{x}, y') \rangle}{\Phi(\overline{x})} = \frac{\frac{1}{D} \int_0^D \Phi(\overline{x}, y')\,dy'}{\Phi(\overline{x})} = \frac{\lambda}{D} \frac{\cosh(D/\lambda) - 1}{\sinh(D/\lambda)}. \tag{6.69}$$

The MR head efficiency versus MR element height D is plotted in Fig. 6.13. To achieve a high read efficiency, it is very important to control the MR element height to $D \leq \lambda$ during the mechanical lapping process.

6.4 PRACTICAL ASPECTS OF MAGNETORESISTIVE HEADS

Although MR head provides great advantages over inductive read head, it has proven to be a new art that is more difficult to master than many

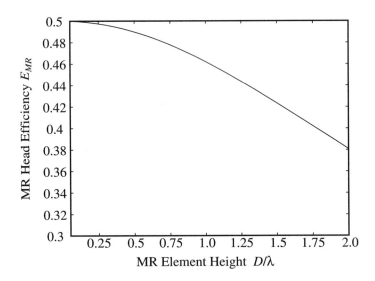

FIGURE 6.13.   Efficiency of shielded MR head vs. normalized MR element height.

people had anticipated. In addition to proper biasing the MR element and controlling MR sensitivity as discussed previously, there are many engineering issues associated with MR heads, some of which will be presented here.

6.4.1 Side reading asymmetry

Side reading effect and read trackwidth of a magnetic recording head is often characterized by microtrack profiles. A microtrack is prepared by symmetrically erasing both edges of a full written track until the remaining track is substantially narrower than the read head to be analyzed. The width of the microtrack is limited by the mechanical precision of positioning head across the track and by the read signal amplitude needed to acquire meaningful track profiles. As shown in Fig. 6.14, a microtrack profile is obtained by measuring the readback amplitude as the read head is stepped across the microtrack while the fly height and disk speed are kept constant.

It has been recognized that side reading and the microtrack profile of MR heads tends to be asymmetric, mainly due to the anisotropic nature of flux propagation in the MR element.[15-17] As shown in Fig. 6.14a, when a transition is next to the MR element at air bearing surface, the flux propagation paths (thick dashed lines) are perpendicular to the magnetization biased at 45°, because the flux along the magnetization is already saturated. According to the transmission line model of MR head, the flux almost decays linearly if the MR element height is much shorter than the characteristic length.

If the signal flux and the change in magnetization angle $\Delta\theta$ are small, then we can apply a simple small signal model to obtain the microtrack profile. The change in resistivity is

$$\Delta\rho \approx -\Delta\rho_{\max} \sin(2\theta_0)\, \Delta\theta, \qquad (6.1)$$

where θ_0 is the quiescent bias angle. The resulting change in resistance of the MR element is

$$
\begin{aligned}
\Delta R &= \langle\Delta\rho\rangle \frac{W}{Dt} = \frac{1}{DW} \iint_{\text{flux path}} \Delta\rho\, dy'\, dz' \cdot \frac{W}{Dt} \\
&= -\frac{\Delta\rho_{\max} \sin(2\theta_0)}{D^2 t} \iint_{\text{flux path}} \Delta\theta\, dy'\, dz' \qquad (6.70) \\
&\approx -\frac{\Delta\rho_{\max} \sin(2\theta_0)}{D^2 t} \frac{\Delta\theta_{\text{air}}}{2}\, wL(z).
\end{aligned}
$$

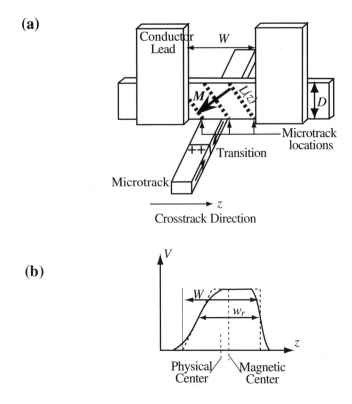

FIGURE 6.14. (a) Microtrack measurement, and anisotropic signal flux propagation in MR element. (b) Schematic microtrack profile and magnetic center.

where $\Delta\theta_{\text{air}}$ is the value of $\Delta\theta$ at the ABS edge of the MR element, $L(z)$ is the length of the flux propagation path varying with the microtrack location z, and w is the width of flux paths, which should be somewhat larger than the microtrack width.

When the microtrack is beyond the active region of the MR element, the flux propagation path disappears. Strictly speaking, this active region is defined by hard bias or exchange tabs rather than by the conductor leads, if they do not flush. Therefore, it is a good idea to flush conductor leads and hard bias (exchange tabs) to reduce side reading. In Fig 6.14a, as the microtrack moves to left from the right edge of the MR element active region, the flux path length is constant at first and then diminishes linearly when it is interrupted by the hard bias layer or exchange tab under the conductor. Consequently, the resulting microtrack profile is shown as the dashed line in Fig. 6.14b, where the solid line is the actual microtrack profile observed. The broadening is due to the finite width of microtrack, the finite magnetic spacing between head and disk, etc.

Having explained why side reading of MR head is asymmetrical, it is useful to introduce a number of important concepts. The *physical center* of the MR head is the center of its active region. In contrast, the *magnetic center* is defined as the equal-area line which divides the microtrack profile. In other words, either side of the magnetic center generates equal voltage signal. Both magnetic center and physical center are denoted in Fig. 6.14b. It is very important to know the offset of magnetic center with respect to the physical center of the MR sensor and inductive writer to achieve an optimum track density (Chapter 14).

From the microtrack profile, we can also find the read width, w_r, defined as the microtrack width at half-amplitude. As shown in Fig. 6.14b, the read width is

$$w_r = W - (D/2)\tan\theta_0 < W, \tag{6.71}$$

which is valid only if $D \tan \theta_0 < W$. This equation indicates that the effective read width of AMR head is smaller than the physical trackwidth defined by the conductor leads. Since the magnetic center should be at the middle of the read width, the physical-magnetic center offset is then

$$W/2 - w_r/2 = (D/4)\tan\theta_0. \tag{6.72}$$

The magnetic center approaches the physical center as the MR element height is reduced.

For MR head designs with exchange-tab longitudinal biasing, microtrack profiles display a *cross-talk null*, as shown in Fig. 6.15a. The dip in the microtrack profile is due to the negative contribution of the *x*-component of signal flux superimposed on the *y*-component of signal flux, which

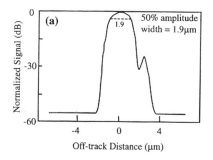
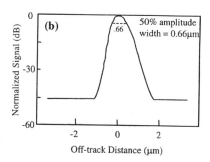

FIGURE 6.15. Microtrack profiles (schematic) of (a) 1 Gb/in.[2] AMR head (read trackwidth ~1.9 μm)[18] and (b) 5 Gb/in.[2] AMR head (read trackwidth ~0.66 μm).[19]

may be picked up through the tail regions of the MR element under the conductor leads. Although the MR signals from the tail regions are shorted out by the leads, any magnetic rotation there can propagate to the central active region of the MR element because it is highly permeable. The undesirable side read, however, can be reduced by improving the exchange-tab biasing design, as shown in Fig. 6.15b, where the cross-talk null disappears. Alternatively, abutted junction longitudinal bias can eliminate the cross-talk null by replacing the soft permalloy tails under the conductor leads with hard magnets, which are designed not to respond to magnetic excitations from the medium. In this sense, the longitudinal bias design is not only important for eliminating Barkhausen noise, but also critical for high-track density applications.

6.4.2 Readback nonlinearity

The transfer curve of MR element is intrinsically nonlinear. This nonlinearity does not show up if the dynamic range of the signal flux is small such that the transfer curve is effectively linear. However, when an MR head is driven into saturation or near-saturation, the MR head read response is no longer linear, as illustrated in Fig. 6.16. One of the consequences of MR read nonlinearity is that the positive peaks and negative peaks have

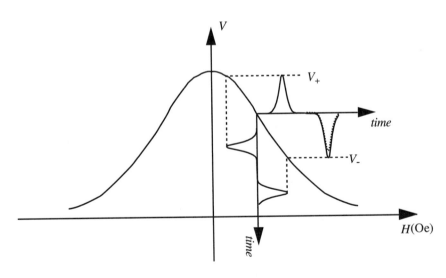

FIGURE 6.16.   Amplitude asymmetry of MR head readback pulses. (The dashed negative pulse has no amplitude asymmetry.)

different amplitudes. The normalized peak asymmetry (PA) can be defined as

$$\%PA = \frac{|V_+| - |V_-|}{|V_+| + |V_-|} \times 100. \tag{6.73}$$

Another definition is

$$\%PA = \frac{|V_+| - |V_-|}{\max\,(|V_+|,|V_-|)} \times 100. \tag{6.74}$$

The latter is a factor of ~2 larger than the former definition. In both conventions, a positive peak asymmetry means a smaller negative-going peak.

In order to limit the peak asymmetry of MR head, we should choose the medium $M_r\delta$ product properly such that the MR element is not driven into saturation. In addition, we should neither overbias nor underbias (transversely) an MR head. Elaborate simulation and experiments have shown that asymmetric placement of MR element within the shield-to-shield gap can significantly affect peak asymmetry.[20]

Generally speaking, MR readback nonlinearity is more serious for isolated pulses than for high-density pulses because the former is generated by the maximum signal flux level from a magnetic medium. Even though MR readback nonlinearity may not be a vital problem in high-density recording, it does interfere with the measurement of other write nonlinearities such as partial erasure and nonlinear transition shift. Therefore, several methods have been devised to characterize MR read nonlinearity.[21,22] An in-depth discussion will be given in Chapter 10.

6.4.3 Baseline shift

Readback voltage waveforms normally display a flat baseline, i.e., the quiescent voltage level in the absence of magnetic transitions is expected to be zero. However, isolated transitions written with an untrimmed thin film inductive head and read with an MR sensor may exhibit a shift in baseline between readback pulses,[23] as shown in Fig. 6.17a. The normalized baseline shift (BLS) is defined as

$$\%BLS = \frac{V_2 - V_1}{V_{p-p}} \times 100, \tag{6.75}$$

where V_1 is the baseline voltage level of positively magnetized domains, V_2 is the baseline voltage level of negatively magnetized domains, and V_{p-p} is the peak-to-peak amplitude of the isolated pulses.

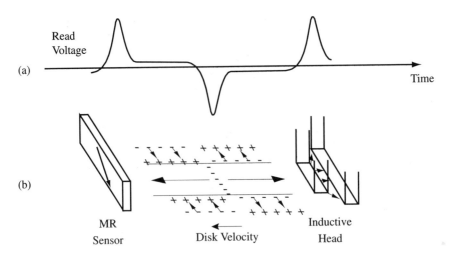

FIGURE 6.17. (a) Baseline shift of MR head readback voltage waveform. (b) Media-induced baseline shift interpretation.

The observed BLS can be interpreted based on the track edge effect during the write process, as shown in Fig. 6.17b. For the untrimmed write head with unequal pole tip widths, the side fringing field will have a transverse component which changes sign alternately as the write gap field is switched. Consequently, magnetic dipoles with alternating directions are written at track edges. These track edge dipoles will contribute no signals to an inductive read head because there is no flux derivative with respect to time, but it will contribute signals to an MR head because the latter is a flux-sensing device. The magnetic dipoles before the transition will increase the readback voltage, while those after the transition will decrease the readback voltage, resulting in a negative baseline shift. At very high recording densities, the magnetic dipoles at the track edges diminish, so the baseline shift also disappears accordingly. Properly trimming the write head pole tips reduces the transverse component of side erase field, thus results in a smaller baseline shift. Using an oriented magnetic disk whose easy axis (EA) is along circumferential direction also helps to reduce the magnetic dipoles at the track edge and thus decreases baseline shift.

    Magnetic dipoles at track edges can also be formed when magnetic transitions are written at skew angles, leading to an asymmetrical charge distribution across the data track and a baseline shift in MR readback waveform.[24]

In addition to the medium-induced baseline shift discussed above, there can also be head-induced baseline shift.[25] If the MR element experiences hysteretic behavior, then the MR transfer curve is no longer a single-value curve. In other words, the quiescent operating point of the MR head will shift slightly, resulting in baseline shift.

6.4.4 Thermal asperity

Like many materials, the resistivity of MR sensor is a function of temperature. The temperature dependence of metallic resistivity is nearly linear, and is often characterized by the temperature coefficient of resistivity (TCR):

$$\alpha = \frac{\rho - \rho_0}{\rho_0 \, (T - T_0)}, \tag{6.76}$$

where ρ and ρ_0 are the resistivity at temperatures of T and T_0, respectively. For example, a NiFe sensor has a positive TCR of $\sim 0.24\%/°C$, which means that a temperature fluctuation of about $\pm 10°C$ can completely overwhelm the MR sensor response, because MR ratio of NiFe is only $\sim 2\%$.

Due to Joule heating of the sense current passing through the MR element, the MR sensor generally operates at 20–50°C above the ambient temperature of a hard disk drive. When a MR sensor moves at a high speed relative to a magnetic disk and hits a high spot on the disk such as sputter splash, debris, mechanical scratch, and corrosion spot, its temperature can increase almost instantaneously upon the contact, then drop gradually to its normal operating temperature. Such a heating event is often called a *thermal asperity*. As a result, the MR head read voltage waveform changes accordingly, as shown in Fig. 6.18, where the readback waveform from a square wave recording pattern is greatly distorted by a large broad voltage peak due to a thermal asperity.

The temperature of the MR sensor is kept constant by the balance between Joule heating and cooling mainly through metallic shields. Sometimes MR head-disk contact can cause a modulation of MR element-to-shield spacing, which in turn causes excessive cooling or heating of MR sensor, giving rise to baseline shifts in MR read voltage waveform. This is called "baseline wander."

For MR head to reliably read data at a magnetic spacing as low as possible, thermal noise spikes such as thermal asperity and baseline wander need to be detected and corrected. To this end, electronic counter-

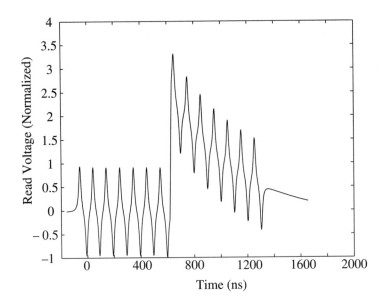

FIGURE 6.18.   Simulated read voltage waveform of all 1s data pattern with thermal asperity.

measures have been devised, and an excellent review has recently been given in the literature.[26]

6.4.5 Electrostatic discharge sensitivity

It has been recognized that MR heads can be easily damaged by electrostatic discharge (ESD) during head fabrication, testing, and assembly.[27,28] Transient ESD current through MR sensor may cause a permanent increase in MR sensor resistance, and eventually severe melting damages of MR sensors. In addition, dielectric breakdown between MR sensor and shields, and between shields and substrate, can also cause the MR heads to fail.

Thin-film inductive heads are also susceptible to ESD damages, although not to the extent of MR sensors. Spurious electrostatic discharges between magnetic pole tips and an adjacent conductor will likely damage the pole tips. The insulating layers surrounding write coils may break down due to ESD, causing the failure of the inductive heads.

To prevent ESD damages, we should properly ground operators, table tops and other equipment which may be in contact with heads. In addition, the materials used for storage container and work trays must be carefully selected. Bypass paths to ESD away from the sensor or pole tip area can also be implemented.[29]

6.5 FABRICATION OF MR HEADS

An integrated inductive write, magnetoresistive read head (often simply called MR head) is schematically shown in Fig. 6.19a. The top shield and the bottom pole (P1) of the inductive head are merged to reduce the built-in offset between the MR sensor and inductive write gap due to skew angles in hard disk drive.[30] The *merged design* is also helpful to decrease the alignment error between the MR sensor and the inductive head pole tips due to photolithography steps in head manufacturing. In the first-generation IBM MR head "Corsair," the bottom magnetic pole of the inductive writer is separate and placed above the top shield of the MR reader. This is often referred to as *piggyback design*. A real cross-sectional view of a merged inductive write/MR read head is shown in Fig. 6.19b, where the bottom shield is ~2 μm thick, and the top shield is ~3 μm thick. The actual thickness of the MR sensor is so small compared with the shield thickness that the MR sensor is not visible in Fig. 6.19b, but it can be seen from the magnified photograph shown in Fig. 6.19c.

6.5.1 Magnetoresistive head wafer process

Fabrication of MR heads is similar to inductive thin-film heads in terms of technology, but more complicated in terms of design and fabrication steps. Magnetoresistive heads still use ceramic substrates, and the front-end (substrate preparation) and back-end (lapping, slicing, slider fabrication) processes are almost identical to inductive head technology. During the wafer processing, usually the MR reader is fabricated first, then the inductive writer is processed.

The fabrication steps of the MR reader, with SAL transverse bias and hard longitudinal bias, are shown in Fig. 6.20. The bottom shield is electroplated or sputtered, then the first gap layer (typically alumina) is sputtered. After sputtering SAL/spacer/MR trilayer, photoresist pattern defining MR element is applied. The sidewall of the photoresist pattern must have a reentrant angle or an undercut to facilitate a lift-off process later. The uncovered part of the trilayer is then removed by ion milling (ion beam etching). The angle and exposure of ion beam are controlled such that a slope is formed at the both ends of the MR trilayer. This slope

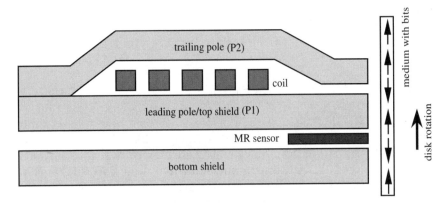

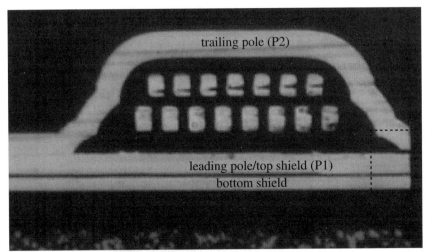

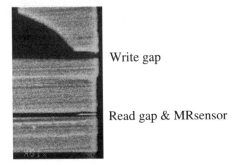

FIGURE 6.19.   (a) Top: schematic of MR head cross section (including the writer). (b) Middle: cross-sectional view of a real merged MR head with inductive writer. (c) Bottom: magnified view of the boxed gap region in (b).

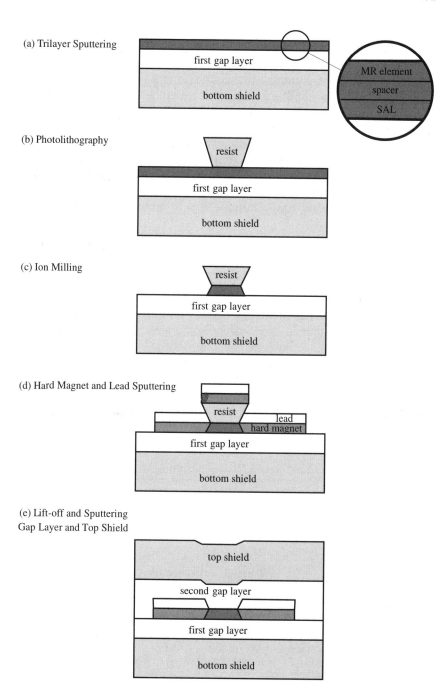

FIGURE 6.20.   Fabrication steps of MR reader with SAL and hard bias.

allows a contiguous junction to form reliably when a hard magnet layer and a conductor layer are deposited, as shown in Fig. 6.20d. The hard magnet layer and the conductor layer deposited on the photoresist are lift off along with the photoresist when the wafer is soaked in a photoresist solvent. Then the second gap layer is sputtered, and the top shield is electroplated or sputtered, as shown in Fig. 6.20e. Steps to complete the inductive writer follows.

Actual processing of MR heads involves more subtlety and complexity, and varies from company to company. The important tricks usually remain trade secrets and are unavailable in the literature. A comprehensive review, but with limited details, on the process complexity of MR heads of various designs and configurations was given in the literature.[31]

6.5.2 Materials used in magnetoresistive head fabrication

MR head fabrication consists of many deposition and lithographic steps, and is truly a "materials-intensive" activity that involves numerous magnetic and nonmagnetic layers. Driven by the tremendous growth rate of the magnetic storage technology, the data storage industry seems to have an insatiable appetite for newer and better materials. In fact, new materials are often the enabling factors for the continued gain in head performance. The materials used or proposed for dual-element inductive write/MR read heads are listed in Table 6.2, and the table is growing fast.

TABLE 6.2. Materials Used in MR Heads (Including Writer and Reader)[a]

Layers	Materials
Undercoat/overcoat/gap	Alumina (Al_2O_3)
Magnetic poles	Permalloy ($Ni_{81}Fe_{19}$,$Ni_{80}Fe_{20}$), $Ni_{45}Fe_{55}$, CoNiFe, CoXZr, FeXN
Coils	Ti/Cu, Cr/Cu
Shields	Permalloy, Sendust (FeAlSi)
MR sensor	Permalloy, spin valves
Soft adjacent layer (SAL)	NiFeRh, CoZrMo, NiFeMo, NiFeCo
Hard bias	CoPt, CoCr, CoCrPt
Exchange bias	FeMn, NiMn, IrMn, PtMn, TbCo, NiO
Spacer between MR & SAL	Ta, Ti, Al_2O_3, SiO_2
Protective coating	Carbon

[a]Ref. 32.

6.6 GIANT MAGNETORESISTIVE (GMR) HEADS

The signal output of MR heads is directly proportional to the MR ratio of the sensor material. No wonder that people have been searching for the materials that display much larger changes in resistivity than permalloy for a long time. It turns out that the match made in haven for the MR head technology is the so-called giant magnetoresistive (GMR) material first discovered in 1988. It took less than a decade to transform this scientific breakthrough into commercial products, both of which will be examined here.

6.6.1 Giant magnetoresistance effect

In 1988 Baibich et al. reported their seminal work on giant magnetoresistance (GMR) effect in Fe/Cr multilayers.[33] In this type of "magnetic/nonmagnetic metal" multilayers, the magnetic layers are antiferromagnetically coupled (or antiparallel) in the absence of external field. When a field strong enough is applied to align the magnetizations of adjacent magnetic layers, the resistivity of the multilayers drop significantly (up to a factor of 2) as shown in Fig. 6.21. The ferromagnetic state, i.e., the saturated state, has a much lower resistivity than that of the antiferromagnetic state, because the former state allows electrons of one spin have a lower average resistance, leading to a so-called "short-circuit" effect. The origin of GMR effect is the spin-dependent scattering either at the interfaces of magnetic/nonmagnetic metals or inside the ferromagnetic metal layers.

The definition of GMR ratio varies somewhat in the literature. To be consistent with the AMR ratio, we adopt the definition of GMR ratio as the ratio of the maximum difference in resistance over the minimum resistance:

$$\text{GMR ratio} \equiv \frac{R_{ap} - R_p}{R_p} = \frac{G_p - G_{ap}}{G_{ap}}, \qquad (6.77)$$

where R_p (or G_p) is the resistance (or conductance) when the magnetizations are parallel, and R_{ap} (or G_{ap}) is the resistance (or conductance) when the magnetizations are antiparallel. This definition gives a GMR ratio of ~85% at 4.2 K for the (Fe 3 nm/Cr 0.9 nm)$_{60}$ multilayers. Since the discovery of GMR effect, it has been observed in numerous ferromagnetic metal/nonmagnetic metal multilayers.[34] In a $Co_{95}Fe_5$/Cu multilayer, GMR ratios of ~220% at 4.2 K and ~110% at 295 K have been reported.[35]

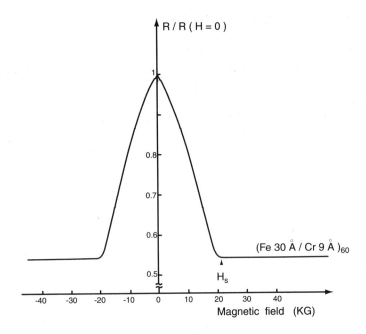

FIGURE 6.21.   Giant magnetoresistance of Fe/Cr superlattice at 4.2 K. The current and applied field are along the [110] axis in the plane of the layers. When the temperature increases to room temperature, the saturation magnetoresistance ratio is reduced by a factor of ~2, and the saturation magnetic field H_s is reduced by about 30%.

The GMR multilayers typically require a rather large saturation field to overcome the antiferromagnetic coupling of magnetic layers to display the large GMR ratios. It was found that the strength of the antiferromagnetic coupling is a periodic function of the nonmagnetic spacer.[36] It was later recognized that the antiferromagnetic (AF) coupling and GMR effect in multilayers are due to entirely different physical mechanisms although the two phenomena often occur at the same time. The origin of oscillatory antiferromagnetic coupling has been explained by the RKKY (Ruderman-Kittel-Kasuya-Yosida) theory[37] and by quantum well models.[38]

The field sensitivity of GMR multilayers is actually lower than the AMR ratio of NiFe because of the large saturation fields in the multilayers. In addition, CoFe/Cu multilayers are difficult to implement in a device like a recording head. These considerations motivated the invention of an alternative GMR structure that consists of NiFe/Cu/NiFe/FeMn, where the FeMn layer pins the magnetization of the top NiFe layer through exchange anisotropy.[39] Between the ferromagnetic NiFe layers, the ferro-

to the right because of a weak ferromagnetic coupling from the pinned layer. The resistance of the spin valve is maximum when bottom and top FM layers are antiparallel, and minimum when parallel. The corresponding MR transfer curve along the EA is shown in Fig. 6.23b.

The GMR ratio of a spin valve can be expressed as

$$\frac{\Delta R}{R_0} = \frac{\Delta R_{max}}{R_0}\left[\frac{1 - \cos(\theta_2 - \theta_1)}{2}\right], \tag{6.78}$$

where θ_1 and θ_2 are the orientation angles of the magnetizations of the pinned and free FM layers in the spin valve, respectively. In contrast, the AMR ratio is expressed as

$$\frac{\Delta\rho}{\rho_0} = \frac{\Delta\rho_{max}}{\rho_0}\cos^2\theta, \tag{6.1}$$

where θ is the angle between the current and the magnetization. Both GMR and AMR effects are shown in Fig. 6.24. As a comparison, the AMR transfer curve is most linear at $\theta = 45°$ or $135°$, while the GMR transfer curve is most linear at $\theta_2 - \theta_1 = 90°$. Also note that the spin valve requires magnetization rotation of $180°$ instead of $90°$ to realize its maximum variation of resistance.

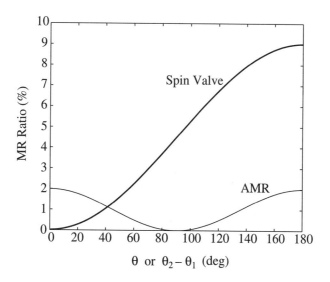

FIGURE 6.24. MR ratios of spin valve and NiFe vs. magnetization orientation. The maximum ratios are 9% and 2%, respectively.

6.6.2 GMR read heads

GMR read heads that are capable of reading areal densities ≥ 5 Gb/in.[2] and are based on spin valve structures have been reported by a number of companies,[42–44] GMR head technology is not a simple extension of AMR technology because of many new requirements, including much stricter control of film thickness and interfaces, better exchange bias materials, good thermal stability of spin valves, etc.

To linearize a spin valve GMR read head, we can design the spin-valve head as shown in Fig. 6.25 such that $\theta_1 = 0°$, $\theta_2 = 90°$, $\theta_2 - \theta_1 = 90°$ in the quiescent state. In this design of spin valve heads, $\theta_2 = 90°$ is fixed, but θ_1 is influenced by the flux signal. In principle the spin valve head is very similar to AMR head shown in Fig. 6.7. However, the free layer magnetization in the former rotates from $-90°$ to $+90°$ (maximum dynamic range), while the magnetization in the latter rotate from $0°$ to $90°$ (if biased at $\sim 45°$).

The readback voltage from a spin valve GMR read head is

$$V_{GMR}(\overline{x}) = I \, \Delta R_{max} \frac{\langle -\cos(90° - \theta_1) + \cos(90° - \theta_{1,0}) \rangle}{2}$$

$$= -\frac{1}{2} I \, \Delta R_{max} \langle \sin \theta_1 - \sin \theta_{1,0} \rangle \qquad (6.79)$$

$$= -\frac{1}{2} I \, \Delta R_{max} \left[\langle \sin \theta_1 \rangle - \langle \sin \theta_{1,0} \rangle \right],$$

where I is the sense current through the spin valve, and $\theta_{1,0}$ is the quiescent magnetization orientation of the free layer. Note that the resistivity and current density in spin valve is not uniform, so we no longer use current

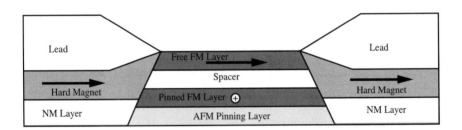

FIGURE 6.25. Schematic cross section of a spin-valve GMR read head with exchange-pinned layer and longitudinal hard bias. The nonmagnetic (NM) layer under the hard magnet is for controlling the coercivity and thickness of the hard magnet.

density and resistivity to express readback voltage. The magnetization orientation of the free layer is determined by minimizing total energy of the free layer (Intrinsic induced anisotropy H_k of the free layer is neglected). This leads to the following relation:

$$ - H_t \cos\theta_1 + H_l \sin\theta_1 + \frac{1}{2} N_{dy}^f M_s^f \sin(2\theta_1) = 0, $$

where H_t is the transverse field (bias plus signal); H_l is the longitudinal bias field, and $N_{dy}^f M_s^f$ is the transverse demagnetizing field of the free layer. To make sure that the free-layer magnetization is oriented at $\theta_1 \sim 0°$, the required transverse bias field is

$$ H_{t,bias} = 0. $$

This means that there is no transverse bias field required to linearize the spin valve head. However, ferromagnetic coupling field from the pinned layer must be canceled by the sense current field. If biased at $\theta_{1,0} = 0°$, the small field sensitivity of the free layer can be found as follows:

$$ -\cos\theta_{1,0} + H_{t,bias}\sin\theta_{1,0}\frac{\partial\theta_1}{\partial H_{sig}} + H_l\cos\theta_{1,0}\frac{\partial\theta_1}{\partial H_{sig}} $$

$$ + N_{dy}^f M_s^f \cos(2\theta_{1,0})\frac{\partial\theta_1}{\partial H_{sig}} = 0, $$

$$ \frac{\partial\theta_1}{\partial H_{sig}} = \frac{\cos\theta_{1,0}}{H_{t,bias}\sin\theta_{1,0} + H_l\cos\theta_{1,0} + N_{dy}^f M_s^f\cos(2\theta_{1,0})} $$

$$ = \frac{1}{H_l + N_{dy}^f M_s^f}. \tag{6.80} $$

For both spin valve head and AMR head, the longitudinal bias field must not be excessive so that the field sensitivity can be maintained.

The spin valve GMR head readback voltage can be related to the flux signal $\Phi(\bar{x},y')$ based on the following relation:

$$ \sin\theta_1 = \frac{M_y^f}{M_s^f} = \frac{\Phi(\bar{x}, y')}{\mu_0 t W M_s^f}. \tag{6.81} $$

where y' is the vertical coordinate of the free layer, t is its thickness, and W is the read trackwidth. Therefore, if $\langle\sin\theta_{1,0}\rangle = 0$, Equation (6.79) becomes

$$ V_{GMR}(\bar{x}) = -\frac{I\Delta R_{max}}{2\mu_0 Dt W M_s^f}\int_0^D \Phi_{sig}(\bar{x}, y')\,dy'. \tag{6.82} $$

According to the transmission line model, the signal flux $\Phi_{sig}(\bar{x}, y')$ in the free layer decays from the ABS ($y' = 0$) to the top of free layer ($y' = D$) due to flux leakage to the shields. Similar to the case of the AMR head, the efficiency of the spin valve GMR head can be defined as

$$F_{GMR} \equiv \frac{\langle \Phi_{sig}(\bar{x}, y') \rangle}{\Phi_{sig}(x, y' = 0)} = \frac{\int_0^D \Phi_{sig}(\bar{x}, y') \, dy'/D}{\Psi_{sig}(\bar{x})}. \tag{6.39}$$

where $\Phi_{sig}(\bar{x}) \equiv \Phi_{sig}(\bar{x}, y' = 0)$ is the signal flux injected into the free layer at the ABS. The spin valve GMR head efficiency can be evaluated by a transmission line model as described earlier. Therefore, Equation (6.82) can be simplified as

$$V_{GMR}(\bar{x}) = -\frac{I \Delta R_{max}}{2\mu_0 t W M_s^f} E_{GMR} \Phi_{sig}(\bar{x}), \tag{6.83}$$

where $\Phi_{sig}(\bar{x}) \equiv \Phi_{sig}(\bar{x}, y' = 0)$ can be found by the reciprocity principle (Section 6.3). When the free-layer-to-shield gaps are small and symmetrical, Equation (6.52) holds, and the GMR head readback voltage from an arctangent transition is approximately

$$V_{GMR}(\bar{x}) = -\frac{E_{GMR} I \Delta R_{max} M_r \delta}{\pi t M_s^f} \frac{g + t}{g} \left(\tan^{-1} \frac{\bar{x} + g/2}{a + d} - \tan^{-1} \frac{\bar{x} - g/2}{a + d} \right), \tag{6.84}$$

where t is the thickness of the free layer, and g is the free-layer-to-shield gap length.

If the flux signal at the ABS and head efficiency are fixed, then the ratio of the readback amplitude of the spin valve (SV) GMR head over that of the AMR head is

$$\frac{V_{0\text{-peak}}^{SV}}{V_{0\text{-peak}}^{AMR}} = \frac{1}{2\sqrt{2}} \frac{t_{MR}}{t} \frac{M_s^{MR}}{M_s^f} \frac{(I \Delta R_{max})^{SV}}{(JW \Delta \rho_{max})^{AMR}}, \tag{6.85}$$

where t_{MR} and M_s^{MR} are the AMR element thickness and magnetization, respectively. For example, a Si/Ta/NiFe(5 nm)/Co(2)/Cu(2)/Co(2)/NiFe(3)/FeMn(15) spin valve displays a GMR ratio of ~9% and a square resistance of $R_{sq} \approx 12 \, \Omega$. If the spin valve total thickness is $t_{sv} \approx 26$ nm, then

$$(I \Delta R_{max})^{SV} \cong J t_{sv} D \cdot \frac{\Delta R_{max}}{R_0} R_{sq} \frac{W}{D} = J W R_{sq} t_{sv} \cdot \frac{\Delta R_{max}}{R_0},$$

where D is the sensor height, and J is the average current density through the spin valve. If $t_{MR} = 10$ nm, $\Delta\rho_{max} = 0.34\ \mu\Omega\cdot$cm, and the sense current density and read trackwidth are the same for the AMR head, then

$$\frac{V^{SV}_{0\text{-peak}}}{V^{AMR}_{0\text{-peak}}} = \frac{1}{2\sqrt{2}}\frac{10}{7}\frac{12 \times 26\ \Omega\cdot\text{nm} \times 9\%}{0.34 \times 10\ \Omega\cdot\text{nm}} \approx 4.$$

The GMR head output is a factor of 4 bigger, which is a significant gain. Note that the ratio above is the ratio of readback voltages under the same flux signal. This ratio is slightly different from the ratio of the maximum readback voltages that can be attained with each head if different flux signals can be applied. In the latter case, if the sense current density and trackwidth are still the same for the SV GMR head and AMR head, then the ratio of the maximum readback voltages is

$$\frac{V^{SV}_{0\text{-peak}}}{V^{AMR}_{0\text{-peak}}} = \frac{(I\,\Delta R)^{SV}_{max}/2}{(JW\,\Delta\rho)^{AMR}_{max}/2} = \frac{12 \times 26\ \Omega\cdot\text{nm} \times 9\%}{0.34 \times 10\ \Omega\cdot\text{nm}} \approx 8. \qquad (6.86)$$

This is a boost of eight times! In short, when properly designed, spin valve GMR heads could generate significantly greater readback voltages than conventional AMR heads. With the help of GMR materials, we can further reduce the trackwidth W while maintaining adequate readback signal. Without the discovery and improvement of GMR materials, magnetic recording at a density of >10 Gb/in.^{2} would be unthinkable.

6.6.3 Tunneling magnetoresistive effect

What is next after GMR? This is anyone's guess. The recent development in room temperature spin-dependent tunneling (SDT) junctions,[45] also called magnetic tunneling junctions, is very exciting and worth our attention. These devices are made of ferromagnetic metal/insulating barrier/ferromagnetic metal sandwiches, as shown in Fig. 6.26. If the barrier is very thin (<2 nm), electrons can tunnel through the barrier from one ferromagnetic electrode to another, or vice versa. Since the electrons in the ferromagnetic electrodes are at least partially spin polarized, the tunneling current through the barrier is larger when the magnetization in the two electrodes is parallel than when they are aligned antiparallel, resulting in a so-called tunneling magnetoresistance (TMR) effect. TMR ratios more than 25% at room temperature have been reported, making it very attractive for applications in magnetic read heads and magnetic random access memory (MRAM, see Chap. 17).[46]

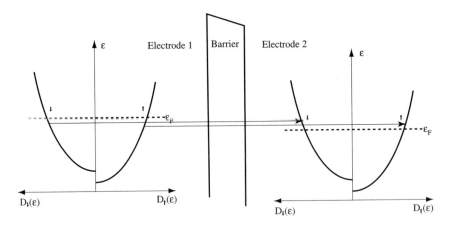

FIGURE 6.26. Spin-dependent (spin polarized) tunneling in ferromagnet/barrier/ferromagnet junctions. The energy bands shown are for free-electron-like tunneling carriers, where $D(\epsilon)$ is the density of states and dashed lines represent the Fermi levels. The spin-up band is split from the spin-down band due to exchange interaction. It is assumed there is no spin flipping.

The *spin polarization* of a ferromagnetic material is defined as

$$p = a_\uparrow - a_\downarrow = 2a_\uparrow - 1, \tag{6.87}$$

where $a_\uparrow$ and $a_\downarrow (= 1 - a_\uparrow)$ are the fractions of spin-up electrons and spin-down electrons, respectively. Related to tunneling phenomena under a small bias voltage, only the electrons near the Fermi energy should be counted because they are the tunneling carriers. A simple model was proposed by Julliere to explain spin-dependent tunneling. It was Julliere who first discovered the effect at low temperatures in Fe/Ge/Co junctions.[47] One can assume that the electrons tunnel through the barrier without spin flipping, i.e., spin-up (or spin-down) electrons tunnel to empty spin-up (or spin-down) electron states only. Therefore, the tunneling current when the magnetizations of the two electrodes are parallel is

$$
\begin{aligned}
I_p &\propto a_{\uparrow 1} a_{\uparrow 2} + a_{\downarrow 1} a_{\downarrow 2} \\
&= (1 + p_1)(1 + p_2)/4 + (1 - p_1)(1 - p_2)/4 \\
&= (1 + p_1 p_2)/2,
\end{aligned} \tag{6.88}
$$

where p_1 and p_2 are the spin polarizations of the two electrodes. When the magnetizations are antiparallel, the spins of one electrode are turned upside down, so the tunneling current becomes

$$
\begin{aligned}
I_{ap} &\propto a_{\uparrow 1} a_{\downarrow 2} + a_{\downarrow 1} a_{\uparrow 2} \\
&= (1 + p_1)(1 - p_2)/4 + (1 - p_1)(1 + p_2)/4 \\
&= (1 - p_1 p_2)/2.
\end{aligned} \tag{6.89}
$$

The TMR ratio is then

$$\frac{R_{ap} - R_p}{R_p} = \frac{G_p - G_{ap}}{G_{ap}} = \frac{I_p - I_{ap}}{I_{ap}} = \frac{2p_1p_2}{1 - p_2p_2}. \tag{6.90}$$

The Julliere model is widely used to predict the TMR ratios attainable in a SDT junction, and ferromagnetic electrodes with larger spin polarizations are sought to achieve higher TMR ratios. For example, for a $Co_{50}Fe_{50}/Al_2O_3/Co$ junction, $p_1 \sim 47\%$, $p_2 \sim 34\%$, and a TMR ratio of $\sim 38\%$ can be predicted, which is close to experimentally observed values.

Consider the band structure shown in Fig. 6.26. Near the Fermi level,

$$a_{\uparrow} \propto D_{\uparrow}(\varepsilon_F) \propto k_{\uparrow F}, \tag{6.91}$$
$$a_{\downarrow} \propto D_{\downarrow}(\varepsilon_F) \propto k_{\downarrow F},$$

where k_F is the Fermi wavevector. Therefore, the spin polarization of an electrode near the Fermi level can be expressed as

$$p = \frac{k_{\uparrow F} - k_{\downarrow F}}{k_{\uparrow F} + k_{\downarrow F}}, \tag{6.92}$$

which is a good approximation if the tunneling electron band is parabolic and the barrier height is much higher than the Fermi level.[48] It has been accepted for a long time that the spin polarization of transition metal ferromagnets is roughly proportional to its saturation magnetization.[49] However, more recent studies indicate that tunneling spin polarization is strongly dependent upon the nature of electrode/barrier bonding at the interfaces, and may vary widely from the traditional accepted values.[50,51]

The major challenge to use SDT junctions in magnetic read heads is that we must lower their junction resistance. The smallest junction resistance-area product reported to date is about 0.24 $K\Omega \cdot \mu m^2$, which is getting close to the required values for MR read head applications.[52] The research of SDT devices is a rapidly developing research field, and more exciting results are anticipated in the future.

References

1. T. R. McGuire and R. I. Potter, "Anisotropic magnetoresistance in ferromagnetic 3d alloys," *IEEE Trans. Magn.*, **11**, 1018, 1975.

2. M. Pinarbasi, private communication, 1998.

3. R. Hunt, "A magnetoresistive readout transducer," *IEEE Trans. Magn.*, **7**, 150, 1971.

4. H. N. Bertram, "Linear signal analysis of shielded AMR and spin valve heads," *IEEE Trans. Magn.*, **31**, 2573, 1995.

5. H. N. Bertram, *Theory of Magnetic Recording*, (Cambridge, UK: Cambridge University Press, 1994, Chapter 2).

6. S. W. Yuan and H. N. Bertram, "Magnetoresistive heads for ultra high density recording," *IEEE Trans. Magn.*, **29**, 3811, 1993.

7. C. Tsang, "Unshielded MR elements with patterned exchange biasing," *IEEE Trans. Magn.*, **25**, 3692, 1989.

8. M. T. Krounbi, O. Voegeli, and P.-K. Wang "Magnetoresistive read transducer having hard magnetic bias," U.S. Patent 5,018,037, 1991.

9. T. Kira, T. Miyauchi, and M. Yoshikawa, "Thin film magnetic head having a magnetized ferromagnetic film on the MR element," U.S. Patent 4,639,806, 1987.

10. M. T. Krounbi, O. Voegeli, and P.-K. Wang, "Magnetoresistive read transducer having hard magnetic shunt bias," U.S. Patent 5,005,096, 1991.

11. R. I. Potter, "Digital magnetic recording theory," *IEEE Trans. Magn.*, **10**, 502, 1974.

12. D. E. Heim, "The sensitivity function for shielded magnetoresistive heads by conformal mapping," *IEEE Trans. Magn*, **19**, 1620, 1983.

13. E. Champion and H. N. Bertram, "The effect of MR head geometry on playback pulse shape and spectra," *IEEE Trans. Magn.*, **31**, 2461, 1995.

14. D. A. Thompson, "Magnetoresistive transducers in high-density magnetic recording," *AIP Conf. Proceeding*, **24**, 528, 1974.

15. N. H. Yeh, "Asymmetric crosstalk of magnetoresistive head," *IEEE Trans. Magn.*, **18**, 1155, 1982.

16. A. Wallash, M. Salo, J. K. Lee, D. Heim, and G. Garfunkel, "Dependence of MR head readback characteristics on sensor height," *J. Appl. Phys.*, **69**, 5402, 1991.

17. D. E. Heim, "On the track profile in magnetoresistive heads," *IEEE Trans. Magn.*, **30**, 1453, 1994.

18. C. Tsang, M. Chen, T. Yogi, K. Ju, "Gigabit density recording using dual-element MR/inductive heads on thin film disk," *IEEE Trans. Magn.*, **26**, 1689, 1990.

19. C. Tsang, T. Lin, S. MacDonald, M. Pinarbasi, N. Robertson, H. Santini, M. Doerner, T. Reith, L. Vo, T. Diola, and P. Arnett, "5 Gb/in.^{2} recording demonstration with conventional AMR dual-element heads and thin film disks," *IEEE Trans. Magn.*, **33**, 2866, 1997.

20. J. S. Feng, "Gap length and sensor height effect on readback signals in shielded magnetoresistive sensors," *IEEE Trans Magn.*, **28**, 1031, 1992.

21. T. C. Arnoldussen, "Harmonic analysis of nonlinear MR head transfer characteristics," *IEEE Trans. Magn.*, **32**, 3449, 1996.

22. B. Wilson, S. X. Wang, A. Taratorin, "A generalized method for measuring readback nonlinearity using a spin stand," *J. Appl. Phys.*, **81**(8), 4828, 1997.

23. J. L. Su and K. Ju, "Track edge phenomena in thin film longitudinal media," *IEEE Trans. Magn.*, **25**, 3384, 1989.

24. T. Chang, M. Salo, H. Fang, M. Madison, J-G. Zhu, and T. Arnoldussen "Affects of skew angle on the written-in baseline shift signal at high track densities," *IEEE Trans. Magn.*, **32**, 3395, 1996.

25. E. Champion and H. N. Bertram, "Hysteresis and baseline shift in permanent magnet stablized MR/SAL heads," *IEEE Trans. Magn.*, **32**, 13, 1996.

26. K. B. Klaassen and J. C. L. van Peppen, "Electronic abatement of thermal interference in (G)MR head output signals," *IEEE Trans. Magn.*, **33**, 2611, 1997.

27. A. J. Wallash, "Standardized ESD test for magnetoresistive recording heads," *IEEE Trans. Magn.*, **33**, 2911, 1997.

28. H. Tian and J. K. Lee, "ESD damage to MR heads," *IEEE Trans. Magn.*, **31**, 2624, 1995.

29. K. H. Elser and R. R. Kerl, "Magnetic head having static discharge means," US Patent 4, 317, 149, 1982.

30. D. Hannon, M. Krounbi, and J. Christner, "Allicat magnetoresistive head design and performance," *IEEE Trans. Magn*, **30**, 298, 1994.

31. R. E. Fontana, Jr., "Process complexity of magnetoresistive sensors: a review," *IEEE Trans. Magn.*, **31**, 2579, 1995.

32. *Semiconductor International*, Feb. 1997, p. 26.

33. M. N. Baibich, I. M. Broto, A. Fert, F. Nguyen Van Dau, F. Petroff, P. Eitenne, G. Creuzet, A. Friederich, and J. Chazelas, "Giant magnetoresistance of (001)Fe/(001)Cr magnetic superlattices," *Phys. Rev. Lett.*, **61**, 2472, 1988.

34. For example, H. Yamamoto and T. Shinjo, "Magnetoresistance of multilayers," *IEEE Translation Journal on Magnetics in Japan*, **7**, 674, 1992.

35. S. S. P. Parkin, private communication, 1997.

36. S. S. P. Parkin, N. More, and. K. P. Roche, "Oscillations in exchange coupling and magnetoresistance in metallic superlattice structures: Co/Ru, Co/Cr, and Fe/Cr," *Phys. Rev. Lett.*, **64**, 2304, 1990.

37. P. Bruno and C. Chappert, "Oscillatory coupling between ferromagnetic layers separated by a nonmagnetic metal spacer," *Phys. Rev. Lett.*, **67**, 1602, 1991.

38. D. M. Edwards, R. P. Erickson, J. Mathon, R. B. Muniz, and M. Villeret, "Spin current and exchange coupling in magnetic multilayers," *Materials Science & Eng.* **B31**, 25, 1995.

39. B. Dieny, V. S. Speriosu, S. S. P. Parkin, D. A. Gurney, D. R. Wilhoit, and D. Mauri, "Giant magnetoresistance in soft ferromagnetic multilayers," *Phys. Rev.* **B43**, 1297, 1991.

40. T. C. Anthony, J. A. Brug, and S. Zhang, "Magnetoresistance of symmetric spin valve structures," *IEEE Trans., Magn.*, **30**, 3819, 1994.

41. W. F. Egelhoff, Jr., P. J. Chen, C. J. Powell, M. D. Stiles, R. D. McMichael, J. H. Judy, K. Takano, and A. E. Berkowitz, "Oxygen as a surfactant in the growth of giant magnetoresistance spin valves," *J. Appl. Phys.*, **82**(12), 6142, 1997.

42. H. Kanai, K. Yamada, K. Aoshima, Y. Ohtsuka, J. Kane, M. Kanamine, J. Toda, and Y. Mizoshita, "Spin valve read heads with NiFe/$Co_{90}Fe_{10}$ layers for 5 Gb/in.^{2} density recording," *IEEE Trans Magn.*, **32**, 3368, 1996.

43. H. Yoda, H. Iwasaki, T. Kabayashi, A. Tsutai, and M. Sahashi, "Dual-element GMR/inductive heads for gigabits density recording using CoFe spin-valve," *IEEE Trans. Magn.*, **32**, 3363, 1996.

44. C. Tsang, R. E. Fontana, T. Lin, D. E. Heim, V. S. Speriosu, D. A. Gurney, and M. L. Williams, "Design, fabrication, and testing of spin-valve heads for high density recording," *IEEE Trans. Magn.*, **30**, 3801, 1994.

45. J. S. Moodera and L. R. Kinder, "Ferromagnetic-insulator-ferromagnetic tunneling: spin-dependent tunneling and large magnetoresistance in trilayer junctions," *J. Appl. Phys.*, **79**(8), 4724, 1996.

46. W. J. Gallagher, S. S. P. Parkin, Y. Lu, X. P. Bian, A. Marley, K. P. Roche, A. Altman, S. A. Rishton, C. Jahnes, T. M. Shaw, and G. Xiao, "Microstructured magnetic tunnel junctions," *J. Appl. Phys.*, **81**(8), 3741, 1997.

47. M. Julliere, "Tunneling between ferromagnetic films," *Phys. Lett.*, **54A**, 225, 1975.

48. J. C. Slonczewski, "Conductance and exchange coupling of two ferromagnets separated by a tunneling barrier," *Phys. Rev. B*, **39**(10), 6995, 1989.

49. M. B. Stearns, "Simple explanation of tunneling spin-polarization of Fe, Co, Ni and its alloys," *J. Magn. Magn. Mater.*, **5**, 167, 1977.

50. E. Y. Tsymbal and D. G. Pettifor, "Modeling of spin-polarized electron tunneling from 3d ferromagnets," *J. Phys.: Condens. Matter*, **9**, L411, 1997.

51. M. Sharma, S. X. Wang, and J. Nickel, "Inversion of spin polarization and tunneling magnetoresistance in spin-dependent tunneling junctions," *Phys. Rev. Lett.*, **82**(3), 616, 1999.

52. H. Tsuge, T. Mitsuzuka, A. Kamijo, and K. Matsuda, "Magnetic tunnel junctions with low resistance, high current density and good uniformity," *Materials Research Society Symp. Proc.*, **517**, 87, 1998.

CHAPTER 7

Magnetic Recording Media

After analyzing magnetic inductive and MR heads, it is natural to revisit the magnetic recording media introduced briefly in Chapter 2. Magnetic recording media contain all the recorded information, with which most users are concerned. No wonder many refer to a magnetic disk drive simply as a "magnetic disk" although it requires a lot of other components to function at all. A magnetic recording layer is composed of either closely packed magnetic particles or continuous magnetic thin films deposited on a substrate. Therefore, magnetic disks are often called *particulate* or *thin-film* disks, respectively.

Magnetic disks are also classified as *flexible* (or floppy) disks and *rigid* (or hard) disks based on the substrates used. Flexible disks commonly use polyester substrates, which are very suitable for removable disk storage such as the 1.4-MB floppy disk and the Iomega zip™ disk. Rigid disks usually use aluminum-magnesium (Al-Mg) alloy, glass, or ceramic substrates. Their rigidity permits higher rotation speeds, lower magnetic spacing, and smaller data track misregistration (see Chapters 14 and 15). Therefore, rigid disks can generally accommodate higher linear data densities and track densities, but they are permanently enclosed in a box and thus remain irremovable. Removable hard disk drives have been designed and commercialized by companies like Iomega and SyQuest, but with somewhat inferior performance compared with irremovable hard disk drives.

Finally, magnetic disks can be classified according to their *form factors* as specified by their physical diameters, as shown in Table 7.1 The first rigid disk drive (IBM 350) commercialized in 1957 employed 24-in.-diame-

TABLE 7.1. Magnetic Disk Form Factors

Form factor (in.)	24	14	8	5.25	3.5	2.5	1.8	1.3
Diameter (mm)	606	355	210	130	95	65	48	33

ter particulate disks. The larger form factors ($>5.25''$) are now out of date, the dominant form factor is 3.5 in. The smaller form factors (≤ 2.5 in.) are common in portable computers. Note that the physical diameters are slightly different from the form factors in inches (1 in. $=$ 25.4 mm). For example, a 3.5 in. form factor disk actually has a diameter of 3.74 in.

This chapter starts with the basic magnetic medium requirements and brief introduction of particular media. The bulk of the chapter is devoted to thin-film media for high-density magnetic recording. Future alternative media such as keepered media and patterned media are covered at the end of the chapter.

7.1 MAGNETIC MEDIUM REQUIREMENTS

The analyses of magnetic read/write processes in previous chapters illustrate that the basic requirements of magnetic recording media for high-density recording are as follows:

1. *High coercivity, H_c.* This is necessary to accommodate very sharp transitions. Coercivities of hard magnetic materials are often determined by magnetocrystalline anisotropy, meaning that high anisotropy materials are required. However, the coercivity cannot exceed the write capability of the corresponding inductive write head.
2. *High remanent magnetization, M_r, but small thickness, δ.* These are needed to provide large enough readback signals with minimum thickness spacing loss. For MR heads, the $M_r\delta$ product must match that of MR element (or spin valve free layer).
3. *Nearly squared hysteresis loop, important to achieve sharp transitions and satisfactory overwrite ratio.*
4. *Very smooth surface and reliable mechanical stability.* These are very important to attaining small magnetic spacing with acceptable tribological performance.
5. *Uniform, small and magnetically isolated magnetic grains.*

The last requirement has not been derived before, but it can be understood considering that magnetic information retrieval is accomplished in the presence of noises, interferences and distortions, which will produce errors in data detection. It will be shown in Chapter 9 that the medium noises in magnetic recording media decrease with grain size. Therefore, magnetic recording layers must consist of *uniform and small grains with little or no intergranular exchange coupling.* This is a very critical constraint in the design and manufacture of magnetic media.

7.2 PARTICULATE MEDIA

As mentioned previously, the first rigid disk drives employed particulate magnetic recording layers because they had already been developed for magnetic tape recording during World War II. Such magnetic media are made of acicular magnetic particles in organic polymer matrix with a volume fraction of 20–45%. These magnetic particles are first prepared through complex chemical reaction process; then they are dispersed in polymer binder, solvents, dispersants, lubricants, etc. The mixture is eventually coated on rotating disk substrates in a fashion similar to painting but at very high speed.[1]

The commonly used particulate magnetic media are listed in Table 7.2. The coercivity of particulate media is mainly due to magnetocrystalline anisotropy and shape anisotropy of the magnetic particles. The γ-iron oxide, the ferromagnetic twin of the nonmagnetic α-iron oxide (rust), is the first widely used magnetic particles. As the required recording densities rise, the coercivities of the particles are also increased. The metal particles and barium ferrite are now the leading edge choices for high-density applications.

The development of thin-film media in late 1980s made apparent that attaining high coercivity and low noise in particular disks was more difficult than in thin-film disks. Therefore, particulate media in rigid disk

TABLE 7.2. Selected Particulate Media[a,b]

Particulate media	Magnetic particles	Coercivity (Oe or 80 A/m)	Saturation magnetization (emu/cc)
Iron oxide	γ-Fe_2O_3	250–400	350–360
Co-modified iron oxide	γ-Fe_2O_3 + Co	380–940	340–360
Chromium dioxide	CrO_2	310–940	360–400
Barium ferrite	$BaFe_{12-2x}Co_xTi_xO_{19}$	630–1900	290–340
Metal particles	Fe_4N, Fe-Ni	690–1130	640–700
	Fe-Co	940–2010	680–920

[a]Ref. 1
[b]1 emu/cc = $4\pi \times 10^{-4}$ T.

drives were quickly replaced by the more superior thin-film media. Now particulate media are exclusively used in floppy drives and magnetic tapes.

7.3 THIN-FILM MEDIA

Thin-film disks have progressed significantly since their introduction in the late 1980s. High-coercivity (2500–3000 Oe) disks are now being employed for high-density recording. As introduced in Chapter 2, magnetic thin disks are complex multilayer structures comprising of aluminum substrate, amorphous NiP undercoat, underlayer (chromium, NiAl, NiP, etc), magnetic film (typically Co-based alloy), overcoat (typically carbon), and lubricant. Here we examine their magnetic properties, microstructures, and fabrication in detail.

7.3.1 Magnetic recording layer

The most important part of a thin-film disk is the magnetic recording layer. The ideal thin-film magnetic recording layer would be composed of high-coercivity and high-anisotropy grains, which are much smaller than recording bit cell, uniform in size, and magnetically isolated (with no exchange coupling). Currently a typical magnetic recording layer in a thin-film disk is Co-based alloy with a hcp crystal structure, including CoCrTa, CoCrPt, CoCrPtTa, CoNiCrPt, etc. The addition of alloying elements[2-7] helps to enhance and control medium coercivity, to minimize the intergranular exchange coupling, and to increase corrosion resistance. It was reported that an ultraclean process can produce higher coercivities in Co-based alloy thin films.[8] High-coercivity alloy films such as SmCo/Cr[9,10] and barium ferrite (BaM)[11-13] have also been proposed as future high-density recording media. The $SmCo_5$ and BaM phase possess higher magnetocrystalline anisotropy field ($H_k = 2K_u/M_s$) than hexagonal close-packed (hcp) Co, where K_u is the uniaxial anisotropy constant and M_s the saturation magnetization.

   The coercivity of a magnetic layer is generally an extrinsic property. On the one hand, it is largely determined by the magnetocrystalline anisotropy and, to a lesser extent, by shape anisotropy of the magnetic grains. On the other hand, the coercivity can be greatly affected by film stress, crystal defects, grain size, grain orientation, grain boundaries, etc. In Chapter 16, we will see that coercivity is also a strong function of switching time and temperature due to the superparamagnetic effect. The correlation between microstructures and coercivity tends to be complicated and qualitative. For example, let us examine how coercivity varies with grain size or medium thickness. As shown in Fig. 7.1, it is common that the coercivity of magnetic thin films increases with decreasing film thickness (or grain size) until it reaches a peak at a critical thickness in the range of 10–30 nm. The decrease in coercivity with increasing film thickness in this range is attributed to a combination of magnetostatic interaction and an increased out-of-plane c-axis orientation of Co grains. For independent magnetic particles, the coercivity peaks at a critical grain size that separates single domain and multidomain regimes. For magnetic thin films, the critical thickness is also dependent on the material and process, and is less understood. Below the critical thickness, the coercivity drops rapidly with decreasing thickness because of the superparamagnetic effect which causes thermally assisted switching of magnetic grains.[14,15]

As linear recording density is scaled up, the remanence-thickness product $M_r\delta$ of the thin-film disk needs to be scaled down. For example, at an areal recording density of ~10 Gb/in^2, the $M_r\delta$ product is expected to be ~0.5 memu/cm^2, and medium thickness to be ~15 nm. This is at the regime where film coercivity may drop quickly with thickness. Therefore, attaining high coercivity in really thin films is an increasingly challenging task.

Exchange coupling, which is responsible for ferromagnetism, originates from the Pauli exclusion principle of electrons, and falls off rapidly with the separation distance between electrons. Exchange coupling between particles is negligible in particulate media because the particles are well separated. In contrast, the grains in a magnetic thin film are so close that magnetostatic and intergranular exchange interaction (coupling) are relatively strong. It was recognized very early that highly coupled mag-

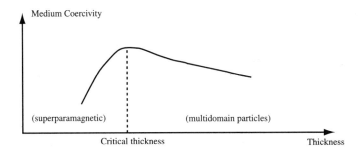

FIGURE 7.1. Typical magnetic thin-film coercivity vs. thickness.

netic thin films with long correlation lengths and zigzag domain walls in recorded transitions tended to display excessive transition noises (jitters).[16,17] In the late 1980s a major breakthrough was achieved by micromagnetic modeling, which clearly elucidated that transition noise was indeed caused by intergranular exchange coupling.[18] Therefore, the reduction of intergranular exchange coupling in thin-film media becomes a starting point in medium design and fabrication.

The interaction among grains can be inferred from remanence magnetization curves.[19–21] The *static remanence curve* $M_r(H)$, also called the *AC-erased isothermal remanence curve*, is measured after AC demagnetization and successive application of a DC magnetic field H. The normalized curve is denoted as $m_r(H) \equiv M_r(H)/M_r(\infty)$. The *DC demagnetizing remanence curve* $M_d(H)$ is measured after DC saturation in one direction and then successive application of a DC field in the opposite direction. The normalized curve is denoted as $m_d(H) \equiv M_d(H)/M_d(\infty)$. For independent particles, the two remanence curves satisfy the Wohlfarth relation:

$$m_d(H) = 1 - 2m_r(H). \qquad (7.1)$$

This relation indicates that the magnetization variation vs. applied field in the DC demagnetizing process should be the same as in AC-erased isothermal magnetization process, but a factor of two arises because the former goes from $-M_r(\infty)$ to $+M_r(\infty)$. The remanence coercivity $H_r [m_d(H_r) = 0]$ is equal to the field $H_{0.5}$ which irreversibly reverses half of the magnetic particles [$m_r(H_{0.5}) = 0.5$]. For interacting particles, we can define that

$$\Delta m \equiv m_d(H) - [1 - 2m_r(H)] = m_d(H) + 2m_r(H) - 1, \qquad (7.2)$$

which indicates the magnitude of interaction qualitatively. A schematic Δm curve for a Co-alloy thin-film medium, which is no longer zero, is shown in Fig. 7.2.

It is believed that a positive peak in the Δm curve is associated with intergranular exchange coupling that favors magnetized state, while a negative peak is associated with magnetostatic interaction that favors demagnetized state. When the applied field is slightly less than coercivity, the "magnetizing" interaction would boost either $m_d(H)$ or $m_r(H)$ compared with noninteracting particles. Near saturation the Δm curve approaches the noninteracting case.

7.3.2 Microstructures of the underlayer and magnetic layer

The microstructures of the magnetic recording layer must be properly designed and well controlled such that desired magnetic and mechanical

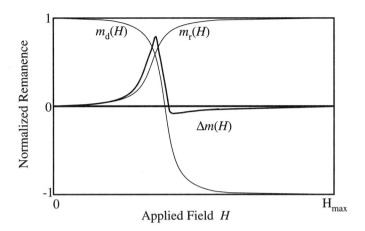

FIGURE 7.2. Schematic remanence magnetization curves.

properties can be attained. In particular, high coercivity and low noise are the primary concerns at a given $M_r \delta$ product. Many techniques and approaches have been developed to accomplish these goals.

Underlayer Design. For hcp Co-based alloy thin films, it is highly desirable to have an in-plane *c*-axis for longitudinal recording because of the resulting high in-plane anisotropy and coercivity. Furthermore, the Co grains need to be uniform and small. These objectives can be achieved by using appropriate underlayers and processes. There have been a variety of underlayers for hcp Co alloy, including Cr, CrV[22] and CrMo.[23] The most popular choice of underlayer is Cr, which has a good epitaxial relation with Co alloy. More recently NiAl is proposed as an alternative underlayer[24,25] because it produces finer grains in magnetic layers.

An underlayer is a polycrystalline film sputter deposited with low-mobility conditions, i.e., low substrate temperatures and high sputtering gas pressures. This corresponds to the "zone 1" structure in the well-known Thornton microstructure zone diagram, and leads to fine columnar grains with voided grain boundaries.[26] The underlayer can be endowed with a desired crystallographic texture. When a magnetic layer is deposited on such an underlayer with similar low-mobility conditions, the resulting magnetic grains tend to have voided boundaries as well, beneficial to reducing intergranular exchange coupling.

It is well known that there are certain epitaxial relations between hcp and bcc crystals, or between fcc and bcc crystals, as listed in Table 7.3. To uniquely specify an epitaxial relation, it is not sufficient to specify the

TABLE 7.3. The hcp/bcc and fcc/bcc Epitaxial Relations[a]

Orientation relationship	Planes (hcp)//(bcc)	Directions [hcp]//[bcc]
Burgers	(0001)//(110)	[11$\bar{2}$0]//[1$\bar{1}$1]
Pitsch-Schrader	(11$\bar{2}$0)//(001)	[0001]//[110]
Potter	(10$\bar{1}$1)//(110)	[1$\bar{2}$10]//[1$\bar{1}$1]

Orientation relationship	Planes (fcc)//(bcc)	Directions [fcc]//[bcc]
Nishiyama-Wasserman	(111)//(110)	[$\bar{1}$01]//[001]
Kurdjumov-Sachs	(111)//(110)	[0$\bar{1}$1]//[$\bar{1}$11]
Bain	($\bar{1}$01)//(001)	[010]//[010]

[a]Ref. 27.

parallel planes alone. The crystallographic directions in the third column, parallel to the corresponding planes, must be aligned to define a unique epitaxial relation. These relations are dictated by the symmetry of crystal structures. Close lattice matching and appropriate processing conditions, such as high-temperature high-mobility growth, are necessary for the epitaxial relation to actually occur.

A common scheme is to exploit the Pitsch-Schrader relation by using a Cr(001) underlayer to orient the c-axis of Co alloy parallel to the disk plane. For a CoCrTa/Cr thin-film medium shown in Fig. 7.3, lattice mismatch is 0.2% and 6.9% along the c-axis and [10$\bar{1}$0], respectively. The epitaxial relation can be written as (11$\bar{2}$0)[0001]//(001)[110]. The c-axis can

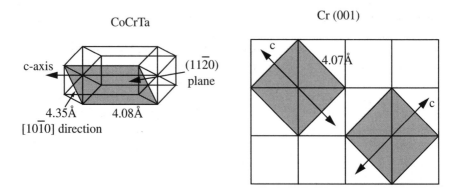

FIGURE 7.3. Pitsch-Schrader epitaxial relation in longitudinal magnetic recording media, CoCrTa(11$\bar{2}$0)[0001] //Cr(001)[110].

be aligned to two equivalent directions: Cr[110] and [1$\bar{1}$0], resulting in a so-called *bicrystal* cluster of Co alloy grains with orthogonally aligned *c*-axes grown on a single Cr(001) columnar grain. This feature may help to reduce medium noise as shown by micromagnetic modeling and experiments. As shown in Fig. 7.4, bicrystal clusters of Co grains are clearly observed by high-resolution transmission electron microscopy (TEM) (HREM) micrographs.[28]

Since Cr thin film grown at a moderate substrate temperature (~260°C) has a (001) *crystallographic texture*, but the [110] axis is randomly oriented in the film plane. Therefore, the *c*-axis of the Co-alloy will also be randomly oriented in the film plane. Then the coercivity of a thin-film disk will be isotropic, and there will be no coercivity anisotropy in circumferential and radial directions. However, magnetic thin-film disks often have circumferential *mechanical texturing* (done at the NiP layer) to prevent stiction between head slider and disk. As a result, the coercivity along the circumferential direction tends to be slightly greater than along the radial direction. The coercivity ratio is sometimes called the *orientation ratio.* An orientation ratio greater than 1 is beneficial to longitudinal magnetic recording. Its origins have been attributed to anisotropic stress,

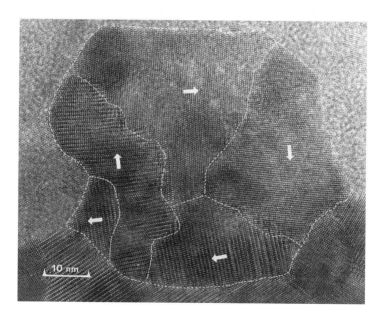

FIGURE 7.4. HREM micrograph showing a bicrystal cluster of Co grains. (Courtesy of Prof. R. Sinclair, Stanford University)

conformal radial roughness, preferential alignment of Co-alloy c-axes along the mechanical texture lines, etc.[29]

Another scheme to align the c-axis in the disk plane is to exploit the Potter relation $(10\bar{1}1)[1\bar{2}10]//(110)[1\bar{1}\bar{1}]$ by using Cr(110) underlayer grown at room temperature to orient the c-axis to *nearly* in-plane, as shown in Fig. 7.5. Note that the $[1\bar{2}10]$ (or [010] in three-axis hexagonal indices) axis can be aligned with $[1\bar{1}\bar{1}]$ in two different ways, but the c-axes (dashed lines) are tilted out of plane. This leads to a so-called *quadcrystal* cluster of Co alloy grains grown on a single Cr columnar grain, with the c-axes pointing at four different directions.[30] So far this scheme has not gained popularity, probably because of the undesirable tilting of c-axes and the complexity involved.

The low-mobility sputtering conditions are applicable to underlayers other than Cr. It was reported that amorphous Ni_xP underlayer instead of Cr leads to reduced intergranular exchange coupling and low medium noise in CoNiPt, CoCrPt, and CoNiCrPt magnetic recording layers.[31] In this case, the Co alloy c-axes are not preferentially oriented; thus such thin film disks are sometimes called *isotropic thin film disks*. If a Co alloy grain has an out-of-plane c-axis, the presence of a strong shape anisotropy will force the magnetization to be in film plane. Therefore, longitudinal recording can still be performed. The main drawback is that the magneto-crystalline anisotropy of Co alloy is not fully utilized and the magnetic hysteresis loop squareness and coercivity orientation ratio may be smaller than those in the thin-film disks with Cr underlayers. Nevertheless, iso-

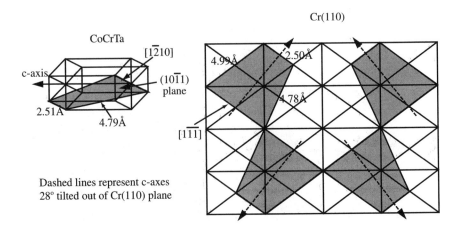

FIGURE 7.5.   Lattice dimensions and epitaxial relations of CoCrTa films and Cr underlayer with $(10\bar{1}1)[1\bar{2}10]//(110)[1\bar{1}\bar{1}]$.

tropic thin-film disks with AlMg substrates have been manufactured by Komag for a long time.

Compositional Segregation. An alternative method to reduce intergranular exchange coupling is to promote compositional segregation in magnetic recording layers. A high substrate temperature ($\sim270°$) during the deposition of the CoCr-based magnetic recording layer can lead to less intergranular exchange interaction and smaller medium noise.[32] Many studies suggest that Cr atoms segregate to Co alloy grain boundaries[33-35] forming Cr-rich phases that diminish intergranular exchange coupling. Tantalum addition into CoCr appears to enhance Cr segregation, resulting in very small medium noise.

Multilayer Magnetic Layer Design. A CoCrPt/CoCrPtSi/Cr dual-magnetic-layer medium with no interlayer[36] was reported in Hitachi 2 Gb/in.^{2} demonstration in 1991. The CoCrPtSi layer has a lower $M_r\delta$ product, a smaller coercivity, and a lower noise, while the CoCrPt layer has a higher $M_r\delta$ product, a higher coercivity, and a slightly higher noise. The appropriate combination of the two layers can produce a higher signal-to-noise ratio than what was obtained with either single magnetic layer.

Multiple magnetic recording layers separated by thin nonmagnetic interlayers have also been proposed. The decoupling between magnetic layers leads to reduced medium noise. However, a laminated recording layer cannot be much thicker than a single magnetic recording layer because of the requirement of small magnetic spacing loss at high-density recording.

7.3.3 Tribological requirements

The magnetic spacing between magnetic recording layer and read/write head tip must be minimized in high-density recording while preserving the reliability of the head-disk interface. This is made possible by the study of friction, wear, and lubrication between surfaces, often called *tribology*. This extremely important subject has been neglected in this book because it emphasizes magnetics and electronics, but it is worthwhile to examine the tribological requirements of thin-film disks here. Experiences tell us that it often takes more effort to meet the tribological requirements than the magnetics requirements.

As mentioned previously, thin-film disks are mechanically textured by polishing NiP surface before depositing subsequent layers. The main function the *mechanical texture* is to diminish the stiction, i.e., the static friction, between head slider and disk when the disk drive is turned off.

The stiction force is proportional to the contact area. Roughening the disk surface reduces the contact area and, thus the stiction force. Otherwise, the head slider may stick and crash upon a stop-and-start operation.

The trade off of the mechanical texture is that the glide height and fly height of the head slider as shown in Fig. 7.6 is increased. The fly height is defined as the distance from the trailing edge of the head slider to the average surface level of the disk. The glide height is the minimum fly height when the slider comes in contact with the disk asperities. The state-of-the-art commercial disks in 1997 have a fly height of ~1 μin. (= 25 nm) and a glide height of ~0.8 μin. (= 20 nm).

The requirements of low glide height and low medium noise dictate that the thin-film disk substrates be supersmooth. For example, at magnetic recording densities ≥ 10 Gb/in^2, the substrate roughness need to be ~1 nm or even smaller. Therefore, the zone texture scheme is devised to get around the conflicting requirements of the smoothness needed for high-density data recording and the roughness needed for head slider landing. As shown in Fig. 7.7, a textured landing zone is prepared typically at the inner radius of the disk where head slider can stop and start as

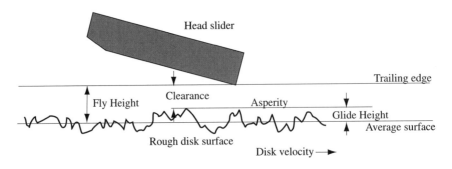

FIGURE 7.6.   Fly height and glide height of a magnetic thin-film disk.

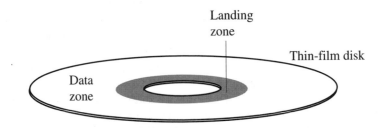

FIGURE 7.7.   Zone texture in thin-film disks. The inner rough zone is for head landing and the outer smooth zone for data storage.

long as the product life. The rest of the disk serves as the data recording zone except for a small fraction near the outer edge of the disk. Clever mechanical load/unload schemes are designed to make sure that the head slider lands only on the landing zone.

It was proposed that disk texturing can be accomplished by a high-power laser beam which burns uniform and small bumps into a selected disk area.[37] This scheme turned out to be a preferred approach for landing zone texture over mechanical polishing. Laser zone texture has gained popularity in the mid-1990s and is now widely used in high-density thin-film disks.[38] The loss of recording area in the laser texture zone is offset by the gain in recording density in the data zone. However, it is possible to make the whole disk supersmooth such that high-density recording can be performed on the whole disk. In this case, load/unload schemes must be designed to ensure that head slider never land on the disk when power is turned off, intentionally or unintentionally.

Generally speaking, the magnetic recording layer (and underlayer) sputtered on disk substrates cannot withstand the friction and wear forces from head slider contact. As a result, overcoat and lubricant must be applied. The most widely used overcoat material is sputtered carbon, often with the addition of H or N. Such carbon films have diamond-like properties including high hardness, good corrosion resistance, and high electrical resistance, which are desirable for rigid magnetic disk applications. Other overcoat materials studied include a-CN_x, ZrO_2, etc. To reduce magnetic spacing between head and disk, the overcoat layers are typically about 10–20 nm thick in the current disk drives. The challenge is to develop overcoat materials that are extremely thin (<10 nm) but still meet all the tribological requirements.

It is a common practice to apply lubricants on contacting surfaces to reduce friction force and wear, and magnetic disk drives are no exception. Disk manufacturers must select a lubricant that can provide adequate wear protection over the life of the disk drives. The most popular choices are perfluoropolyethers (PFPE), a type of polymer with all the hydrogen atoms replaced by fluorine atoms. The desirable attributes of lubricants include high chemical and thermal stability, low vapor pressure, low surface tension, low viscosity, and extremely small thickness (~3 nm or smaller).

7.3.4 Fabrication of low-noise thin-film disks

Efficient large-volume disk manufacturing is essential to the availability of low-cost high-performance thin-film disks. The typical manufacturing process flow can be summarized as follows:

1. The Al-Mg, glass, or ceramic substrates are cut to the appropriate dimensions according the desired form factor. They are then polished abrasively to achieve the desired smoothness.
2. The Al-Mg substrate is plated with a thick NiP undercoat using an electroless plating process. The NiP surface is again polished abrasively, and then textured mechanically or by laser. The glass or ceramic substrate usually does not require a NiP undercoat, and the mechanical textures on the latter substrates are generally isotropic in circumferential and radial directions.
3. Underlayer, magnetic recording layer, and overcoat are deposited on disk substrates using vacuum-deposited processes. This step will be further discussed later.
4. The disks are dipped into a solution of lubricant. The solvent evaporates after the disks are taken out of the solution, leaving a uniform thin lubricant film on disk surfaces.
5. The disks are finally burnished to remove any asperities that could interfere with the flying head slider.

Multichamber sputtering systems emerge as the systems of choice for depositing underlayers, magnetic thin-film layers, and overcoat materials. *Both sides* of the disks are coated simultaneously, as shown schematically in Fig. 7.8, where only one chamber is illustrated. A plasma consisting of argon (Ar) ions and electrons is generated by the applied voltage. The Ar ions bombard the targets and target materials are sputter deposited on the substrates. The deposited film composition is typically similar to the target composition. The sputtering processes have been nearly perfected by engineers over the years. In comparison, electroless plating or evaporation can deposit large quantities of materials rapidly, but cannot meet the versatility and yield requirements for high-performance thin-film disks.

Many scientific and empirical insights have been gained to fabricate low noise high performance thin-film disks. Here is a "recipe" to design your own high performance thin-film disk[39]:

1. Choose a $M_r\delta$ product to match the readback head design.
2. Choose a high-anisotropy alloy *magnetic recording layer* to yield the desired medium coercivity as required by the targeted linear recording density. The coercivity must also be greater than the maximum demagnetizing field, and compatible with the inductive write head capability.
3. Select an *underlayer* that has small grain size, narrow size distribution, right crystal symmetry, and lattice parameters matching the chosen magnetic recording layer.

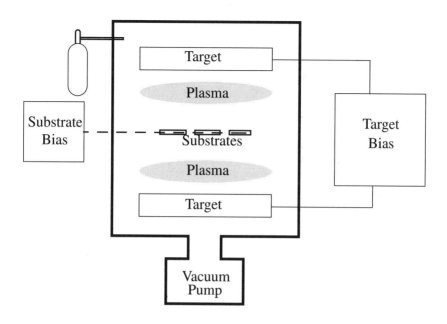

FIGURE 7.8. A schematic of sputtering chamber for magnetic disk coating.

4. If necessary, select a *seed layer* to buffer the underlayer from the substrate and to induce the desired underlayer texture. Select an optional *intermediate layer* to transfer epitaxial growth from the underlayer to the magnetic layer and to provide the magnetic layer with grain-to-grain isolation via diffusion.
5. Choose a sputtering process to induce small magnetic switching units, small intergranular exchange coupling, and desirable crystallographic orientation.

7.3.5 Thermal stability of bits

The grains in a magnetic disk cannot be too small because of thermally assisted switching of magnetization. This is a very complex phenomenon, but a simplified model of thermal switching can be used at least qualitatively: The probability of switching per unit time is

$$f = f_0 \exp\left(-\frac{\Delta E}{kT}\right), \tag{7.3}$$

where f_0 is about 10^9 Hz, k is the Boltzmann's constant, T is the absolute temperature, and ΔE is the energy barrier height associated with the switching process.

For an isolated grain with a volume of V and a uniaxial anisotropy constant of K, the energy barrier in the absence of external field is KV. If the grain is very small, the energy barrier height is also very small. Then, the magnetization will switch very easily, just like an electron. This is called the *superparamagnetic limit* or *superparamagnetic effect*.

How do you estimate the superparamagnetic limit in a magnetic disk? It depends on your criterion. Assume we need to store the bits safely for $t > 10$ year (3×10^8 s), then $f < 3.33 \times 10^{-9}$ Hz. This means that

$$\Delta E > 40kT, \text{ or } V > 40kT/K. \tag{7.4}$$

For current Co-based thin film disks, K is about 2×10^6 erg/cm^3, $T = 300$ K, $kT = 4.14 \times 10^{-14}$ erg, then V should be greater than 8×10^5 Å^3.

More detailed discussion of superparamagnetic limit and magnetic viscosity will be given in Chapter 16.

7.4 KEEPERED MEDIA

Laminated magnetic recording media comprised of a thin high-coercivity recording layer over a *thick* low-coercivity *keeper* layer have been used in perpendicular recording. They are separated by a nonmagnetic interlayer to eliminate the exchange coupling between the recording layer and the keeper. The keeper layer has a high permeability, and thus significantly enhances the efficiency of single pole perpendicular write/read head.[40,41] In this case, the keeper is so thick that it is never saturated. Perpendicular recording can also be implemented with an unkeepered medium[42] or a perpendicular medium with a very thin keeper,[43] which is written by a inductive write head and read by an MR read head.

In longitudinal recording, a keeper is generally not necessary because of the high-efficiency thin-film head and MR head. However, laminated magnetic recording media comprised of a *thin* low-coercivity keeper layer *over* or *under* a high-coercivity recording layer were proposed as an alternative to conventional single-layer longitudinal recording media. Unlike the case of perpendicular recording, the thin keeper layer in longitudinal recording is locally saturated during read/write processes. The addition of a keeper layer is claimed to narrow pulse widths and to improve off-track performance.[44] Most interestingly, it may slow down the thermal decay of recorded bits because of the reduction in demagnetizing field.

However, nonlinear effects such as pulse asymmetry and timing shifts degrade partial-response maximum likelihood (PRML) channel performance. In addition, it is more difficult for an MR head to saturate the keeper than the inductive head, resulting in an amplitude loss. Some of these challenges are discussed in the following.

7.4.1 Effect of keeper on spacing loss

As shown in Fig. 7.9, a highly permeable keeper layer is deposited under the magnetic recording layer. The magnetic recording layer and the interlayer have a relative permeability of ~1, thus can be treated like air in solving the head field. From Section 2.1, the magnetic potential $\phi(x,y)$ in the free space of the 2D keepered head, which is uniform along the z-direction normal to the paper, must obey Laplace's equation. Its Fourier transform (with respect to x) must satisfy Equation (2.9):

$$\frac{\partial^2 \phi(k, y)}{\partial y^2} - k^2 \phi(k, y) = 0. \tag{2.9}$$

The general solution to this equation is

$$\phi(k,y) = \phi_s(k)\frac{A^+e^{ky} + A^-e^{-ky}}{A^+ + A^-},$$

where k (>0) is the wavevector, $\phi_s(x)$ is the surface magnetic potential at the head ABS ($y = 0$), and $A^{\pm}$ are the constants determined by the bound-

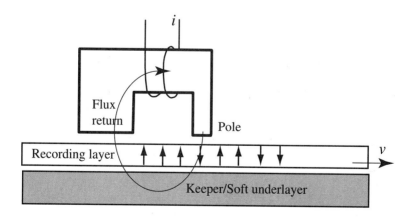

FIGURE 7.9. Schematic single-pole perpendicular head and keepered medium.

ary conditions. If the keeper-head distance is t and the keeper permeability is assumed to be ∞, then $\phi(k,t) = 0$, for the magnetic potential in the keeper must be constant, which can be chosen to be zero. Based on the boundary condition, we obtain that

$$\phi(k,y) = \phi_s(k) \frac{\sinh[k(t-y)]}{\sinh (kt)},$$

$$H_x (k,y) = H_x (k,0) \frac{\sinh[k(t-y)]}{\sinh (kt)}, \qquad (7.5)$$

$$H_y (k,y) = H_y (k,0) \frac{\cosh[k(t-y)]}{\cosh (kt)}.$$

Note that the spacing loss without a keeper layer is exponential:

$$\frac{H_x(k,y)}{H_x(k,0)} = \frac{H_y(k,y)}{H_y(k,0)} = e^{-ky}, \qquad (2.12)$$

which is quite different from the case with a keeper. Consider an example of $t = 0.07$ μm, $y = 0.06$ μm, and disk velocity $v = 10$ m/s. The wavevector is $k = 2\pi f/v$. The spacing loss vs. recording frequency for unkeepered and keepered cases are shown in Fig. 7.10. It is apparent that the exponential spacing loss law is not valid for keepered recording.

The readback amplitude is proportional to $H(k,y)$, thus keepered recording can slow down the signal loss with increased recording frequency. This is quite beneficial for perpendicular recording. However, the signal loss due to the shunting effect of the keeper at DC is so great for longitudinal component that the signal for keepered recording is actually *smaller* than that in unkeepered case. At very high frequencies, all the three cases merge.

7.4.2 Modulated head efficiency model

The trick in longitudinal keepered recording is to locally saturate the keeper such that the shunting effect is removed. However, this may also reduce the beneficial effect of the keeper that slows down signal loss at high frequencies. The schematic of longitudinal keepered recording is shown in Fig. 7.11. During the write process, the thin keeper is most saturated by the write field, so it can be regarded as air. The situation is identical to conventional longitudinal recording. During the read process, however, only a small section of keeper right under the head gap is saturated by a *bias current*, forming a so-called "virtual gap." This virtual gap brings the head gap closer to the recording medium, thus generating

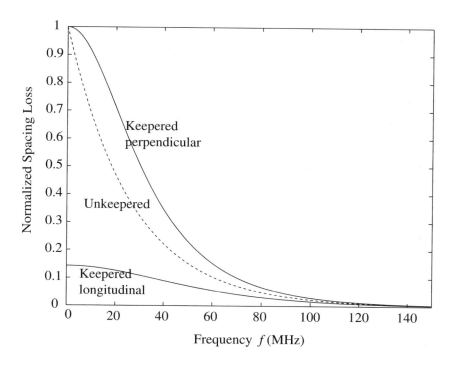

FIGURE 7.10.   Spacing loss vs. recording frequency for keepered and unkeepered cases.

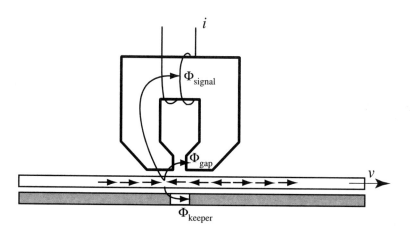

FIGURE 7.11.   Longitudinal keepered recording with thin-film inductive head.

a larger readback signal. If the bias current is not sufficient to generate a virtual gap, the readback amplitude is very small because the flux shunted through keeper does not contribute to the readback signal.

One of the major problems in the above keepered recording scheme is the readback nonlinearity including timing asymmetry and amplitude asymmetry. As shown in Fig. 7.12a, the positive-going peak and negative-going peak are aligned to each other. Not only the peak amplitudes are slightly different, but their peak locations are also shifted. This readback nonlinearity is data pattern dependent, unlike the MR head readback nonlinearity which is memoryless.[45]

The readback nonlinearity can be explained by the *modulated head efficiency model*, also called the *variable reluctance model*, as shown in Fig. 7.12b.[46] The efficiency of a thin-film head is the ratio of the reluctance of the gap to the reluctance of the head magnetic circuit. If head efficiency is high, most of the flux captured by the poles during the read process, $\Phi_{capture}$, will pass through the turns of the coil and thus contribute to the read signal. If head efficiency is low, most of the flux captured by the poles will be shunted across the gap and will not contribute to the read signal. Normally the head efficiency is fixed and flux passing through the turns of the coil, Φ_{signal}, is a linear function of field in the gap. During read-back from keepered media the saturated region of the keeper forms a virtual head gap. The effective gap length is thus the size of the saturated region of the keeper and is a function of the head bias field and the demagnetization field from recorded transitions. When the demagnetization field from the media reinforces the head bias field, more of the keeper layer is saturated and head efficiency increases. Conversely, when the

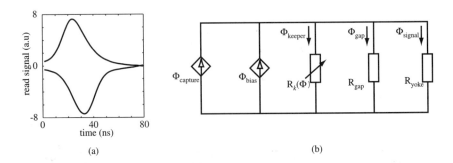

(a)   (b)

FIGURE 7.12. (a) Read-back pulses from keepered media show strong timing asymmetry and amplitude asymmetry (a.u. = arbitrary unit.) (b) An equivalent circuit model to explain nonlinear readback from keepered media. Φ denotes magnetic flux, R denotes reluctance.

demagnetization field from the media opposes the head bias field, less of the keeper layer is saturated and head efficiency decreases. In other words, the reluctance of the keeper, R_k, is now *variable* depending on the fluxes from bias current and recording pattern. This is why the slopes of the isolated peaks are asymmetrical.

7.4.3 Extension to magnetoresistive heads

The concept of the keeper was initially proposed for inductive heads only. With an appropriate combination of inductive head, medium, keeper, and bias current, it is possible to realize a signal gain of ~1.5 dB in keepered inductive readback over its unkeepered counterpart. However, inductive readback has been almost replaced by MR readback by 1998. As a result, a keepered recording scheme must be made compatible with MR read technology.

The bias field generated by the soft adjacent layer (SAL) is generally not enough to saturate the keeper. The total flux from the saturated NiFe SAL is $t_{SAL}WM_s$, and the total flux needed to saturate the keeper (e.g., FeAlSi Sendust, which has the same magnetization as NiFe) is t_kWM_s. If the magnetic coupling efficiency is <30%, then the thickness of the FeAlSi keeper must be

$$t_k < 0.3 t_{SAL} \approx 3 \text{ nm} \quad \text{for a SAL thickness of 10 nm.}$$

However, an effective keeper often need to be ~30 nm thick. It appears that an innovative MR head design, including the addition of an auxiliary bias coil as shown in Fig. 7.13, must be considered to work with keepered media.

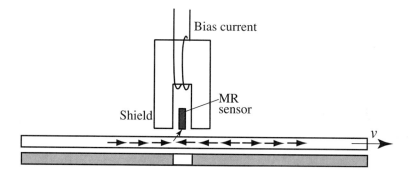

FIGURE 7.13. Schematic MR read head for keepered medium, with additional bias coil.

Note that the keeper shunting problem goes away in perpendicular recording mode, as shown in Figures 7.9 and 7.10. Therefore, an MR head can read effectively from perpendicular keepered media, and no additional bias coil is needed. However, the readback voltage waveform in this case is a square wave, so a differentiator after the MR head is required to generate a voltage pulse for each magnetic transition.

7.5 PATTERNED MEDIA

The idea of patterned media dates back to as early as 1963 when physically etched discrete tracks were first proposed to improve head positioning capability.[47] As will be discussed in Chapters 14 and 15, precisely positioning head sliders on the data track is essential to achieving high data track densities. In the late 1980s an IBM research group investigated submicron discrete tracks using lithographic patterning. The approach reduced medium noises from the track edge and from the guardband between data tracks, which would lead to enhanced offtrack performance of recording head.[48] Applications of patterned media for head servomechanism and read-only memory have been explored by a number of groups.[49,50]

There are many discussions on rewritable patterned magnetic media with a single-bit-per-island recording methodology.[51-53] Both longitudinal and perpendicular recording schemes have been considered, as shown in Fig. 7.14. Note that the shape anisotropy along the long axis of the small islands shown in Fig. 7.14a is quite small such that additional anisotropy such as stress anisotropy can be induced to align the magnetization along the short axis, which is the downtrack direction.[54]

The potential advantages of patterned media include elimination of nonlinear transition shift and track edge noise, reduction of medium noise, and extension of the superparamagnetic limit. The last attribute is probably the most attractive one for patterned media, and will be further discussed next.

(a)   (b)

FIGURE 7.14.   Schematic patterned magnetic islands for (a) longitudinal, (b) and perpendicular recording.

7.5.1 Single-domain particles and superparamagnetic limit

Generally speaking, sufficiently small magnetic particles tend to be in a single-domain state, i.e., the magnetization is uniform throughout the particle and there is no domain wall. The critical volume below which the particle may become a single-domain particle depends on the geometry of the particle. For example, consider a spherical Co particle with a domain wall energy of $\gamma \sim 0.01$ J/m^2. The particle has a uniaxial anisotropy and a radius of r, as shown in Fig. 7.15. The magnetostatic demagnetizing energy of the single-domain state is

$$E_d = \frac{1}{2}\, \mu_0 N_d M_s^2 \cdot \frac{4}{3}\, \pi r^3, \tag{7.6}$$

where N_d is the demagnetizing factor, which is 1/3 for spheres. The total energy for the two-domain state, neglecting the demagnetizing energy, is

$$E_w = \gamma \cdot \pi r^2. \tag{7.7}$$

Therefore, the critical radius r_c is given by[55]

$$E_d = E_w, \, r_c = \frac{3\gamma}{2\mu_0 N_d M_s^2} = \frac{9\gamma}{2\mu_0 M_s^2} \approx 1.8 \times 10^{-8}\text{m} = 18\text{nm}, \tag{7.8}$$

where $M_s = 1.424 \times 10^6$ A/m ($B_s = 1.79$ T) for Co.

The above argument is not general because it compared only two specific domain configurations. Brown proved that the single-domain state for a sufficiently small magnetic particle has a lower energy than any possible multidomain states.[56] It can be derived from the thermodynamics principle that the state of lowest energy for a single crystal homogeneous spherical magnetic body is a single-domain state of uniform magnetization if its radius (r) is smaller than a certain value (r_1), and a multidomain

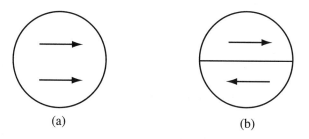

(a)   (b)

FIGURE 7.15. Small ferromagnetic particle: (a) single domain, (b) two-domain. (Easy axis is horizontal.)

state if r is greater than a certain value (r_2). This is called Brown's theorem. However, it made no statement in the intermediate range $r_1 < r < r_2$.

Although Brown's theorem was only rigorously proven for single-crystal spheres or prolate spheroids[57] with uniaxial or cubic anisotropy, it is often applied to small magnetic particles in general. This needs to be cautioned, especially for polycrystalline magnetic particles. It should also be noted that single-domain particles observed in experiments are only approximately uniformly magnetized. Researches on the subject are presently active.

The concept of patterned media offers to push out the superparamagnetic limit several orders of magnitude. In conventional recording a bit must contain ~100 exchange-decoupled magnetic grains as required by signal-to-noise ratio (SNR), each of these magnetic grains must be thermally stable. In contrast, a patterned bit can contain one or any number of exchange-coupled magnetic grains because there is no longer transition noise due to exchange coupling. If only the superparamagnetic limit is concerned, patterned magnetic densities can be ~100 times that of conventional recording if the same medium material is used. In practice, the limit of patterned media will be constrained by patterning defects, signal-to-noise ratio, head design, and servo requirements.

7.5.2 Read/write processes and noises in patterned media

Read/write experiments on patterned nickel (Ni) islands have been performed by magnetic force microscope (MFM) tips and by quasi-static MR heads. All these approaches suffer low data rates. It is an open question what the read/write heads for patterned media will be like.

Imaging that a scaled-down version of the conventional read/write head with its trackwidth and gap length matching the patterned islands can be used instead of MFM tips, then a high data rate can be maintained. In this case the readback signals from patterned media are surprisingly simple. For longitudinal patterned islands, the edge of the island is ideally a sharp step transition (from 0 to $\pm M_r$), which will generate a voltage pulse similar to that from a step transition (from $-M_r$ to M_r), but with half of the amplitude. If the read head is Karlqvist-head-like, and the islands are very thin, then the isolated pulse is

$$V_{sp}(x) \propto \tan^{-1}\left(\frac{x + g/2}{d}\right) - \tan^{-1}\left(\frac{x - g/2}{d}\right), \qquad (3.4)$$

where g is the gap length, and d is the magnetic spacing.

For example, consider longitudinal patterned bits with a duty cycle (island length/island gap) of 50%. The island length is l, $d = g/4$, and the recorded magnetization pattern is "$\rightarrow \rightarrow \leftarrow \leftarrow \leftarrow \rightarrow \rightarrow \rightarrow \leftarrow \rightarrow \leftarrow \rightarrow$," where the arrows represent the remanence magnetization orientation of each island. The data can be coded as "01001001111" with the NRZI rule: 1 if neighboring magnetic islands oppose each other, 0 if they do not. Then the readback signal can be derived by the linear superposition of isolated pulses, as shown in Fig. 7.16. Linear bit shifts still exist in patterned media. However, if $g \approx 2l =$ bit length, then the recorded bits can be recovered readily like conventional longitudinal recording with peak detection (Chapter 11). With more advanced detection schemes such as PRML channels (Chapter 12), even higher recording densities at $g \approx 5l$ $= 2.5 \times$ bit length may be realized.

The main noise sources in patterned media may be from patterning defects such as island irregularities, missing islands, merging islands, etc. A simple model considering island merging/vanishing errors and edge

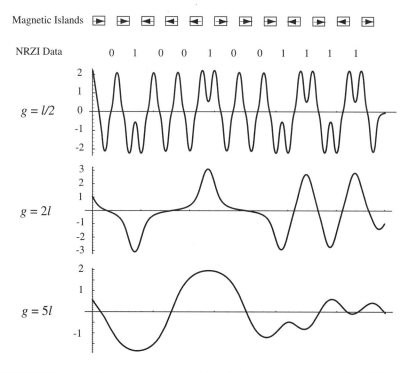

FIGURE 7.16. Readback waveforms of longitudinal patterned media. (Without any equalization.)

roughness errors gives an optimum duty cycle of ~43%.[58] However, the presence of electronic noise will make the optimum duty cycle to be <43%.

Although conventional nonlinear transition shift no longer exists in patterned media because the location of transition is now defined by lithography, several new issues must be resolved to write patterned bits. First, the timing of write current switching must coincide with the period of the islands, which may shift due to mechanical disturbances. Potential solutions include read-while-write for clock generation, and "under-clocked" asynchronous writing which intentionally allows the write timing to be out of phase with island period.[59] The second issue is that how the single-domain particle will behave under write field at high frequencies. If domain switching does not coincide with the write current, or multidomain states arise, additional errors will occur. So far little is known on these subjects.

7.5.3 Challenges in patterned media

The main advantage of magnetic disk drives over semiconductor memories is their low cost per megabyte. This partially derived from the fact that magnetic disks can be manufactured without patterning, and very short bits (<0.1 μm) can be written just by switching write current at high frequencies. The critical dimension in read/write head patterning steps is greater than that in VLSI industry, so the lithographic technology developed there can be leveraged. However, adopting patterned media would change the whole landscape.

The major challenge in patterned media is to find a low cost pattern generation technology. Although the critical dimension required is smaller than that in the VLSI technology, but the patterns in magnetic media are periodic and much simpler than semiconductor memory cells. The proposed concepts include nano-imprinting, laser interference lithography, self-assembly patterning, etc. The subject promises to develop and evolve in the years to come.

References
1. E. Koster, "Particulate media," in *Magnetic Recording Technology,* edited by C. D. Mee and E. D. Daniel, (New York, NY: McGraw-Hill, 1996), p. 3.37.
2. T. Chen and T. Yamashita, "Physical origin of limits in the performance of thin-film longitudinal recording media," *IEEE Trans. Magn.* **24**, 2700, 1988.
3. R. D. Fisher, J. C. Allan and J. L. Pressesky, "Magnetic properties and longitudinal recording performance of corrosion-resistant alloy films," *IEEE Trans. Magn.,* **22**, 352, 1986.

4. K. E. Johnson, L. R. Ivett, D. R. Timmons, M. Mirzamaani, S. E. Lambert, and T. Yogi, "The effect of Cr underlayer thickness on magnetic and structural properties of CoPtCr thin films," *J. Appl. Phys.*, **67**, 4686, 1990.

5. S. Miura, T. Yamashita, G. Ching, and T. Chen, "Noise and bit jitter performance of CoNiPt thin film longitudinal recording media and its effect on recording performance," *IEEE Trans. Magn.*, **24**, 2718, 1988.

6. B. B. Lal, H. Tsai, and A. Eltoukhy, "A new series of quaternary Co-alloys for high density rigid-disk applications," *IEEE Trans. Magn.*, **27**, 4739, 1991.

7. K. Utsumi, T. Inase, and A. Kondo, "Magnetic properties and crystal structures of CoCrTaPt thin film for longitudinal recording media," *J. Appl. Phys.*, **73**, 6680, 1993.

8. M. Takahashi, A. Kikuchi, and J. Nakai, "The ultra-clean sputtering process and high density magnetic recording media," *IEEE Trans. Magn.*, **33**, 2938, 1997.

9. E. M. T. Velu and D. N. Lambeth, "CoSm-based high coercivity thin films for longitudinal recording," *J. Appl. Phys.*, **69**, 5175, 1991.

10. Y. Okumura, O. Suzuki, H. Morita, X. Yang, and H. Fujimori, "Microstructure and magnetic properties of sputtered CoSm/Cr thin films with high coercivities," *J. Magn. Magn. Mater.*, **146**, 5, 1995.

11. T. L. Hylton, M. A. Parker, M. Ullah, K. R. Coffey, R. Umphress, and J. K. Howard, "Ba-ferrite thin film media for high-density longitudinal recording," *J. Appl. Phys.*, **75**, 5960, 1994.

12. X. Sui and M. H. Kryder, "Magnetic easy axis randomly in-plane oriented barium hexaferrite thin film media," *Appl. Phys. Lett.*, **63**, 1562, 1993.

13. J. Li, Preparation and Characterization of Sputtered Barium Ferrite Magnetic Thin Films, Ph.D. dissertation, Department of Materials Science and Engineering, Stanford University, 1995.

14. T. Yogi and T. A. Nguyen, "Ultra high density media: gigabit and beyond," *IEEE Trans. Magn.*, **29**, 307, 1993.

15. B. D. Cullity, *Introduction to Magnetic Materials,* (Medford, MA: Addison-Wesley, 1972), p. 387.

16. R. A. Baugh, E. S. Murdock, and B. R. Nararajan, "Measurement of noise in magnetic media," *IEEE Trans. Magn.*, **19**, 1722, 1983.

17. N. R. Belk, P. K. George, and G. S. Mowry, "Noise in high performance thin-film longitudinal magnetic recording media," *IEEE Trans. Magn.*, **21**, 1350, 1985.

18. J-G Zhu and H. N. Bertram, "Micromagnetic studies of thin metallic films," *J. Appl. Phys.*, **63**, 3248, 1988; J-G Zhu and H. N. Bertram, "Recording and transition noise simulations in thin film media," *IEEE Trans. Magn.*, **24**, 2739, 1988.

19. P. E. Kelly, K. O'Grady, P. I. Mayo, and R. W. Chantrell, "Switching mechanism in Co-P thin films," *IEEE Trans. Magn.*, **25**, 3881, 1989.

20. I. A. Beardsley and J-G Zhu, "Significance of ΔM measurements in thin film media," *IEEE Trans. Magn.*, **27**, 5037, 1991.

21. K. O'Grady, R. W. Chantrell, and I. L. Sanders, "Magnetic characterization of thin film recording media," *IEEE Trans. Magn.*, **29**, 286, 1993.

22. J. K. Howard, "Magnetic recording medium with a chromium alloy underlayer and a cobalt-based magnetic layer," U.S. Patent 4, 652, 499, 1987.

23. I. Okamoto, C. Okuyama, K. Sato, M. Shinohara, "Rigid disk medium for 5 Gb/in^2 recording," paper AB-03, Intermag, Seattle, 1996. (Only digest abstract published.)

24. L-L Lee, D. E. Laughlin, and D. N. Lambeth, "Seed layer induced (002) crystallographic texture in NiAl underlayers," *J. Appl. Phys.*, **79**, 4902, 1996.

25. C. A. Ross, T. P. Nolan, R. Ranjan, and D. Lu, "The role of NiAl underlayers in longitudinal thin-film recording media," *J. Appl. Phys.*, **81**, 7441, 1997.

26. J. A. Thornton, "The microstructure of sputter-deposited coatings," *J. Vac. Sci. Technol.*, **A4**, 3059, 1986.

27. B. Y. P. Wong, *The Effect of Chromium Underlayer on the Structures and Properties of CoNiCr Magnetic Thin Films*, Ph.D. Dissertation, Dept. of Metallurgical Engineering and Materials Science, Carnegie Mellon University, 1991.

28. Courtesy of R. Sinclair, C. Habermeier, and N. Fussing, Stanford University.

29. J. A. Bain, B. M. Clemens, S. M. Brennan, and H. Kataoka, "X-ray diffraction characterization of stress and crystallographic texture in thin film media," *IEEE Trans. Magn.*, **29**, 300, 1993.

30. Y. Deng, *Crystallographically-Oriented CoCrTa Thin-Film Media for High Areal Density Recording*, Ph.D. Dissertation, Department of Electrical and Computer Engineering, Carnegie Mellon University, 1993.

31. T. Yamashita, L. H. Chan, T. Fujiwara, and T. Chen, "Sputtered Ni_xP underlayer for CoPt-based thin film magnetic media," *IEEE Trans. Magn.*, **27**, 4727, 1991.

32. M. Lu, T. Min, Q. Chen, and J. H. Judy, "Effects of morphology on the magnetic properties and transition noise of longitudinal CoCrTa/Cr thin film media," *IEEE Trans. Magn.*, **28**, 3255, 1992.

33. D. J. Rogers, L. N. Chapman, J. P. C. Bernards, and S. B. Luitjens, "Determination of local composition in CoCr films deposited at different substrate temperatures," *IEEE Trans. Magn.*, **25**, 4180, 1989.

34. K. Tang, M. B. Schabes, C. A. Ross, L. He, R. Ranjan, T. Yamashita, and R. Sinclair, "Magnetic clusters, intergranular exchange interaction and their microstructural basis in thin film longitudinal media," *IEEE Trans. Magn.*, **33**, 4074, 1997.

35. Y. Hirayama, M. Futamoto, K. Kimoto, and K. Usami, "Compositional microstructures of CoCr-alloy perpendicular magnetic recording media," *IEEE Trans. Magn.*, **32**, 3807, 1996.

36. M. Futamoto, F. Kugiya, M. Suzuki, H. Takano, Y. Matsuda, N. Inaba, Y. Miyamura, K. Akagi, T. Nakao, and H. Sawaguchi, "Investigation of 2 Gb/in^2 recording at a track density of 17 KTPI," *IEEE Trans. Magn.*, **27**, 5280, 1991.

37. R. Ranjan, D. N. Lambeth, M. Tromel, P. Golia, and Y. Li, "Laser texturing for low flying height media," *J. Appl. Phys.*, **69**, 5745, 1991.

38. P. Baumgart, D. J. Krajnovich, T. A. Nguyen, and A. C. Tam, "A new laser texturing technique for high performance magnetic disk drives," *IEEE Trans. Magn.*, **31**, 2946, 1995.

39. D. N. Lambeth, "The role and control of crystalline texture, isolation, and orientation in magnetic media performance," MRS (Materials Research Society) Spring Meeting, paper L8.1, 1998.

40. S. Iwasaki, Y. Nakamura, S. Yamamoto, and K. Yamakawa, "Perpendicular recording by a narrow track single pole head," *IEEE Trans. Magn.*, **19**, 1714, 1983.

41. H. Hamilton, R. Anderson, and K. Goodson, "Contact perpendicular recording on rigid media," *IEEE Trans. Magn.*, **27**, 4921, 1991.

42. Y. Hirayama, M. Futamoto, K. Ito, Y. Honda, and Y. Maruyama, "Development of high resolution and low noise single-layered perpendicular recording media for high density recording," *IEEE Trans. Magn.*, **33**, 996, 1997.

43. S. Iwasaki, K. Ouchi, and N. Honda, "Gbit/in^2 perpendicular recording using double layer medium and MIG head," *IEEE Trans. Magn.*, **32**, 3795, 1996.

44. B. Gooch, R. Niodermeyer, R. Wood, and R. Pisharody, "A high resolution flying magnetic disk recording system with zero reproduce spacing loss," *IEEE Trans. Magn.*, **27**, 4549, 1991; T. Coughlin, Y. S. Tang, E. Velu, and B. Lairson, "Magnetic recording performance of keepered media," *J. Appl. Phys.*, **81**(8), 4693, 1997.

45. B. Wilson, S. X. Wang, and A. Taratorin, "A generalized method for measuring read-back nonlinearity using a spin stand," *J. Appl. Phys.*, **81**(8), 4828, 1997.

46. B. Wilson, S. X. Wang, and T. Coughlin, "Read-back nonlinearity in longitudinal keepered recording," *IEEE Trans. Magn.*, **34**, 51, 1998; J. Loven, Y. Guo, K. Sin, J. Judy, and J. G. Zhu, "Modeling study of isolated read-back pulses from keepered longitudinal thin film media using the boundary element method," *IEEE Trans. Magn.*, **30**, 4275, 1994.

47. L. F. Shew, "Discrete tracks for saturation magnetic recording," *IEEE Trans. Broadcast & TV Rec.*, **9**, 56, 1963.

48. S. E. Lambert, I. L. Sanders, T. D. Howell, D. McCown, A. M. Patlach, and M. T. Krounbi, "Reduction of edge noise in thin film metal media using discrete tracks," *IEEE Trans. Magn.*, **25**, 3381, 1989.

49. S. E. Lambert, I. L. Sanders, A. M. Patlach, and M. T. Krounbi, "Beyond discrete tracks: other aspects of patterned media," *J. Appl. Phys.*, **69**(8), 4724, 1991.

50. Y. Ohtsuka, T. Koshikawa, K. Aikawa, H. Maeda, and Y. Mizoshita, "A new magnetic disk with servo pattern embedded under recording layer," *IEEE Trans. Magn.*, **33**, 2620, 1997.

51. S. Y. Chou, M. S. Wei, P. R. Krauss, and P. B. Fischer, "Single-domain magnetic pillar array of 35 nm diameter and 65 Gbits/in^2 density for ultrahigh density quantum magnetic storage," *J. Appl. Phys.*, **76**(10), 6673, 1994; S. Y. Chou

"Patterned magnetic nanostructures and quantized magnetic disks," *Proc. of IEEE,* **85**(4), 652, 1997.

52. R. L. White, R. M. H. New, and R. F. W. Pease, "Patterned media: a viable route to 50 Gbit/in^2 and up for magnetic recording?" *IEEE Trans. Magn.,* **33**, 990, 1997.

53. S. Y. Yamamoto, R. O. Barr, and S. Schultz, "MR head response from arrays of lithographically patterned perpendicular nickel columns," *IEEE Trans. Magn.,* **33**, 3017, 1997.

54. R. M. H. New, Patterned Media for High Density Magnetic Recording, Ph.D. Dissertation, Department of Electrical Engineering, Stanford University, 1995.

55. R. M. White, *Introduction to Magnetic Recording,* (New York: IEEE Press, 1985), p. 11.

56. W. F. Brown, Jr., "The fundamental theorem of the theory of fine ferromagnetic particles," *Ann. NY Aad. Sci.,* **147**, Art. 12, 1969; W. F. Brown, Jr., "The fundamental theorem of fine-ferromagnetic-particle theory," *J. Appl. Phys.,* **39**, 993, 1968.

57. A. Aharoni, "Elongated single-domain ferromagnetic particles," *J. Appl. Phys.,* **63**, 5879, 1988.

58. S. K. Nair and R. M. H. New, "Patterned media recording: noise and channel equalization," *IEEE Trans. Magn.,* **34**, 1916, 1998.

59. B. A. Wilson and S. X. Wang, unpublished.

CHAPTER 8

Channel Coding and Error Correction

So far we have only discussed readback voltage waveforms and the means to write/read them, but the question remains how they represent digital information reliably. In this chapter we will discuss the principles of the information encoding for magnetic recording channels: channel encoding and error detection/correction codes. Channel encoding is required in order to satisfy specific requirements of the magnetic recording channel: provide a certain separation between transitions, eliminate continuous regions of uniform magnetization. Error-correction encoding is used to provide additional robustness of the magnetic recording channel to possible errors in the output data stream. All encoding schemes require a certain overhead, i.e., the encoded data contains more bits than the original user data. However, this overhead is the inevitable price, which is paid for by the efficiency and reliability of the magnetic storage devices.

8.1 ENCODING FOR MAGNETIC RECORDING CHANNELS

Prior to writing information on the magnetic medium, this information is encoded. There are several layers of encoding which are schematically illustrated in Fig. 8.1.

The actual bits of information, which have to be stored on the magnetic medium are usually referred to as "user bits" or "user data." The stream of the user bits has to be encoded prior to recording them in the form of changing magnetization pattern.[1,2] After several encoding steps, the stream of the user bits is converted into "channel bits" which are actually written on the magnetic medium. When the recorded information is read from the disk drive, the channel bits are detected in a detector and then decoded to convert them back into decoded user data.[2]

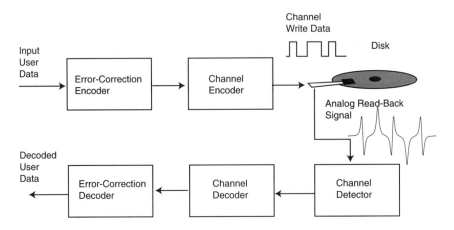

FIGURE 8.1.   Typical magnetic recording channel.

The first step of encoding user bits is usually error-correction encoding.[4,5] The error-correction code (ECC) introduces additional bits of information into the stream of the user bits. While this operation requires some extra disk storage, it improves error rate performance of the disk drive and increases the reliability of the storage device.

The ECC encoding is followed by the channel encoding. This step is necessary due to different system requirements: channel synchronization, data detection and recording density considerations. When the data bits are converted into magnetization pattern written on the magnetic medium, it is important to ensure that the magnetization changes at least once within a certain number of channel bits. In other words, it is necessary to avoid long segments of uniformly magnetized medium. The read-back signal from the uniformly magnetized medium equals to zero, it does not contain any transitions, which creates numerous problems, related to channel synchronization and detection.[2]

First consider the synchronization problem. Magnetic disk drives have special clock recovery system, which synchronizes the writing and reading processes with the incoming data pattern. The clock recovery system is required to compensate for the rotational instability of the motor. This instability is unavoidable due to mechanical limitations. For example, if a magnetic disk rotates at 3600 RPM, then one revolution takes ≈16.7 ms. At a reasonably high density, approximately 500,000 transitions are written on a single track of the disk. The best available motors have instability of rotational speed on the order of about 0.1%. If the clocking signal in the disk drive is made independent of this rotational instability, the phase error at the end of the revolution can be on the order of several thousands of transitions. For example, if the disk at the particular revolution rotates slightly faster, the actual distance between transitions on the magnetic medium will be increased and less transitions will be written on the track. This is unacceptable and phase-locked loops (PLL) are used to provide clock recovery in the magnetic disk drives.

A clock recovery circuit generates the clock signal which tracks and adjusts its frequency to the current variations of the disk rotational speed. PLL adjusts the phase of the oscillator based on the value of the phase error. The phase error of the PLL circuit is calculated by comparing the signal from the current transition with the prior values of the PLL clock. If, for example, the current transition comes a little earlier than expected, and this "early arrival" situation holds for some number of transitions, it means that the disk rotates slightly faster and the clock frequency should be slightly increased. Obviously, when the long stream of zeros is encountered, phase update signals cannot be generated. This makes tracking difficult, resulting in phase shifts and detection errors.

The second reason that transitions should not be separated by long segments of uniformly magnetized medium is related to data detection. For example, the maximum likelihood (ML) detector used in PRML systems makes its decisions based on nonzero transitions data. If a long string of zeroes is met by the ML detector, its path memory may be not enough to determine which sequence is correct. Therefore, when a long zero sequence is processed, the ML detector may produce a burst of errors.

Another purpose of channel encoding is to constrain the distance between adjacent transitions on the magnetic medium in order to avoid intersymbol interference (ISI) and nonlinear distortions. At high densities, linear ISI results in amplitude loss and linear bit shift. In addition, closely spaced transitions are subject to a number of nonlinear distortions, such as nonlinear transition shift and partial erasure, which need to be reduced by means such as channel encoding.

8.2 CHANNEL BIT REPRESENTATION BY NRZ AND NRZI DATA

When actual channel bits of information are written on the magnetic disk, each "0" or "1" is transformed into write current waveform. When write current takes a positive value, it magnetizes the magnetic medium into one direction. A negative write current forces the medium to remagnetize into an opposite direction. Two different conventions are used to describe states of magnetization on the recording medium: NRZ and NRZI.[1,2]

The nonreturn to zero (NRZ) scheme transforms channel bits directly into different directions of the medium magnetization. For example, channel bit "1" means one direction of magnetization and channel bit "0" means the opposite direction. The NRZ scheme is straightforward. A simple example of NRZ encoding for a sequence of channel bits {0100110. . .} is shown in Figure 8.2.

The NRZ scheme is very convenient to describe the write current waveform. However, the read head detects only transitions between different directions of magnetization, i.e., read data constitutes an absolute value of the derivative of the NRZ waveform. Each change of magnetization from "0" to "1" or from "1" to "0" will result in a pulse of voltage in the readback waveform, which will be detected as a transition by a peak

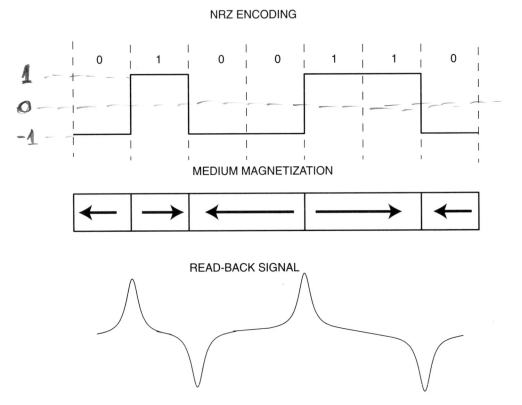

FIGURE 8.2.   Example of NRZ encoding.

detection channel regardless of its sign. Therefore, it is often convenient to represent read waveform in the following manner: denote each "1" as a transition and each "0" as "no transition." This representation is usually called the modified NRZ, or NRZI scheme. An example of the NRZI encoding is shown in Fig. 8.3.

Both NRZ and NRZI conventions are commonly used. The translation of NRZ encoding scheme into NRZI is easily provided by differentiating NRZ data and discarding the transition sign. Consider the following example, from which we can see that the NRZI bit is the absolute value of the difference between the corresponding NRZ bit and its previous NRZ bit.

NRZ data:	0	1	0	0	1	1	0
NRZ derivative:		1-0	0-1	0-0	1-0	1-1	0-1
NRZI data:		1	1	0	1	0	1

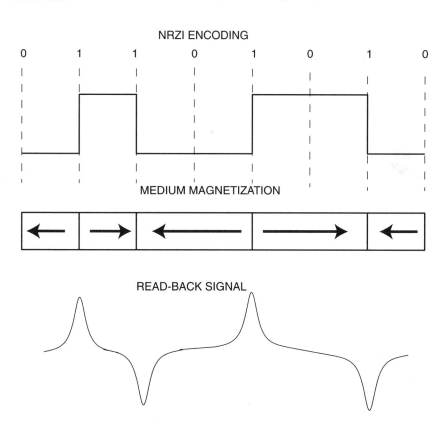

FIGURE 8.3. Example of NRZI encoding.

The difference between NRZ and NRZI schemes can be summarized as follows: In the NRZ scheme, "1" means one direction of medium magnetization, "0" means opposite direction of medium magnetization; in the NRZI (modified NRZ) scheme, "1" means transition, "0" means no transition.

8.3 RUN-LENGTH LIMITED (RLL) CODES

As we mentioned previously, channel encoding is required for synchronization and data detection. One of the early codes used in magnetic recording, which deals with long sequences of zeros in the user data pattern was the frequency modulation (FM) code.[2] This code is based on the following simple encoding rule (in NRZI notation): Each "1" or "0" is always followed by an extra "1":

$$1 \Rightarrow 11,$$
$$0 \Rightarrow 01.$$

For example, the user data 110000 will be FM encoded as 111101010101. Note that now we have many adjacent transitions "11", and, instead of 1 user bit we now have to write 2 bits of information, i.e., we lose 50% of the disk space to channel encoding.

To improve FM code, a modified frequency modulation (MFM) code was proposed. The MFM code maps the user data into the encoded pattern according to the following rule:

$$1 \Rightarrow 01,$$
$$0 \Rightarrow x0,$$

where x is the complement of the preceding bit in the coded pattern. For example, user data 1100011 will be MFM encoded as 01010010100101. We still need to write twice as many bits as we have in the user data. However, if the minimum distance between transitions is fixed, the MFM encoded bit rate can be twice as high as the FM encoded bit rate because there is at least a "0" between two "1"s, which offsets the disk space lost to MFM encoding. Although there is no net gain in data storage capacity over the NRZI user data, at least the synchronization problem has been solved by avoiding long streams of "0"s.

The FM and MFM codes are the simplest realizations of a more general concept: *run-length-limited* (RLL) *codes* or *modulation codes*. RLL codes limit the minimum and maximum number of consecutive zeros allowed in the encoded pattern. It is often more convenient to describe RLL codes using the NRZI convention, i.e., "1" means a transition and "0" means absence of transition. RLL codes are typically characterized by four parameters: m, n, d, k, and are referenced as (m/n)(d,k) codes[2,3]:

1. The modulation code maps m user bits into n encoded bits, and $n \geq m$. The encoded pattern always contains more bits than the user pattern.
2. The minimum allowed number of consecutive "0"s between two "1"s is d, and $d \geq 0$.
3. The maximum allowed number of consecutive "0"s between two "1"s is k, and $k \geq 0$.

For example a popular (1,7) code will encode an arbitrary user pattern so that two transitions will always be separated by at least one "0" and the maximum length of the zero string which will be met in the encoded pattern is 7. The MFM code is a particular case of RLL code which has the following parameters: $m = 1$, $n = 2$, $d = 1$, $k = 3$. In other words, MFM code is a (1/2)(1,3) RLL code.

The *code rate* of a RLL code ($= m/n$) describes the ratio of the user bits to the encoded channel bits. For MFM code, the code rate equals to 1/2, meaning that twice as much information as the user data is written on the magnetic medium, which is not that efficient. The higher the code rate, the lower the loss of disk space to encoding. It is desirable to have a code rate as close to 1 as possible. There are several popular and efficient RLL codes used in magnetic recording,[1,2] which will be discussed next.

The 2/3(1,7) code has at least one "0" between two adjacent transitions and at most seven consecutive zeros. It encodes each 2 user bits into 3 bits, therefore it has a rate of 2/3. Several realizations of 2/3(1,7) code are possible. A typical encoding table for 2/3(1,7) code is shown in Table 8.1.

TABLE 8.1. An Encoding Table for 2/3(1,7) Code

User bits	Encoded bits
00	101
01	100
10	001
11	010
0000	101000
0001	100000
1000	001000
1001	010000

The first four rows of this table provide a simple translation rule which substitutes 3 channel bits for 2 user bits. However, a certain extension of this 2-to-3 rule is required, which is given by the last four rows, called the substitution rows or substitution table. For example, if we use the first row of the table to encode user bits "0000", the encoded data will be "101101", which violates the $d = 1$ constraint because of a string of "11" in the encoded bits. Therefore, the substitution encoding table is required to satisfy the $d = 1$ constraint.

Another popular modulation code is the 1/2(2,7) code, which has at least two consecutive zeros between transitions, but requires writing two channel bits for each user bit as shown in Table 8.2.

A popular code widely used in digital sampling (PRML) disk drives is (8/9)(0,4/4) code. In this code each 8 user bits of data are encoded into 9 channel bits.[5] At the same time, there is no constraint on transition separation, i.e., two transitions may be written without additional zeros between them. The notation (4/4) appears because of the specific realization of PRML channel when a stream of user bits is split into odd and even bits. It indicates that no more than four consecutive zeros can be present in either the odd or even data sequence. The encoding table for (8/9) (0,4/4) code is complicated. It consists of 256 8-bit words which are translated into 9-bit sequences, e.g., the word "00000000" is translated into "011101110" and the word "11111111" into "111101111".

Another popular code that is used in PRML channels is the (16/17)(0,6/6) code.[5] This code translates 2 bytes of user data (16 bits) into 17 bits of channel data, has $d = 0$ constraint, and allows no more than six consecutive zeros in either odd or even interleave channels. This code has a better code rate than the 8/9 code. Only 1 extra channel bit is required for 16 user bits in the 16/17 code compared with 1 extra bit per 8 user bits in the 8/9 code.

TABLE 8.2.   An Encoding Table for 1/2(1,7) Code

User bits	Encoded bits
10	0100
11	1000
000	000100
010	100100
011	001000
0010	00100100
0011	00001000

There is a certain theoretical limit on code rate which can be achieved under a specific (d, k) constraint. This limit is called the *capacity* of the modulation code. For example, the capacity of (1,7) codes is 0.679. The 2/3(1,7) code has a code rate of 0.67, which is very close to the optimum.

8.4 USER DATA RATE, CHANNEL DATA RATE, AND FLUX FREQUENCY

Since user data pattern is encoded prior to writing it on a magnetic disk, the parameters of user data pattern and encoded channel data pattern are to be treated separately. One must be very careful to distinguish channel data from user data.[2] Most important is to distinguish the *user data rate* (Mbit/s) from the *channel data rate* (Mbit/s), or the *flux change frequency* (Mflux/s) in the magnetic media.

The user data rate (UDR) is defined as the number of user data bits per second that are recorded in the magnetic disk. Encoded extra channel bits are not counted in the UDR definition. In today's commercial disk drives UDR can be 100 Mbit/s or more.

The channel data rate (CDR) is defined as the number of encoded bits per second that are recorded in the disk (after encoding). When RLL codes are used, m user bits are encoded into n channel bits. Therefore,

$$CDR = (n/m) \times UDR > UDR. \qquad (8.1)$$

If the 2/3(1,7) code is used, UDR = 100 Mbit/s, then CDR = 150 Mbit/s. It means that the disk drive electronics encodes the user data and writes to magnetic media at a channel data rate that is 1.5 times as high as the user data rate.

The *flux frequency* in magnetic media is usually defined as the maximum rate of magnetic flux change, which is equal to the inverse of the minimum distance between two transitions multiplied by the channel data rate. Therefore,

$$\text{Flux frequency} = \frac{CDR}{(d + 1)} \qquad (8.2)$$

A simple relation between the user data rate and the flux frequency can be derived from Equations (8.1) and (8.2):

$$\frac{UDR}{\text{Flux frequency}} = (d + 1)\frac{m}{n}. \qquad (8.3)$$

If the UDR is 100 Mb/s and the 2/3(1,7) code is used for channel encoding, then the flux frequency is (3/4) × UDR = 75 Mflux/s. Equation (8.3) demonstrates that the ratio between the user data rate and the flux frequency is higher for $d = 1$ constraint codes than for $d = 0$ constraint codes. Let us consider what will happen if we encode the same user pattern using (8/9)(0,4/4) code. UDR = 100 Mbit/s, so CDR = (9/8) × 100 = 112.5 Mbit/s. At first glance, the (8/9) code appears to be much better than the (2/3) code since we can write channel data at a lower rate (112.5 Mbit/s vs. 150 Mbit/s) while attaining the same user data rate. However, an examination of the flux frequency reveals the disadvantage of the (8/9) code. Since $d = 0$, the flux frequency when using the 8/9(0,4/4) code is 112.5 Mflux/s, which is significantly larger than 75 Mflux/s for the 2/3(1,7) code. We should not forget pulse interactions in the magnetic medium, which will limit the minimum transition distance that can be allowed. Therefore, it is not a simple matter to judge which code is better. The 2/3(1,7) code is more advantageous in terms of lowering flux frequency, while the 8/9(0,4/4) code is more advantageous in terms of lowering the channel data rate.

Other important parameters include *channel density, channel bit period, user density, user bit period, flux density,* and *flux change period.* The channel density is defined as

$$D_{ch} \equiv \frac{PW_{50}}{T_{ch}},$$

where PW_{50} is the width of an isolated pulse at 50% of its amplitude and T_{ch} is the channel bit period. The channel density describes how many channel bits of information are stored on a single pulse from the medium. The channel bit period T_{ch} is the distance between two adjacent channel bits, so the channel data rate is CDR = $1/T_{ch}$.

The user density is defined as

$$D_u \equiv \frac{PW_{50}}{T_u} = D_{ch}\frac{m}{n}, \tag{8.4}$$

where T_u is the user bit period and equal to $(n/m) × T_{ch}$. The user density measures how many user bits of information are stored on a single pulse from the medium.

Similarly, the flux density is defined as

$$D_f \equiv \frac{PW_{50}}{B} = \frac{D_{ch}}{d+1} = \frac{nD_u}{m(d+1)},$$

where B is the flux change period, i.e., the minimum distance between transitions. The flux density measures how close transitions are written with respect to the width of an isolated pulse, which is often limited by signal-to-noise considerations, linear ISI, nonlinear distortion, etc. The flux density is most relevant to head and medium engineers before considering channel encoding. However, we must remember that the relation among the flux density, channel density, and user density depends on the modulation code chosen for channel encoding.

8.5 PRINCIPLES OF ERROR DETECTION AND CORRECTION

Prior to RLL encoding, the user pattern is encoded with special error correction codes (ECC) to ensure the highest fidelity of recorded information. ECC codes are capable of correcting errors that may occur during the detection of a data pattern.[3–5] The errors may arise from noises, nonlinear distortions, and/or interferences, which will be discussed in Chapters 9, 10, and 14, respectively.

There are generally two types of errors that occur in magnetic recording: single-bit errors and bursts of errors. Single-bit errors typically occur due to a single short-duration noise event, which results in an extra pulse or a missing pulse. Burst errors occur when a group of bits is detected erroneously. Burst errors may occur due to defects of magnetic medium such as a scratch or a defective spot spanning over many bit periods. Burst errors may also occur at the output of the ML detector in a PRML channel. The ML detector decision contains a group of channel bits. Any error event in the ML detector will result in detecting a wrong sequence of bits. Figure 8.4 illustrates a single-bit error and a burst of three errors that occur in a peak detection channel when the signal amplitude is reduced.

Losing the information in magnetic recording devices should be an extremely low-probability event. Typically magnetic recording systems must provide a probability of error less than 10^{-9}. The reliability of data recovery may be greatly boosted by using ECCs. Error correction works by adding some special bits to the user data pattern. Different ECC are designed to provide the correction of a finite number of corrupted bits. Like any encoding schemes, ECC encoding requires more bits of information to be stored on the medium. However, once we are guaranteed that a certain number of errors are corrected by ECC, we can afford to increase the recording densities until reaching a required bit error rate (BER). Therefore, the storage capacity with ECC can in fact be the same as or

Single bit error

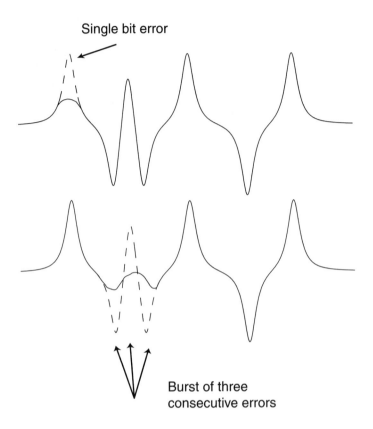

Burst of three
consecutive errors

FIGURE 8.4.   Illustration of single bit error and a burst of errors.

even more than that without ECC. This will be clearly seen from the
following example.

Consider a simple error correction code capable of correcting a
single error in a group of 16 bits, which requires the transmission of
extra 8 bits of information for each 16-bit block. Therefore, to provide
this error correction capability we have to write 50% more data on
the disk. However, now an error will occur only if 2 or more bits out
of 16 are in error. If the probability of a single-bit error is p_0, then the
probability of 2 bits being in error equals p_0^2 and we have a total of
120 possible 2-bit errors in a 16-bit sequence. Therefore, the probability
of error using ECC is $P \approx 120p_0^2(1 - p_0)^{14}$. If we fix the error rate at
$P = 10^{-9}$, then p_0 is $\sim 3 \times 10^{-6}$. If the error rate is determined by

random noise, it is possible to calculate the SNR which is required to achieve a specific BER for different detection methods. For example, to achieve BER $= 10^{-9}$ without ECC in a typical peak detection channel, the SNR of about 22 dB is required. To achieve BER $= 3 \times 10^{-6}$, it is enough to provide SNR $= 19$ dB, which gives us a gain of ~3 dB to spare. The reduction in the SNR (dB) required to achieve a specified error performance due to ECC is called the *coding gain*. In this example, the coding gain is ~3 dB.

The coding gain in SNR allows increased densities under the same target error performance. If the main source of noise is the medium noise, then a reduction of track width by a factor of 2 will result in losing SNR by 3 dB (Chapter 9). In the above example, using ECC need 50% more disk space for extra bits, but the coding gain of 3 dB allows us to record 100% more information by doubling the track density. The net gain from ECC is 50% more storage capacity.

Of course, the calculation above did not take into account of cross-track interference, head positioning accuracy, etc. However, it is good enough to show that the disk space given to ECC can be more than gained back from the coding gain in SNR.

8.5.1 Error detection by parity checking

Most error detection and correction codes can be traced back to the principle of parity checking.[3,4] A simple parity checking scheme can be implemented by using an exclusive OR (XOR) circuit as shown in Fig. 8.5.

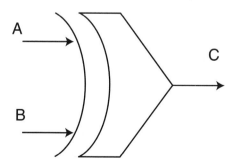

FIGURE 8.5.   XOR circuit.

The XOR circuit shown in Fig. 8.5 performs an exclusive OR operation on the input bits A and B. The output of this circuit is $C = A \oplus B$: "0" if the two input bits A and B are the same, and "1" if the two input bits are different. The table of XOR operation is shown in the following table:

A	B	C = A $\oplus$ B
0	0	0
0	1	1
1	0	1
1	1	0

If {A, B} is the word consisting of two user bits, then an extra bit $C = A \oplus B$ can be added to form a codeword {A, B, C}. Note that due to the property of XOR operation, there must be an even number of 1s in each row of the above table, meaning that the *parity* is even. If we now transmit three bits and the number of 1s is not even, it means that at least one of the detected bits contains an error. Similarly if a four bit sequence {A, B, C, D} is transmitted, a parity bit P can be added to form a codeword {A, B, C, D, P}, where $P = A \oplus B \oplus C \oplus D$. Again, every codeword {A, B, C, D, P} has an even number of 1s. If a codeword {10100} is detected with an error as {10000}, then the number of 1s in a codeword is odd. This is shown in the following table:

A	B	C	D	P
1	0	1	0	0
1	1	1	0	1
0	1	1	0	0
0	1	0	0	1

$$P \oplus A \oplus B \oplus C \oplus D = \begin{cases} 1, & \text{error;} \\ 0, & \text{no error.} \end{cases}$$

The error above is detected but can not be corrected, i.e., we do not know which bit was in error. To detect which bit was in error, we can construct a "crossword" table[3] and perform a parity check on both rows and columns as shown in the following table for a 16-bit word {A1, . . . ,A16}:

A1	A2	A3	A4	P1
A5	A6	A7	A8	P2
A9	A10	A11	A12	P3
A13	A14	A15	A16	P4
CP1	CP2	CP3	CP4	CP5

$$P1 \quad = A1 \oplus A2 \oplus A3 \oplus A4,$$

. . .

$$P4 \quad = A13 \oplus A14 \oplus A15 \oplus A16,$$
$$CP1 = A1 \oplus A5 \oplus A9 \oplus A13,$$

. . .

$$CP4 = A4 \oplus A8 \oplus A12 \oplus A16,$$
$$CP5 = P1 \oplus P2 \oplus P3 \oplus P4.$$

Now, if one bit is in error, for example, A10, then P3 and CP2 are wrong and we can locate an error and correct it. However, if two bits are in error, e.g., A6 and A14, this scheme will be "fooled" and no error can be corrected. Therefore, this "crossword" type of table is capable of detecting and correcting a single error in data bits.

In the preceding discussions we have assumed that errors occur at data bits rather than the parity bits. In reality, the parity bits can be wrong as well, so more practical ECC should assume that errors can occur at any encoded bits.

8.5.2 Hamming distance and error detection/correction theorems

Different error detection and correction codes can be constructed. To understand what can and what can not be done with ECC, it is important to introduce the concept of *Hamming distance* between two codewords. It is defined as the number of bits that differ from each other in the two codewords.[3] For example, consider two codewords $\mathbf{A} = \{1, 0, 1, 0\}$ and $\mathbf{B} = \{1, 1, 1, 0\}$. The Hamming distance is $d(\mathbf{A}, \mathbf{B}) = 1$ because only the second bit in $\mathbf{A}$ and $\mathbf{B}$ is distinct.

The formal definition of the Hamming distance between two code-words $\mathbf{A} = \{a_1, \ldots, a_k, \ldots, a_N\}$ and $\mathbf{B} = \{b_1, \ldots, b_k, \ldots, b_N\}$ is the sum of the XOR operation between the corresponding bits in each codeword:

$$d(\mathbf{A}, \mathbf{B}) = \| \mathbf{A} - \mathbf{B} \| = \sum_{k=1}^{N} (a_k \oplus b_k). \tag{8.5}$$

Since XOR operation equals "1" when bits are different, this sum will result in the Hamming distance.

Hamming distance can be considered as the geometric distance between two vectors in an N-dimensional space, where the calculations are based on *modulo-2 addition*:

$$1 + 1 = 0,$$
$$0 + 0 = 0,$$
$$0 + 1 = 1,$$
$$1 + 0 = 1,$$

where the arguments can only be 0 or 1. Obviously, modulo-2 addition of **A** and **B** is equivalent to XOR operation $\mathbf{A} \oplus \mathbf{B}$. Sequences of user bits (messages) are encoded into codewords so that the Hamming distances between the codewords are larger than the original Hamming distances between messages. This means that the size of the codeword should usually be larger than that of the message. The encoding process is illustrated in Fig. 8.6. Among all the codewords we can find the two closest codewords having the minimum Hamming distance, which is called the *minimum distance* (d_{min}) of the ECC code. Error detection and correction capabilities of any code are determined by the minimum distance, as formulated by the error detection/correction theorems.[4]

The *error detection theorem* states that all the errors having a Hamming distance $d < d_{min}$, where $d_{min} > 1$, can be detected. If the distance between codewords is at least 2, then any single-bit error will have a distance from the corresponding codeword equal to 1 and this error can be detected.

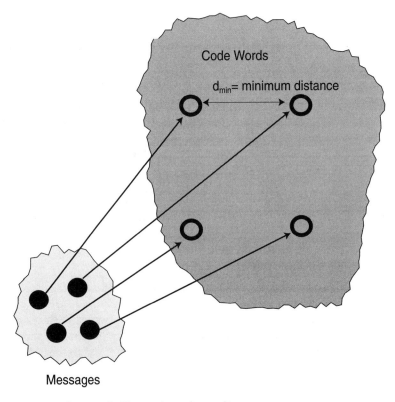

FIGURE 8.6.   Geometric illustration of encoding process.

Let us revisit the parity check example, where the codeword is {A, B, C, D, P}, with P as the parity bit. The Hamming distance between any two codewords listed in the table is 2. For example,

Codeword 1:	1 0 1 0 0,
Codeword 2:	1 1 1 0 1,
Hamming distance:	0+1+0+0+1 = 2.

It means that a single error in this codeword can be detected.

The *error correction theorem* states that all the errors with a Hamming distance $d < d_{min}/2$, where $d_{min} > 2$, can be corrected. For example, a single bit error will result in a corrupted codeword having a Hamming distance of $d = 1$ from the correct codeword. If $d_{min} = 3$, then the corrupted codeword is closer to the correct codeword ($d = 1$) than to any other codewords ($d = 2$). Therefore, the correct codeword can be identified based on the simple distance consideration. In comparison, multiple-bit errors can not be corrected reliably here. A 2-bit error will result in a corrupted codework having a Hamming distance of 2 from the correct codeword, but we can not tell which is the correct codeword based on the distance consideration anymore.

The error detection and correction theorems show that we can construct one-dimensional ECC codes to provide error detection and correction instead of the crossword type of coding. There are many ways to implement ECC codes, and we will consider the most popular ECC in magnetic recording: cyclic codes.

8.5.3 Cyclic codes

Let $\mathbf{U} = \{u_0, u_1, \ldots, u_{n-1}\}$ be an encoded sequence of bits, often called a *code vector*, in an ECC code. The code is called a *cyclic code* if any code vector obtained from $\mathbf{U}$ by cyclic (end-around) shifts is also a code vector. This means that an end-around-shifted code vector such as $\mathbf{U}^{(1)} = \{u_1, u_2, \ldots, u_{n-1}, u_0\}$ can also be obtained by applying the same encoding procedure to a different sequence of input bits. The cyclic codes are based on a well-formalized mathematical theory.[4,5] Encoding and error correction devices for cyclic codes are based on relatively simple circuits with digital shift registers.

A convenient representation of the code vector $\mathbf{U} = \{u_0, u_1, \ldots, u_{n-1}\}$ is a polynomial with the corresponding coefficients[4]:

$$U(X) = u_0 + u_1 X + u_2 X^2 + \cdots + u_{n-1} X^{n-1} \qquad (8.6)$$

Any cyclic shift of the code vector is equivalent to multiplying $U(X)$ by X^i, where i is an integer. For example,

$$XU(X) = u_0X + u_1X^2 + u_2X^3 + \cdots + u_{n-1}X^n = u_{n-1} + u_0X + u_1X^2$$
$$+ u_2X^3 + \cdots + u_{n-2}X^{n-1} + u_{n-1}(X^n + 1).$$

Note that $u_{n-1} + u_{n-1} = 0$ according to modulo-2 addition. More generally speaking, if $U(X)$ is a code vector of a cyclic code, then any remainder $U^{(i)}(X)$ resulting from dividing $X^iU(X)$ by $(X^n + 1)$ is also a codeword:

$$\frac{X^iU(X)}{X^n + 1} = q(X) + \frac{U^{(i)}(X)}{X^n + 1'} \qquad (8.7)$$

or

$$U^{(i)}(X) = X^iU(X) \text{ modulo } (X^n + 1), \qquad (8.8)$$

where the quotient $q(X)$ is any polynomial of X, and "x modulo y" is defined as the remainder obtained from dividing x by y.

Any cyclic code can be generated by a generator polynomial.[4,5] If a message vector has a length of $k = 4$ bits and is encoded to a code vector with a length of $n = 7$: {A, B, C, D, E, F, G}, then the cyclic code is called a (7,4) cyclic code and can be generated by a unique generator polynomial of degree $r = n - k = 3$. In general, for a (n, k) cyclic code, the generator polynomial is

$$g(X) = g_0 + g_1X + \cdots + g_rX^r. \qquad (8.9)$$

where $r = n - k$. Note that the generator polynomial is always a factor of $X^n + 1$, i.e., it can he divided by $X^n + 1$ without a remainder.

The message polynomial of degree $k - 1$ for a sequence of message bits $\{m_0, m_1, \ldots, m_{k-1}\}$ is

$$m(X) = m_0 + m_1X + m_2X^2 + \cdots + m_{k-1}X^{k-1} \qquad (8.10)$$

Encoding from the generator polynomial is given by the product of the two polynomials:

$$U(X) = u_0 + u_1X + u_2X^2 + \cdots + u_{n-1}X^{n-1} = m(X)g(X). \qquad (8.11)$$

Now it is simple to understand the principle of error detection. When the code vector $U(X)$ is received, it is divided by generating polynomial $g(X)$. When the received vector is not corrupted, the result of this division equals the ideal message: $U(X)/g(X) = m(X)$. However, when the received code vector is corrupted, it becomes $R(X) = U(X) + e(X)$, where $e(X)$ is the error term. The division operation in this case results in a nonzero remainder:

$$\frac{R(X)}{g(X)} = \frac{m(X)g(X) + e(X)}{g(X)} = m(X) + \frac{e(X)}{g(X)} \qquad (8.12)$$

The remainder of $e(X)$ divided by $g(X)$ is called the *syndrome*,[5] which can be expressed as

$$S(X) = e(X) \text{ modulo } g(X).$$

Zero syndrome means that no errors are detected, so $e(X) = 0$ or is dividable by $g(X)$ without a remainder.

Encoding is usually performed in such a way that the code vector **U** consists of $(n - k)$ parity bits and k message bits: $\mathbf{U} = \{r_0, r_1, \ldots, r_{n-k-1}, m_0, m_1, \ldots, m_{k-1}\}$. The parity bits must satisfy the cyclic code property that polynomial $U(X)$ can be divided by a generator polynomial $g(X)$ without a remainder. Therefore,[4]

$$U(X) = r(X) + X^{n-k} m(X) = q(X)g(X), \qquad (8.13)$$

or

$$r(X) = X^{n-k} m(X) \text{ modulo } g(X). \qquad (8.14)$$

where $q(X)$ is a quotient polynomial. In other words, to find the parity bits, the message polynomial is first "up-shifted" to occupy the rightmost positions in a coadeword, and then divided by the generator polynomial. The remainder is the parity bits.[4,5]

The beauty of the cyclic codes is that two polynomials can be divided using a simple feedback circuit consisting of modulo-2 adders (XOR gates) and shift registers.[5] The example of a circuit for dividing any arbitrary polynomial $V(X)$ by another polynomial $g(X)$ is shown in Fig. 8.7. All the

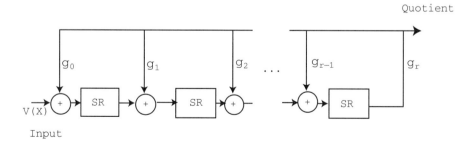

FIGURE 8.7. Shift register for dividing polynomial $V(X)$ by $g(X)$.

shift registers are initially reset to zero. The first k shifts propagate k most significant coefficients of $V(X)$ until the last one appears at the feedback loop. This is the first serial value of the quotient. The feedback loop value is added (modulo-2) to the inputs of each register with a feedback tap, and the result is kept at the corresponding register. At each step, the operation is shifted one stage. After all the coefficients of the polynomial $V(X)$ pass through the first shift register (SR), all the serial values of quotient will have passed the last SR, and the remainder coefficients are represented by the states of all the SRs. This procedure finds a quotient polynomial $q(X)$ and a remainder polynomial $r(X)$:

$$V(X) \, / \, g(X) = q(X) + r(X) \, / \, g(X).$$

For example, consider a simple circuit shown in Fig. 8.8, which divides input polynomial $V(X) = X^3 + X^5 + X^6$, or code vector $\mathbf{V} = \{0,0,0,1,0,1,1\}$, with the generator polynomial $g(X) = 1 + X + X^3$. The sequence of operations is illustrated in the following table:

Input bits	Shift steps	Content of registers	Output bits
0 0 0 1 0 1 1	0	000	—
0 0 0 1 0 1	1	100	0
0 0 0 1 0	2	110	0
0 0 0 1	3	011	0
0 0 0	4	011	1
0 0	5	111	1
0	6	101	1
—	7	100	1

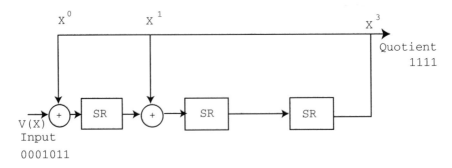

FIGURE 8.8.   Shift register for dividing by $g(X) = 1 + X + X^3$.

If we divide these two polynomials, we get $q(X) = 1 + X + X^2 + X^3$ and $r(X) = 1$. Therefore, this circuit indeed performs polynomial division.

Based on Equation (8.14), it is necessary to divide the "up-shifted" message polynomial by the generating polynomial to find parity bits.[4,5] This can be done in an encoding shift register shown in Fig. 8.9. Switch 1 is closed during first k shifts to allow transmission of the message bits into the $n - k$ stage encoding register. During that time switch 2 is "down" to allow transmission of the message bits to the register output. After k shifts switch 1 is open and switch 2 is moved to the "up" position to allow transmission of parity bits to the register output. As a result, the output encoded vector equals $U = \{r_0, r_1, \ldots, r_{n-k-1}, m_0, m_1, \ldots, m_{k-1}\}$.

Decoding is based on Equation (8.12). The content of the shift register shown in Fig. 8.7 is the syndrome after the entire received vector has entered the shift register.[5] A syndrome polynomial corresponds to a particular error pattern $e(X)$, which can be corrected if the corrupted code vector is within the half-minimum-Hamming-distance of the correct code vector.

The error detection and correction capabilities of the cyclic codes are determined by its design in each particular case. In general, the higher the degree of the generator polynomial, the better the error-detection capability. For example, a class of Hamming codes is characterized by the following (n,k) structure: $(n,k) = (2^m - 1, 2^m - 1 - m)$, where $m = 2, 3, \ldots$. These codes have a minimum distance of 3 and are capable to correct single-bit errors and detect two or fewer errors within a block.

An important class of ECC codes are *Reed-Solomon codes*,[4,5] which are especially powerful in correcting burst errors. A Reed-Solomon code is

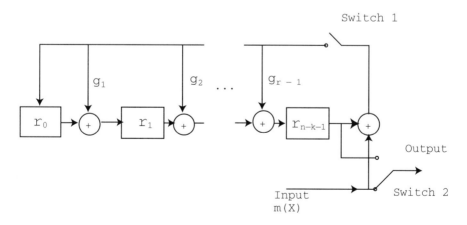

FIGURE 8.9. Encoding shift register for cyclic codes. ($g_o = g_{n-k} = 1$).

the most efficient code among the (n,k) cyclic codes, i.e., it achieves the largest possible minimum distance:

$$d_{min} = n - k + 1.$$

Therefore, a (n,k) Reed-Solomon code can correct any burst of errors having a burst length of $(n - k)/2$. For example, if $k = 15$, $n = 31$, then a burst of eight errors can be corrected.

Another way to correct burst errors is based on interleaving to spread the effect of the burst error over many codewords.[3,5] This can be done by interleaving m codewords to create an m-times-long codeword. An example of interleaving is shown in Fig. 8.10. A bit sequence is split into three bit sequences so that the burst of three errors C1, A2, and B2 is dispersed into single-bit errors in each sequence.

8.5.4 Matrix formulation of ECC

This section contains some additional mathematical formalism of ECC.[4] When the user pattern is encoded, the encoder transforms k user bits of data into n encoded bits, $n > k$. The extra $n - k$ bits are called the redundant bits, parity bits, or check bits. The *code redundancy* is defined as the ratio of the number of redundant bits to the number of user bits: $(n - k)/k$. The typical ECC code redundancy in hard disk drive is ~8% at present. In general, any k user bits can form 2^k message sequences, and any n encoded bits can form 2^n sequences. An encoding procedure assigns one of the 2^n

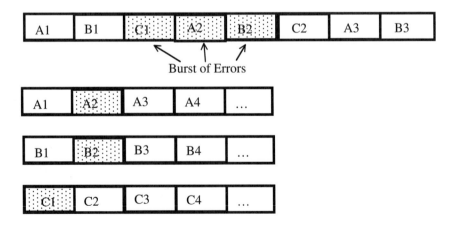

FIGURE 8.10.   Example of interleaved code.

sequences to each of 2^k messages to enhance the distance between the encoded messages. The encoding procedure may be described as a linear matrix operation using a *generator matrix*.

Let the message vector of the message bits $\{m_1, m_2, \ldots, m_k\}$ be

$$\mathbf{m} = [m_1 m_2 \ldots m_k],$$

then the encoded data can be represented by the code vectors generated as follows:

$$\mathbf{U} = \mathbf{m}\, \mathbf{G},$$

where **G** is the generator matrix. For example, the general expression for the parity generator matrix is

$$\mathbf{G} = \begin{bmatrix} 1 & 0 & \cdots & 0 & p_{11} & p_{12} & \cdots & p_{1,(n-k)} \\ 0 & 1 & \cdots & 0 & p_{21} & p_{22} & \cdots & p_{2,(n-k)} \\ & \vdots & & & & & & \\ 0 & 0 & \cdots & 1 & p_{k1} & p_{k2} & \cdots & p_{k,(n-k)} \end{bmatrix} = [\mathbf{I}_k \ \mathbf{P}],$$

where

$$\mathbf{I}_k = \begin{bmatrix} 1 & 0 & \cdots & 0 \\ 0 & 1 & \cdots & 0 \\ & \vdots & & \\ 0 & 0 & \cdots & 1 \end{bmatrix}, \quad \mathbf{P} = \begin{bmatrix} p_{11} & p_{12} & \cdots & p_{1,(n-k)} \\ p_{21} & p_{22} & \cdots & p_{2,(n-k)} \\ & \vdots & & \\ p_{k1} & p_{k2} & \cdots & p_{k,(n-k)} \end{bmatrix},$$

and $p_{i,j} = 0$ or 1, designed to generate the parity bits. $\mathbf{I}_k$ is the $k \times k$ identity matrix with 1s on the main diagonal and 0s elsewhere. It is readily derived that the general expression of the code vector is

$$\mathbf{U} = [m_1 m_2 \cdots m_k \ p_1 \ p_2 \cdots p_{n-k}],$$

where the parity bits are

$$p_1 = m_1 p_{11} + m_2 p_{21} + \cdots + m_k p_{k1},$$
$$p_2 = m_1 p_{12} + m_2 p_{22} + \cdots + m_k p_{k2},$$
$$\cdots$$
$$p_{n-k} = m_1 p_{1,(n-k)} + m_2 p_{2,(n-k)} + \cdots + m_k p_{k,(n-k)}.$$

Let us define a *parity-check matrix,* **H**, which allows us to decode the received vectors. It is such a matrix that $\mathbf{G} \cdot \mathbf{H}^T = 0$, where $\mathbf{H}^T$ is the transpose of **H**. In other words, the rows of **G** are *orthogonal* to the rows of **H**. The parity-check matrix can be constructed from the generator matrix:

$$\mathbf{H} = [\mathbf{P}^T \ \mathbf{I}_{n-k}],$$

$$\mathbf{H}^T = \begin{bmatrix} \mathbf{P} \\ \mathbf{I}_{n-k} \end{bmatrix} = \begin{bmatrix} p_{11} & p_{12} & \cdots & p_{1,(n-k)} \\ p_{21} & p_{22} & \cdots & p_{2,(n-k)} \\ \vdots & & & \\ p_{k1} & p_{k2} & \cdots & p_{k,(n-k)} \\ 1 & 0 & \cdots & 0 \\ 0 & 1 & \cdots & 0 \\ 0 & 0 & \cdots & 1 \end{bmatrix}.$$

It is verified that

$$\mathbf{G} \cdot \mathbf{H}^T = \begin{bmatrix} p_{11} + p_{11} & \cdots & p_{1,(n-k)} + p_{1,(n-k)} \\ p_{21} + p_{21} & \cdots & p_{2,(n-k)} + p_{2,(n-k)} \\ \cdots & & \\ p_{k1} + p_{k1} & \cdots & p_{k,(n-k)} + p_{k,(n-k)} \end{bmatrix} = 0,$$

where the modulo-2 addition rule is applied. It follows that the multiplication of any correct code vector and the transposed parity-check matrix must also vanish:

$$\mathbf{U} \cdot \mathbf{H}^T = \mathbf{m} \cdot (\mathbf{G} \cdot \mathbf{H}^T) = 0.$$

The concepts of error detection and correction are now easily expressed as matrix operations. For example, let **R** be the received code vector, which is equal to the source code vector **U** and error vector **E**:

$$\mathbf{R} = \mathbf{U} + \mathbf{E},$$

then the syndrome vector **S** is calculated as

$$\mathbf{S} = \mathbf{R} \cdot \mathbf{H}^T.$$

or

$$\mathbf{S} = (\mathbf{U} + \mathbf{E}) \cdot \mathbf{H}^T = \mathbf{E} \cdot \mathbf{H}^T$$

Therefore, we can predict the syndrome **S** from the corresponding error **E**, or vice versa. Of course, the mapping between **E** and **S** should be a one-to-one correspondence.

References

1. J. C. Mallinson, *The Foundations of Magnetic Recording*, (New York, NY: Academic Press, 1993).

2. P. H. Siegel and J. K. Wolf, "Modulation and coding for information storage," *IEEE Communications Magazine*, December 1991, pp. 68–86.

3. J. Watkinson, *Coding for Digital Recording*, (Boston, MA: Focal Press, 1990).

4. B. Sklar, *Digital Communications. Fundamentals and Applications*, (Englewood Cliffs, NJ: Prentice Hall, 1988).

5. A. M. Patel, "Signal and error-control coding," in *Magnetic Storage Handbook*, edited by C. D. Mee and E. D. Daniel (New York, NY: McGraw-Hill 1996, p. 10.1–10.93).

CHAPTER 9

Noises

Magnetic recording would be effortless if what we obtain from a recording channel were clean signals as presented earlier. In reality, signals always coexist with noises. The higher the recording density, the more serious the noise. This is clearly illustrated in Fig. 9.1, where signal peaks are no longer clearly distinguishable at the high recording density. As a matter of fact, usually it is the noise level in a magnetic recording system that limits the achievable recording density.

Generally speaking, *noises are due to uncertainties in physical phenomena and need to be treated statistically.* Noises in digital magnetic recording arise from mainly three sources: recording medium, readback head, and preamplifier (or readback amplifier). Usually a magnetic recording system is designed to be medium noise limited.

In addition to noises, readback signals coexist with unwanted *interferences* and *nonlinear distortions.* Interferences are the reception or reproduction of signals other than those intended. They are *deterministic,* and in principle, may be reduced to an arbitrarily small level by proper design. The remaining interference signals add to the noises in the magnetic recording system. Nonlinear distortions cause the linear superposition principle to fail. They include any nonlinear effects during the write and read process, such as the partial erasure effect between closely spaced magnetic transitions and peak asymmetry of MR read heads. Nonlinear distortions are also deterministic.

In order to achieve as high recording densities as possible, we must understand the mechanisms of noises, nonlinear distortions, and interferences in magnetic recording. This chapter will focus on noise mechanisms of recording media, heads and electronics, while the latter topics will be addressed in Chapters 10 and 14. The signal-to-noise ratio (SNR) considerations and measurements in magnetic recording systems will also be covered here.

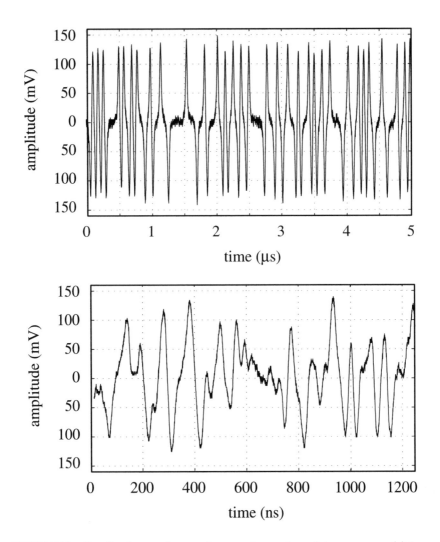

FIGURE 9.1. Readback waveforms of a pseudo-random data sequence: (a) top: low density, minimum bit length = 40 ns, (b) bottom: high density, minimum bit length = 10 ns.

9.1 NOISE FORMULATION

Noise are universal in engineering systems, not unique to magnetic recording system at all. Standard methods have been developed to rigorously treat noises, as found in numerous textbooks in signal processing.[1]

Here we briefly describe the basics of noise formulation that will be essential for the understanding of noise mechanisms and SNR in magnetic recording. Detailed theoretical treatment of noises in magnetic recording can be found in more advanced texts.[2,3]

9.1.1 Power spectral density

Noises and signals coexist in time or spatial domain, and it is difficult to separate them. However, we usually know the frequencies of the signals, so we can better separate noises from signals in frequency (or spatial frequency) domain. Consequently, power spectral density of signals and noises discussed in the following are central concepts in noise formulation.

Let a function $s(x)$ represent the readback voltage from a magnetic recording channel, then $s(x)$ contains all the recorded information including signal and noise. If $s(x)$ is a nonperiodic signal, it may not have a Fourier transform $s(k)$. However, we can always form a *truncated* version of $s(x)$ by observing it only in the interval $[-L/2, L/2]$, i.e., we only read the recorded information of length L, as expressed in the following equation:

$$S_L(x) = \begin{cases} s(x), & \text{for } -L/2 \le x \le L/2, \\ 0, & \text{for } |x| > L/2. \end{cases} \tag{9.1}$$

The energy of the above function is finite: $\int_{-\infty}^{\infty} dx |S_L(x)|^2 < \infty$. Therefore, $S_L(x)$ will always have a proper Fourier transformation $S_L(k)$. From Parseval's theorem, we have

$$\int_{-\infty}^{+\infty} dx |S_L(x)|^2 = \int_{-L/2}^{+L/2} dx |S_L(x)|^2 = \frac{1}{2\pi} \int_{-\infty}^{+\infty} dk |S_L(k)|^2. \tag{2.7}$$

The average power within length L is then

$$\langle P \rangle_L = \frac{1}{L} \int_{-L/2}^{+L/2} dx |S_L(x)|^2 = \frac{1}{2\pi L} \int_{-\infty}^{+\infty} dk |S_L(k)|^2. \tag{9.2}$$

Since

$$s(x) \equiv \lim_{L \to \infty} S_L(x), \tag{9.3}$$

then the average power of $s(x)$ is

$$P_s = \lim_{L \to \infty} \langle P \rangle_L = \frac{1}{2\pi} \int_{-\infty}^{+\infty} dk \lim_{L \to \infty} \frac{1}{L} |S_L(k)|^2. \tag{9.4}$$

The *power spectral density in spatial frequency domain* is defined as

$$PSD_s(k) \equiv \lim_{L \to \infty} \frac{1}{L} |S_L(k)|^2. \qquad (9.5)$$

Substituting Eqn. (9.5) into (9.4), we get the average power of $s(x)$:

$$P_s = \frac{1}{2\pi} \int_{-\infty}^{+\infty} PSD_s(k)dk. \qquad (9.6)$$

The power spectral density is the density of power in a signal in the spatial frequency domain. If the average power P_s is in the unit of volt^2, then the power spectral density $PSD_s(k)$ is in the unit of volt$^2 \cdot$ meter.

Strictly speaking, the above definition of power spectral density is true only for an *ergodic* random process which has zero probability that any state will never recur. For a *nonergodic* process, the power spectral density is defined as

$$PSD_s(k) \equiv \lim_{L \to \infty} \frac{1}{L} \langle |S_L(k)|^2 \rangle. \qquad (9.7)$$

where $\langle \, \rangle$ denotes an *ensemble average* over the possible random variables that comprise $S_L(x)$. For an ergodic process, its ensemble average is equal to its length or time average, and the statistical properties of the process can be determined by length or time averaging over a single sample function of the process, so $\langle \, \rangle$ can be dropped.

Experimentally we always deal with a signal in time domain (such as with oscilloscope) or in frequency domain (such as with spectrum analyzer). If we replace x by time t, L by observation time period T, and k by frequency f, then we can define *the power spectral density of a signal $s(t)$ in frequency domain* as

$$PSD_s(f) \equiv \lim_{T \to \infty} \frac{1}{T} |S_T(f)|^2, \qquad (9.8)$$

where $S_T(f)$ is the Fourier transform of $S_T(t)$:

$$S_T(t) = \begin{cases} s(t), & \text{for} \quad -T/2 \le t \le T/2, \\ 0, & \text{for} \quad |t| > T/2. \end{cases} \qquad (9.9)$$

$$S_T(f) = \int_{-\infty}^{\infty} dt \, S_T(t) e^{-i2\pi ft}. \qquad (9.10)$$

Note that Parseval's theorem for time and frequency domains is

$$\int_{-\infty}^{+\infty} dt |S_T(t)|^2 = \int_{-\infty}^{+\infty} df |S_T(f)|^2. \qquad (2.7)$$

Therefore, the average power of a signal in time domain: $s(t) = s(x)$, where $x = vt$, is

$$P_s \equiv \lim_{T\to\infty} \frac{1}{T} \int_{-T/2}^{T/2} dt |S_T(t)|^2 = \lim_{T\to\infty} \frac{1}{T} \int_{-\infty}^{\infty} df |S_T(f)|^2 = \int_{-\infty}^{\infty} PSD_s(f) df. \qquad (9.11)$$

Obviously the power spectral density in frequency domain has the unit of volt^2/Hz. Since $k = 2\pi f/v$, we can get the following relation from Equations (9.6) and (9.11):

$$PSD_s(k) = v \cdot PSD_s(f). \qquad (9.12)$$

The power spectral density in frequency domain, $PSD_s(f)$, can be experimentally measured by a spectrum analyzer (Section 9.1.3), from which we can obtain the power spectral density in spatial frequency domain, $PSD_s(k)$. Theoretically, both can be derived from the autocorrelation of the signal as detailed next.

9.1.2 Autocorrelation

When dealing with random variables, it is very useful to look at their *correlation*. The *autocorrelation function* of a signal $s(x)$ is defined as

$$R_s(\eta) \equiv \lim_{L\to\infty} \frac{1}{L} \int_{-L/2}^{L/2} s(x)s^*(x+\eta)dx$$
$$= \lim_{L\to\infty} \frac{1}{L} \int_{-L/2}^{L/2} S_L(x)S_L^*(x+\eta)dx, \qquad (9.13)$$

where η represents an arbitrary displacement of $s(x)$, and * denotes conjugate. It can be shown that the autocorrelation function $R_s(\eta)$ is an even function, and that it is maximum at $\eta = 0$.

The *statistical autocorrelation function* is the ensemble average of $s(x)s^*(x + \eta)$:

$$R^{st}(\eta) \equiv \langle s(x)s^*(x+\eta) \rangle. \qquad (9.14)$$

If the random process is *ergodic,* its ensemble averages are the same as the length (or time) averages. Then

$$R^{st}(\eta) = R(\eta)$$

An ergodic process must be *stationary:* its statistics are not affected by a shift in the length or time origin. (The converse is not necessary.)

The Fourier transform of the autocorrelation function is

$$\int_{-\infty}^{\infty} R_s(\eta)e^{-ik\eta}d\eta = \lim_{L\to\infty}\frac{1}{L}\int_{-L/2}^{L/2} dx S_L(x)\int_{-\infty}^{\infty} d\eta S_L^*(x+\eta)e^{-ik\eta}$$

$$= \lim_{L\to\infty}\frac{1}{L}\int_{-L/2}^{L/2} dx S_L(x)e^{ikx}\int_{-\infty}^{\infty} d\eta S_L^*(x+\eta)e^{-ik(x+\eta)} \tag{9.15}$$

$$= \lim_{L\to\infty}\frac{1}{L} S_L(-k)S_L^*(-k) = PSD_s(-k) = PSD_s(k).$$

This indicates that *the power spectral density of a signal is the Fourier transform of its autocorrelation function.* The relation is called the Wiener-Kinchin theorem. Conversely, the autocorrelation function of a signal is the inverse Fourier transform of the power spectral density of the signal:

$$R_s(\eta) = \frac{1}{2\pi}\int_{-\infty}^{\infty} PSD_s(k)e^{ik\eta}dk. \tag{9.16}$$

Thus *the average power of a signal is equal to its autocorrelation at the origin* $\eta = 0$:

$$P_s = \frac{1}{2\pi}\int_{-\infty}^{\infty} PSD_s(k)dk = R_s(\eta=0). \tag{9.17}$$

The *autocorrelation function* of a signal in time domain $s(t)$ is defined as

$$R_s(\zeta) \equiv \lim_{T\to\infty}\frac{1}{T}\int_{-T/2}^{+T/2} s(t)s*(t+\zeta)dt, \tag{9.18}$$

where ζ represents an arbitrary time displacement of $s(t)$. The corresponding Wiener-Kinchin theorem is

$$PSD_s(f) = \int_{-\infty}^{\infty} R_s(\zeta) e^{-i2\pi f\zeta} d\zeta. \tag{9.19}$$

Now we have two ways to calculate power spectral density, either by using the definition in Section 9.1.1 [Equations (9.5) and (9.7)] directly, or by taking the Fourier transform of the autocorrelation function (the Wiener-Kinchin theorem).

9.1.3 Power spectral density of noise and signal

A spectrum analyzer measures an average (mean) signal power in a narrow bandwidth called *resolution bandwidth* Δf that is centered on a given frequency f. It can display *slot noise power* $P_n(f)$ of a noise signal $n(t)$ as a function of frequency f at a resolution bandwidth Δf:

$$P_n(k) = P_n(f) = 2\ PSD_n(k)\frac{\Delta k}{2\pi} = 2\ PSD_n(f)\ \Delta f, \qquad (9.20)$$

where the factor 2 arises from the sum of the power density at both positive and negative k. Obviously, $P_n(k)$ or $P_n(f)$ is in the unit of volt^2. If the noise power spectral density is fixed, *the measured noise power is dependent on the resolution bandwidth setting.* However, a spectrum analyzer can also display a *normalized noise voltage:*

$$\sqrt{P_n(f)/\Delta f} = \sqrt{2\ PSD_n(f)}, \qquad (9.21)$$

which is in the unit of volt/$\sqrt{Hz}$ and is independent of the resolution bandwidth setting.

What is the power spectral density of a signal in time domain such as $s(t) = a\ \cos(2\pi f_0 t)$? Well, its autocorrelation function is

$$
\begin{aligned}
R_s(\zeta) &\equiv \lim_{T\to\infty} \frac{1}{T}\int_{-T/2}^{T/2} s(t)s^*(t+\zeta)dt \\
&= a^2 \lim_{T\to\infty} \frac{1}{T}\int_{-T/2}^{T/2} \cos(2\pi f_0 t)\cos[2\pi f_0(t+\zeta)]dt \\
&= a^2 \lim_{T\to\infty} \frac{1}{2T}\int_{-T/2}^{T/2} \{\cos(2\pi f_0\zeta)+\cos[2\pi f_0\ (2t+\zeta)]\}dt \\
&= \frac{a^2}{2} \cos(2\pi f_0\zeta).
\end{aligned}
$$

Therefore, its power spectral density is the Fourier transform of $R_s(\zeta)$:

$$
\begin{aligned}
PSD_s(f) &= \int_{-\infty}^{\infty} R_s(\zeta)\ e^{-i2\pi f\zeta}d\zeta \\
&= \frac{a^2}{2}\ \frac{\delta(f-f_0)+\delta(f+f_0)}{2}.
\end{aligned}
\qquad (9.22)
$$

The *signal power* measured by a spectrum analyzer at a resolution bandwidth Δf is

$$
\begin{aligned}
P_s(k_0) &= P_s(f_0) = 2PSD_s(f_0)\ \Delta f \\
&= 2\cdot\frac{a^2}{2}\cdot\frac{\delta(f-f_0)}{2}\ \Delta f = \frac{a^2}{2},
\end{aligned}
\qquad (9.23)
$$

which is the average power of the cosine wave, *independent of the resolution bandwidth setting.* A spectrum analyzer can be set up to display the root mean square (rms) amplitude of the cosine signal:

$$\sqrt{P_s(f_0)} = a/\sqrt{2}. \qquad (9.24)$$

For a square wave recording signal at a frequency f_0 (period $T = 1/f_0$), made of alternating single pulses $V_{sp}(t)$, the fundamental Fourier series term is

$$V_{f_0}(t) = a\cos(2\pi f_0 t),$$

where

$$
\begin{aligned}
a &= \frac{2}{T}\int_{-T/2}^{T/2} V(t)e^{-i2\pi f_0 t}dt = \frac{2}{T}\int_{-T/2}^{T/2}[V_{sp}(t) - V_{sp}(t - T/2)]e^{-i2\pi f_0 t}dt \\
&= 2f_0[V_{sp}(f_0) - V_{sp}(f_0)e^{-i\pi}] = 4f_0V_{sp}(f_0) \tag{9.25}\\
&= 4\frac{vk_0}{2\pi}\frac{V_{sp}(k_0)}{v} = \frac{2k_0V_{sp}(k_0)}{\pi},
\end{aligned}
$$

and v is the disk velocity. Therefore, the spectrum analyzer displays a rms fundamental harmonic of $2k_0|V_{sp}(k_0)|/\pi\sqrt{2}$. This is consistent with Equation (3.22) in Section 3.3.2.

A spectrum analyzer displays both signal and noise power spectra, as shown in Fig. 9.2, where the factor of 2 arises from the fact that we only display the positive frequency range. Since signal peaks occur at certain known frequencies, they can be separated from noises. The power spectral density spectrum with the signal peaks removed is the noise

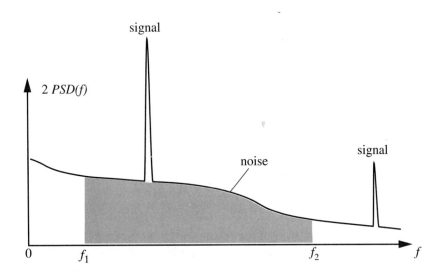

FIGURE 9.2.   Power spectral density in frequency domain.

power spectral density, $PSD_n(f)$. Then, the *wide-band noise power* in the frequency range $[f_1, f_2]$ or the spatial frequency range $[k_1, k_2]$ is

$$N = 2 \int_{f_1}^{f_2} PSD_n(f) df = 2 \int_{k_1}^{k_2} PSD_n(k) dk / 2\pi, \qquad (9.26)$$

which is the shaded area in Fig. 9.2. The *average noise power density* in the same band can be defined as

$$\langle PSD \rangle = \frac{N}{f_2 - f_1} = \frac{2 \int_{f_1}^{f_2} PSD_n(f) df}{f_2 - f_1}. \qquad (9.27)$$

9.2 MEDIUM NOISE MECHANISMS

A main source of noises in magnetic recording is the magnetic recording medium itself. Equipped with the concept of power spectral density and the noise formulation developed in the last section, we are now ready to tackle the mechanisms of magnetic medium noises. We will discuss various noise mechanisms and analyze them in somewhat detail.

Readback signals in magnetic recording are directly related to the medium magnetization distributions, as revealed by the reciprocity principle. Therefore, medium noises are due to the fluctuations (or uncertainties) in the medium magnetization, which can be classified into three types:

1. *Transition noise* due the magnetization fluctuation concentrated near the recorded transition centers. This is the dominant noise source in thin film disks. The transition noise is *nonstationary* by definition, because the transition noise depends on recording pattern.
2. *Particulate or granularity noise* due to the random dispersion of magnetic particles or grains in magnetic medium. This is the dominant noise source in magnetic tapes, floppy disks and particulate thin film hard disks. The particulate noise in magnetic media is usually *stationary*, meaning that the noise is independent of the location along the data track.
3. *Modulation noise* due to the magnetization fluctuation proportional to recorded magnetization between magnetic transitions (nontransition areas). This noise source can be observed in both particulate or continuous thin-film disks. The modulation noise in magnetic media is also *nonstationary*.

Noise power spectral densities are often measured by analyzing square wave recording waves at interested linear recording densities. In this case signal peaks can be readily removed from noise spectra, and nonlinear transition shifts can be avoided. It has been recognized from experiments that particulate noise power spectral density is generally independent of linear recording density because both transition areas and nontransition areas equally contribute to particulate noise. By contrast, transition noise power density generally increases with linear recording density because there are more transition areas at higher densities. On the contrary, modulation noise power spectral density tends to decrease with increasing linear recording density because there are less nontransition areas contributing to modulation noise at higher recording densities.[2–5] These characteristics will be explained in the following sections.

9.2.1 Transition noise

Thin film hard disks are quite uniform and closely packed in which the transition noise tends to play a dominant role. Transition noise can be analyzed by numerical *micromagnetic modeling.*[6,7] However, we will adopt a simple *analytical model* to gain some valuable insights into transition noise. Due to the "zigzag" nature of the magnetic transitions in thin film disks, the *transition center,* defined as the average transition center across data track, becomes uncertain. This uncertainty, usually called transition position *jitter,* will lead to transition noise. The zigzag nature also causes fluctuations in read pulse shape, resulting in additional transition noises.

Transition Noise Spectral Density.   For simplicity, let us first consider that only the transition center fluctuates while the readback pulses keep the same shape. Then a magnetic transition can be expressed as:

$$M(x) = M_r f\left(\frac{x - x_0 - w}{a}\right),\qquad(9.28)$$

where f represents the average shape of magnetic transition, and x_0, a and w are the mean transition center, mean transition parameter, and transition center fluctuation, respectively. The readback voltage for square wave recording can be expressed as

$$V(x) = \sum_{-\infty}^{\infty}(-1)^n V_{sp}(x - nB - w_n),\qquad(3.12)$$

where w_n is the position fluctuation of the nth transition, and B is the bit length. Assume that w_n is small, then the voltage can be expanded into a first-order Taylor series:

$$V(x) = \sum_{-\infty}^{\infty} (-1)^n V_{sp}(x - nB) + \sum_{-\infty}^{\infty} (-1)^n w_n \frac{\partial V_{sp}(x - nB)}{\partial x}, \quad (9.29)$$

where the first term is the signal and the second term is the noise.

The *transition noise power spectral density* is

$$PSD_n(k) = \lim_{L \to \infty} \frac{1}{L} \left\langle \left| \sum_{-L/2B}^{L/2B} (-1)^n w_n \, FT \left[\frac{\partial V_{sp}(x - nB)}{\partial x} \right] \right|^2 \right\rangle, \quad (9.30)$$

which can be simplified by using the following relations:

$$FT \left[\frac{\partial V_{sp}(x - nB)}{\partial x} \right] = ik V_{sp}(k) e^{-iknB},$$

$$\left\langle \left| \sum_{-L/2B}^{L/2B} (-1)^n w_n ik V_{sp}(k) e^{-iknB} \right|^2 \right\rangle = k^2 |V_{sp}(k)|^2 \left\langle \left| \sum_{-L/2B}^{L/2B} (-1)^n w_n e^{-iknB} \right|^2 \right\rangle.$$

Assume that the position fluctuations are *uncorrelated*, then

$$\left\langle \left| \sum_{-L/2B}^{L/2B} (-1)^n w_n e^{-iknB} \right|^2 \right\rangle = \sum_{-L/2B}^{L/2B} \langle w_n^2 \rangle = \frac{L}{B} \sigma_x^2,$$

where σ_x is the standard deviation of the transition center, i.e., the mean transition jitter. Simplifying Equation (9.30), the transition noise power spectral density becomes

$$PSD_n(k) = \frac{\sigma_x^2 k^2 |V_{sp}(k)|^2}{B} \quad (9.31)$$

Several observations can be made regarding the transition noise power spectral density:

1. Uncorrelated transition noise power spectral density is proportional linear recording density ($PSD_n(k) \propto 1/B$). For recording medium dominated by transition noises, as the recording density is increased, the noise regions increase relative to the bit length, so the total noise power increases.
2. Uncorrelated transition noise power spectral density is proportional to the variance of transition positions ($PSD_n(k) \propto \sigma_x^2$). The more the transition centers fluctuate, the higher the transition noise.

3. The dependence on the spatial frequency of the uncorrelated transition noise power spectral density is similar to that of the fundamental harmonic signal power $(PSD_n(k) \propto k^2|V_{sp}(k)|^2)$. In other words, the power spectral density diminishes either at low spatial frequencies or at high spatial frequencies.

Now let us consider a more general case in which not only the transition centers fluctuate but readback pulses also change shapes. If a single pulse in the absence of transition noise can be approximated by a Lorentzian pulse:

$$V_{sp}(x) = V_{sp}^{peak}\frac{(PW_{50}/2)^2}{(x - x_0)^2 + (PW_{50}/2)^2}, \tag{3.8}$$

then all fluctuations can be described by the standard deviations of the transition center x_0, the half-amplitude pulse width PW_{50}, and the zero-to-peak amplitude V_{sp}^{peak}, which are denoted as σ_x, σ_{PW50}, and σ_V, respectively. Using the first-order Taylor series approximation as above, we can derive that the total transition noise power spectral density is[8]

$$PSD_n(k) = \frac{|V_{sp}(k)|^2}{B}\left[k^2\sigma_x^2 + \frac{k^2\sigma_{PW50}^2}{4} + \left(\frac{\sigma_V}{V_{sp}^{peak}}\right)^2\right], \tag{9.32}$$

where

$$V_{sp}(k) = \frac{\pi}{2}V_{sp}^{peak}PW_{50}e^{-kPW_{50}/2}. \tag{9.33}$$

The wide-band noise power over all spatial frequencies $[-\infty, \infty]$ is

$$N = \int_0^{+\infty}2PSD_n(k)\frac{dk}{2\pi} \tag{9.34}$$

$$= \frac{\pi}{2}(V_{sp}^{peak})^2\frac{\sigma_x^2}{BPW_{50}}\left(1 + \frac{\sigma_{PW50}^2}{4\sigma_x^2}\right) + \frac{\pi\sigma_V^2PW_{50}}{4B},$$

where we have used the following integrals:

$$\int_0^{+\infty}x^ne^{-ax}dx = \frac{n!}{a^{n+1}} \quad (n = 0, 1, 2, 3, \ldots ; a > 0).$$

Defining an effective transition jitter as

$$\sigma_n \equiv \sigma_x\sqrt{1 + \frac{\sigma_{PW50}^2}{4\sigma_x^2} + \frac{\sigma_V^2PW_{50}^2}{2(V_{sp}^{peak})^2\sigma_x^2}}, \tag{9.35}$$

and substituting Equation (9.35) into (9.34), then the wide-band transition noise power becomes

$$N = \frac{\pi}{2}(V_{sp}^{peak})^2 \frac{\sigma_n^2}{BPW_{50}}. \tag{9.36}$$

In experiments, we cannot measure noise power over infinite frequency ranges. If the frequency range of measurement is $[0,f_2]$, and f_2 is sufficiently large (e.g., 30 MHz), then the measured wide-band noise power is approximately represented by the above expression because the noise power spectral density beyond f_2 is negligible, Therefore, the average transition noise power density is

$$\langle PSD_n \rangle = \frac{N}{f_2 - 0} = \frac{\pi}{2} \frac{(V_{sp}^{peak})^2}{f_2} \frac{\sigma_n^2}{BPW_{50}}. \tag{9.37}$$

As shown in Fig 9.3, the average transition noise power density of the low noise Co alloy thin-film disk is indeed proportional to the linear density $(1/B)$. By contrast, that of the high-noise disk is only proportional to the linear density at low densities (≤ 4 μm^{-1}). Note that 1 μm^{-1} = 25.4 KFCI (1000 flux changes per inch). At certain high recording densities, the transition noise becomes *correlated*, and the transition noise power spectral density shows a *supralinear* behavior as a function of recording

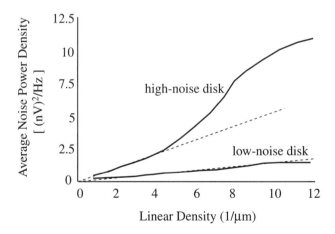

FIGURE 9.3. Average transition noise power density vs. linear density of square wave recording, measured on two disks (solid lines). The dashed lines are the theoretical average uncorrelated transition noise power density.

density. For the low-noise disk, the supralinear behavior is probably beyond the highest linear density in Fig. 9.3.

Signal to Transition Noise Ratio.   An important figure of merit in magnetic
recording is the signal to noise ratio (SNR). The higher the SNR is, the
easier the data detection. There are many definitions of SNR in the literature (Section 9.5). For example, SNR can be defined as the ratio of single-
pulse peak signal power over integrated transition noise power:

$$SNR_{sp} \equiv \frac{(V_{sp}^{peak})^2}{N} = \frac{2BPW_{50}}{\pi\sigma_n^2}, \tag{9.38}$$

which is almost the square of the ratio of the geometric mean of bit length
and pulse width to the effective transition jitter. Alternatively, SNR can
be defined as the ratio of peak signal power at a linear density $D (= 1/B)$
to the transition noise power:

$$SNR \equiv \frac{(V^{peak})^2}{N} \equiv SNR_{sp} \cdot R^2$$
$$\approx \frac{2BPW_{50}}{\pi\sigma_n^2} \cdot \left[\frac{\pi PW_{50}/2B}{\sinh(\pi PW_{50}/2B)}\right]^2 \tag{9.39}$$

where R is the resolution for square-wave recording at a bit length of B.
    If the transition parameter is a, the cross-track correlation width is l_c,
and the read track width is W, then the transition jitter can be approximated
as[9]

$$\sigma_n \approx \frac{\pi^2 a}{4}\sqrt{\frac{l_c}{3W}}. \tag{9.40}$$

Then,

$$SNR_{sp} \approx \frac{96BPW_{50}W}{\pi^5 a^2 l_c} \approx 0.3D_{ch}\frac{B^2W}{a^2 l_c}, \tag{9.41}$$

where $D_{ch} = PW_{50}/B$ is the channel density for a PRML channel using the
8/9(0,4) code (Chapter 8). Note that *the SNR is proportional to the read
trackwidth, and proportional to the square of the bit length.*
    For a very rough estimate, we may assume that

$$\sigma_n \approx \frac{2 \times \text{grain size}}{\sqrt{\text{number of grains across track}}} \approx \frac{2D}{\sqrt{W/D}} = 2\sqrt{\frac{D^3}{W}}, \tag{9.42}$$

where D is the grain size. This leads to

$$SNR_{sp} \approx \frac{D_{ch}B^2W}{2\pi D^3}.$$ (9.43)

For the 10 Gb/in.[2] recording design specifying $B \sim 0.0625\ \mu$m, $W \sim 0.5$ μm, track pitch $\sim 1\ \mu$m, $D_{ch} \sim 2$, $SNR_{sp} \sim 400$ (26 dB), then the maximum grain size should be $\sim0.012\ \mu$m.

9.2.2 Particulate medium noise

For a DC-erased medium, if the magnetic film is continuous, no readback voltage will arise. In reality, however, the medium is of course discontinuous and is made of magnetic particles (in magnetic tapes) or magnetic grains (in thin-film disks). Each particle will produce a voltage and add to a total voltage based on linear superposition. This readback voltage is determined by the granularity of the medium and the distribution of grains (or particles), and therefore is a source of noise in addition to signals resulting from a magnetic transition.

Readback Voltage of a Single Particle.   Consider a longitudinal particle near a recording head with a head field $H_x(x,y)$, as shown in Fig. 9.4. In general,

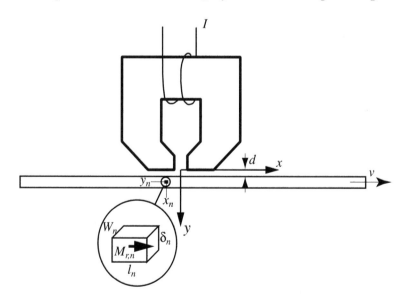

FIGURE 9.4.   Readback from a rectangular magnetic particle.

the readback voltage due to the particle can be readily derived from the reciprocity principle. However, in this case, we can make a shortcut by using the linear superposition principle.

The magnetic particle is essentially composed of two step transitions in the both ends: one changes from 0 to $M_{r,n}$, the other change from $M_{r,n}$ to 0. Therefore, according to Section 3.2, the readback voltage from an arbitrary particle is

$$V_n(x) = 2\mu_0 v W_n \frac{M_{r,n}}{2} \delta_n \left[\frac{H_x(x + x_n + l_n/2, y_n)}{I} - \frac{H_x(x + x_n - l_n/2, y_n)}{I} \right] \cdot \alpha$$

where, n is an integer denoting an arbitrary particle, I is the imaginary write current, v is the head-disk relative velocity, W_n, $M_{r,n}$, δ_n, and l_n are the width, remanent magnetization, thickness and length of the arbitrary magnetic particle in the disk, respectively, and $\alpha = \pm 1$, corresponding to whether the magnetic particle has positive or negative magnetization. If the horizontal magnetization and the volume of the particle are

$$\mu_n = M_{r,n} \cdot W_n \delta_n l_n, \quad v_n = W_n \delta_n l_n, \tag{9.44}$$

then the readback voltage can be rewritten as

$$V_n(x) = \frac{\alpha \mu_0 v}{I} \frac{\mu_n}{l_n} [H_x(x + x_n + l_n/2, y_n) - H_x(x + x_n - l_n/2, y_n)]. \tag{9.45}$$

The Fourier transform of the nth particle readback voltage is

$$\begin{aligned}
V_n(k) &= \frac{\alpha \mu_0 v}{I} \frac{\mu_n}{l_n} [H_x(k, y_n) e^{ik(x_n + l_n/2)} - H_x(k, y_n) e^{ik(x_n - l_n/2)}] \\
&= \frac{\alpha \mu_0 v}{I} \frac{\mu_n}{l_n} H_x(k, y_n) e^{ikx_n} \cdot i2\sin(kl_n/2) \tag{9.46} \\
&= i \frac{\alpha \mu_0 v}{I} \mu_n k H_x(k, y_n) e^{ikx_n} \cdot \frac{\sin(kl_n/2)}{kl_n/2},
\end{aligned}$$

where the *particle length-loss* term $\sin(kl_n/2)/(kl_n/2)$ is similar to the gap-loss term for a Karlqvist head. It is ~1 if we only consider the case that the length scale of the particles is much smaller than magnetic spacing. Since the *spacing loss* term due to Laplace's equation is

$$H_x(k, y_n) = H_s(k) e^{-|k|y_n}, \tag{2.12}$$

then

$$V_n(k) \approx i \frac{\alpha \mu_0 v}{I} \mu_n k H_s(k) e^{ikx_n - |k|y_n} \tag{9.47}$$

If the read head is a Karlqvist head, if follows from Equation (2.19) that

$$\frac{H_s(k)}{I} = \frac{gH_g}{I}\frac{\sin(kg/2)}{kg/2} = NE_{ind}\frac{\sin(kg/2)}{kg/2},$$ (9.48)

where N and E_{ind} is the number of coils and the head efficiency. If the read head is an anisotropic magnetoresistive (AMR) head, the readback voltage expression should be multiplied by a constant (Section 6.3.2):

$$\frac{V_{MR}}{V_{ind}} = \frac{\sqrt{2}J\,\Delta\rho_{max}E_{MR}}{Nv\mu_0 M_s E_{ind}}\frac{g+t}{t}.$$ (6.58)

This constant changes the expression of noise spectral density, noise power, and signal power, but it does not change the particulate SNR.

Particulate Noise Spectral Density. The readback voltage from all the particles under the read head can be expressed by the linear superposition principle:

$$V(x) = \sum_n V_n(x),$$ (9.49)

where the sum is over all particles. This is the particulate noise because it carries no recorded information, and it would be zero if the magnetic particles are all connected to one another or the magnetic medium is not granular! The power spectral density of the particulate noise is

$$PSD_n(k) \equiv \lim_{L\to\infty}\frac{1}{L}\langle|V(k)|^2\rangle = \lim_{L\to\infty}\frac{1}{L}\left\langle\left|\sum_n V_n(k)\right|^2\right\rangle.$$ (9.50)

Now that

$$\left|\sum_n V_n(k)\right|^2 = \sum_n V_n(k)\sum_j V_j^*(k) = \sum_n |V_n(k)|^2 + \sum_{n\neq j} V_n(k)V_j^*(k),$$

then

$$PSD_n(k) = \lim_{L\to\infty}\frac{1}{L}\left(\left\langle\sum_n|V_n(k)|^2\right\rangle + \left\langle\sum_{n\neq j}V_n(k)V_j^*(k)\right\rangle\right).$$ (9.51)

Assume that the particle readback voltages are *uncorrelated*, then the second term vanishes, so

$$PSD_n(k) = \lim_{L\to\infty}\frac{1}{L}\left\langle\sum_n|V_n(k)|^2\right\rangle.$$ (9.52)

If the average particle packing fraction is p, then in the volume $LW\delta$ of the magnetic disk (where L, W and δ are the length, trackwidth, and the thickness of the disk, respectively), the average number of particles is $pLW\delta/\langle v_n \rangle$, where $\langle v_n \rangle$ is the average volume of magnetic particles. Then, Equation (9.52) becomes

$$
\begin{aligned}
PSD_n(k) &= \lim_{L \to \infty} \frac{1}{L} \sum_n^{pLW\delta/\langle v_n \rangle} \langle |V_n(k)|^2 \rangle \\
&= \lim_{L \to \infty} \frac{1}{L} (pLW/\langle v_n \rangle) \int_d^{d+\delta} dy' \langle |V_n(k)|^2 \rangle \qquad (9.53) \\
&= \frac{pW}{\langle v_n \rangle} \int_d^{d+\delta} dy' \langle |V_n(k)|^2 \rangle .
\end{aligned}
$$

From Equation (9.47), we have

$$
\langle |V_n(k)|^2 \rangle = \left(\frac{\mu_0 v}{I} \right)^2 \langle \mu_n^2 \rangle |kH_s(k)|^2 e^{-2|k|y'} . \qquad (9.54)
$$

Substituting Equation (9.54) into (9.53), we get

$$
PSD_n(k) = \frac{pW}{\langle v_n \rangle} \left(\frac{\mu_0 v}{I} \right)^2 \langle \mu_n^2 \rangle |kH_s(k)|^2 \frac{e^{-2|k|d}(1 - e^{-2|k|\delta})}{2|k|} . \qquad (9.55)
$$

If the remanence magnetization of the magnetic disk is M_r, then

$$
\begin{aligned}
\mu_n &= M_{r,n} v_n = \frac{M_r}{p} v_n , \\
\langle v_n^2 \rangle &= \langle v_n \rangle^2 + \sigma_v^2 ,
\end{aligned} \qquad (9.56)
$$

where σ_v is the standard deviation of the magnetic particle volume. The particulate noise power spectral density can be rewritten as

$$
\begin{aligned}
PSD_n(k) &= \frac{W}{2p} (\mu_0 v M_r)^2 |k| \left| \frac{H_s(k)}{I} \right|^2 e^{-2|k|d}(1 - e^{-2|k|\delta}) \frac{\langle v_n^2 \rangle}{\langle v_n \rangle} \\
&= \frac{W \langle v_n \rangle}{2p} (\mu_0 v M_r)^2 |k| \left| \frac{H_s(k)}{I} \right|^2 e^{-2|k|d} (1 - e^{-2|k|\delta}) \left(1 + \frac{\sigma_v^2}{\langle v_n \rangle^2} \right) .
\end{aligned} \qquad (9.57)
$$

This result indicates that there is particulate noise even if the particles are all uniform ($\sigma_v = 0$). It diminishes only when the particle sizes are very small because the particulate noise power spectral density is proportional to the average particle size.

The *slot noise power* in the bandwidth Δk at a spatial frequency k is

$$
\begin{aligned}
P(k > 0) &= 2PSD_n(k)\frac{\Delta k}{2\pi} \\
&= \frac{W\langle v_n \rangle}{2\pi p}(\mu_0 v M_r)^2 k \left|\frac{H_s(k)}{I}\right|^2 e^{-2kd}(1 - e^{-2k\delta})\left(1 + \frac{\sigma_v^2}{\langle v_n \rangle^2}\right)\Delta k \qquad (9.58) \\
&= \frac{W\langle v_n \rangle}{2\pi p}(\mu_0 v NE_{ind}M_r)^2 k \left(\frac{\sin(kg/2)}{kg/2}\right)^2 e^{-2kd}(1 - e^{-2k\delta})\left(1 + \frac{\sigma_v^2}{\langle v_n \rangle^2}\right)\Delta k.
\end{aligned}
$$

In the last step, we have assumed a Karlqvist read head and used Equation (9.48). It should be noted that the slot noise power is proportional to the trackwidth and average particle size. Furthermore, the spacing and gap loss terms are identical to those of readback voltage power, so the slot noise power will drop at high frequencies.

Although the above relation is derived for DC erased media, it should also be valid for the media recorded with square wave patterns. If the frequency range of measurement is $[0, f_2]$, then the average noise spectral density is

$$
\langle PSD_n \rangle = \frac{N}{f_2} = \frac{\int_0^{f_2} 2PSD_n(f)df}{f_2} = \frac{\int_0^{k_2} PSD_n(k)dk}{\pi f_2}, \qquad (9.27)
$$

where $k_2 = 2\pi f_2/v$, and v is the head-disk relative velocity. Obviously, the average noise spectral density is independent of the square wave recording linear density $(1/B)$.

For example, the average noise power spectral density of a barium ferrite thin film hard disk,[10] measured over a 40-MHz bandwidth, is shown in Fig. 9.5. The noise spectral density shows a very weak increase with the linear density, indicating that the transition noise jitter is very small. The medium noise appears to be dominated by particulate noise, averaging ~ 0.25 $(nV)^2/Hz$. The small transition noise is attributed to the absence of intergranular exchange interaction in the barium ferrite thin-film disk.

Signal-to-Particulate-Noise Ratio. It has been shown in Section 3.3 that the *fundamental harmonic* for a square wave recording of arctangent magnetic transitions as read by a spectrum analyzer is

$$
V_{rms}(k_0) = \frac{2k_0 V_{sp}(k_0)}{\sqrt{2\pi}}. \qquad (3.23)
$$

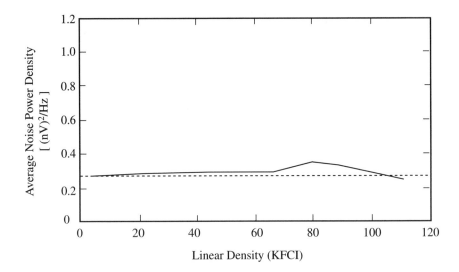

FIGURE 9.5.   Average noise power spectral density of a barium ferrite disk.

At high densities, the square wave readback waveform is approximately sinusoidal, so the zero-to-peak signal voltage amplitude is

$$V_{0-p}(k_0) \approx \sqrt{2} V_{rms}(k_0) \tag{9.59}$$

$$= \frac{4W\mu_0 v NE_{ind}M_r}{\pi} \frac{\sin(k_0 g/2)}{(k_0 g/2)} e^{-k_0(d+a)} (1 - e^{-k_0\delta}).$$

Note that $V_{0-p}(k_0)$ is not the same as the peak amplitude of a single pulse V_{sp}^{peak}, and their ratio is the readback resolution, R. The *slot signal-to-noise ratio* for a spatial frequency slot Δk_0 at k_0 is defined as

$$SNR_{slot}(k_0) \equiv \left[V_{0-p}(k_0)\right]^2 / P(k_0) \tag{9.60}$$

$$= \frac{32Wp(1 - e^{-k_0\delta})^2 e^{-2k_0 a}}{\pi \langle v_n \rangle k_0 \Delta k_0 (1 - e^{-2k_0\delta}) \left[1 + \sigma_v^2 / \langle v_n \rangle^2\right]}.$$

Note that the slot SNR is proportional to the trackwidth, but inversely proportional to the particle size.

Assume that the transitions are very sharp ($a = 0$), the total medium thickness is very small ($k_0\delta \to 0$), and $\sigma_v = \langle v_n \rangle$, $p/\langle v_n \rangle \equiv n$, where n is the particle number density, then

$$1 - e^{-k_0\delta} \approx k_0\delta, \quad SNR_{slot}(k_0) \approx \frac{8Wp\delta}{\pi \langle v_n \rangle \Delta k_0} = \frac{8nW\delta}{\pi \Delta k_0}. \tag{9.61}$$

The slot SNR does not depend on k_0 in this case, and is inversely proportional to the bandwidth.

For a *wide-band* with a minimum wavelength of $\lambda_{min} = 2B$, where B is the minimum bit length, the wide bandwidth is

$$\Delta k \cong 2\pi / \lambda_{min} = \pi/B,$$

and the *wide-band signal to particulate noise ratio* is

$$SNR_{wb} = \frac{8nW\delta}{\pi(\pi/B)} \approx nWB\delta \equiv N, \qquad (9.62)$$

where N is the number of magnetic particles in a single bit. This simple result states that *the wide-band signal to particulate noise ratio is approximately equal to the number of particles in the smallest bit. In the log scale,*

$$SNR_{wb} (dB) = 10\log N. \qquad (9.63)$$

The SNR_{wb} is 20 dB if N is 100. It is often quoted that the required SNR for sufficiently low BER in a magnetic recording channel is $\geq$20 dB, meaning that a bit must contain at least 100 particles or grains.

At recording densities ~1 Gb/in.^{2}, a single bit contains thousands of magnetic particles, giving a wide-band SNR of >30 dB. In other words, particulate noise is not as severe as other noise sources such as transition noise and modulation noise. However, for magnetic recording at an areal density of ~10 Gb/in.^{2} and higher, the number of magnetic grains or particles in a single bit may approach 100 such that the particulate noise becomes not negligible.

As an example, consider that a 10 Gb/in.^{2} recording design specifies $B \sim 0.0625\ \mu m$, $\delta \sim 15$ nm, $W \sim 0.5\ \mu m$, and the particulate noise is the only noise source. Assume that there must be at least 100 grains in a single bit, and the grains extend through the whole thickness of the magnetic disk. Then the maximum grain area is

$$A \sim \frac{0.0625\ \mu m \times 0.5\ \mu m}{100} \approx 313\ nm^2,$$

The maximum lateral grain size is

$$\sqrt{A} \sim 18\ nm = 180\ \text{Å}.$$

9.2.3 Modulation noise

Another source of noise in magnetic recording media (magnetic tapes and textured thin-film disks in particular) is due to magnetization fluctuations associated with finite packing cluster sizes or mechanical substrate texture. This form of the noise is proportional to recorded signal, so it is called modulation noise.

From the reciprocity principle, the readback voltage in longitudinal recording is

$$V(x) = -v\mu_0 \int_{-\infty}^{\infty} dx' \int_{d}^{d+\delta} dy' \int_{-W/2}^{W/2} dz' \frac{H_x(x', y') \, dM_x(x' - x, y', z')}{I}$$

$$= v\mu_0 \int_{-\infty}^{\infty} dx' \int_{d}^{d+\delta} dy' \int_{-W/2}^{W/2} dz' \frac{H_x(x' + x, y') \, dM_x(x', y', z')}{I} \frac{1}{dx'}, \qquad (3.3)$$

where I is the imaginary head current. Its statistical autocorrelation function is

$$R^{st}(\eta) = \left\langle V(x)V * (x + \eta) \right\rangle$$

$$= (v\mu_0)^2 \int_{-\infty}^{\infty} dx' \int_{d}^{d+\delta} dy' \int_{-W/2}^{W/2} dz' \int_{-\infty}^{\infty} dx'' \int_{d}^{d+\delta} dy'' \int_{-W/2}^{W/2} dz'' \qquad (9.64)$$

$$\left[\frac{H_x(x' + x, y')H_x^*(x'' + x + \eta, y'')}{I^2} \times \frac{d^2}{dx'dx''} \left\langle M(x', y', z')M(x'', y'', z'') \right\rangle \right],$$

where only the magnetization is random, while the head field is not random. The magnetization can be expressed as

$$M_x(x, y, z) = M_r m(x, y) \cdot p(x, y, z), \qquad (9.65)$$

where $m(x,y)$ is the normalized recorded magnetization $(-1 \le m \le 1)$, $p(x,y,z)$ is the normalized packing fraction of the magnetic particles, which is the real random variable due to fluctuations in the particle density (or particle clustering).

Assume that $p(x,y,z)$ is a Gaussian random variable, then,

$$\langle p \rangle = 1, \quad \langle p^2 \rangle = 1 + \sigma_p^2. \qquad (9.66)$$

where σ_p is the standard deviation of p. The *cross-correlation* of $p(\vec{r}')$ and $p(\vec{r}'')$ can be expressed as

$$\left\langle p(\vec{r}')p(\vec{r}'') \right\rangle = 1 + \sigma_p^2 R_p(\vec{r}' - \vec{r}''), \qquad (9.67)$$

where $R_p(\vec{r}' - \vec{r}'')$ is the *cross-correlation coefficient* between $p(\vec{r}')$ and $p(\vec{r}'')$, which should be zero if $p(\vec{r}')$ and $p(\vec{r}'')$ are *uncorrelated*. Obviously,

$$R_p(0) = 1, \quad R_p(\infty) = 0.$$

The statistical autocorrelation of the readback voltage becomes

$$R^{st}(\eta) = (v\mu_0 M_r)^2 \int_{-\infty}^{\infty} dx' \int_{d}^{d+\delta} dy' \int_{-W/2}^{W/2} dz' \int_{-\infty}^{\infty} dx'' \int_{d}^{d+\delta} dy'' \int_{-W/2}^{W/2} dz''$$

$$\left[\frac{H_x(x'+x,y')H_x^*(x''+x+\eta,y'')}{l^2} \right.$$

$$\left. \times \frac{d^2}{dx'dx''} m(x',y')m(x'',y'')\left(1 + \sigma_p^2 R_p(\vec{r}'-\vec{r}'')\right) \right]$$

From which the power spectral densities of signal and noise can be derived by the Wiener-Kinchin theorem. The term containing $R_p(r'-r'')$ is the modulation noise, while the rest is the signal. It can be seen that the modulation noise is indeed proportional to the recorded signal.

Assume that the correlation coefficient of magnetic medium can be approximated by

$$R_p(\vec{r}) \approx \frac{1}{1+(x/l_x)^2} \cdot \frac{1}{1+(y/l_y)^2} \cdot \frac{1}{1+(z/l_z)^2}, \tag{9.68}$$

where l_x, l_y, and l_z are the *correlation lengths* in the x-, y-, and z-directions, respectively, in magnetic media. Its spatial Fourier transform with respect to x is

$$R_p(k, y, z) = \pi l_x e^{-|k|l_x} \frac{1}{1+(y/l_y)^2} \frac{1}{1+(z/l_z)^2}. \tag{9.69}$$

Consider a square wave recording pattern of arctangent transitions with a bit length of $B = \pi/k_0$. Usually $l_y \ll \delta, l_z \ll W$, then it can be derived that the *modulation noise power spectral density* is[11]

$$PSD_n(k) = 2\pi l_x l_y l_z \sigma_p^2 W(NE_{ind} v\mu_0 M_r)^2 e^{-k_0 a}$$

$$\times \frac{\sin(kg/2)}{kg/2} e^{-2|k|d}|k|(1 - e^{-2|k|\delta})(e^{-|k-k_0|l_x} + e^{-|k+k_0|l_x}),$$

which is proportional to the read trackwidth like the transition noise and particulate noise. The modulation noise decreases with increasing linear density $(1/B)$ as a function of $\exp(-\pi a/B)$, opposite to transition noise. The k dependence of the modulation noise spectral density is somewhat similar to that of signal power. The last term in Equation (9.70) is a function exponentially decaying from $k = \pm k_0$, indicating that there is a side noise band at the recording frequency, as shown in Fig. 9.6. In this figure we assumed that the square-wave recording bit length is g, the magnetic spacing is $d \sim g/3$, the medium thickness is $\delta \sim g/10$, and the correlation length is $l_x \approx g$. The constant coefficient is ignored in the plot.

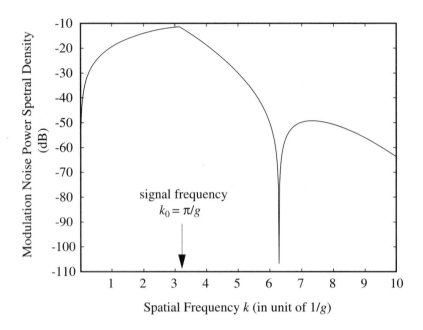

FIGURE 9.6.   Modulation noise power spectral density versus k.

The side noise band at the signal frequency in Fig. 9.6 is not very significant because the noise correlation length is assumed to be very small. The shape of the side band depends on the correlation length and the exact form of correlation function. For thin-film hard disks, an important form of modulation noise is the so-called *texture noise*, which is associated with the mechanical texture of disk substrate. This type of modulation noise tends to display a more prominent side noise band at the signal frequency.[12]

9.3 HEAD AND ELECTRONICS NOISES

When a magnetic read head is electrically activated, but not near a magnetic disk, it can still generate appreciable noises, including noises from head itself and from head preamplifier which magnifies read voltage by several hundreds typically.

Head noise arises from fluctuations of magnetic domain walls of the head core material (*Barkhausen noise*), or from the resistive dissipation in the head (*Johnson noise*). Large domain wall jumps can cause significant flux changes in read heads, leading to intolerable high Barkhausen noise peaks. Such noise peaks have to be avoided by the careful design and manufacturing of read heads. The other head noise can usually be described by the Nyquist noise theorem: any device that dissipates energy will generate noise power as a passive device, and the rms Johnson noise voltage is

$$V_{rms,n} = \sqrt{4kT(\text{Re}Z)\,\Delta f}, \tag{9.71}$$

where k is the Boltzmann's constant, T is the absolute temperature, $\text{Re}Z$ is the real part of the impedance, and Δf is the frequency range within which the noise is measured.

Electronics noise is caused by the random fluctuations in time of the electric carriers. The dominant electronics noise is generated by the preamplifier (the first stage amplifier) in data-detection circuitry. *Shot noise and thermal noise* usually dominates electronics noise in magnetic recording, which are more or less *white noise,* i.e., independent of frequency. In other words, white noise power spectral density is a constant in frequency domain. For a good preamplifier, the normalized white noise voltage is on the order of 0.5–2 nV/$\sqrt{\text{Hz}}$.

Another form of electronics noise is *flicker noise* or *excess noise* (due to device imperfection). Its power spectral density is proportional to $1/f^{\alpha}$, where $0.5 < \alpha < 2$. As an example, if the flicker noise power spectral density is $2000/f$ (nV)^2/Hz, and the white electronics noise (shot and Johnson noise) power spectral density is ~ 1 (nV)^2/Hz, then the total noise spectra are shown in Fig. 9.7. Apparently, the flicker noise is only important at low frequencies (<50 KHz) and can be neglected in magnetic recording.

9.4 SNR CONSIDERATIONS FOR MAGNETIC RECORDING

So far we have analyzed medium noises and head/preamplifier noise. They must be taken into consideration along with nonlinear distortions and interferences. The noise, distortion, and interference sources in magnetic recording are listed in Table 9.1, where the distortions and interferences will be considered in Chapters 10 and 14. There are so many noise,

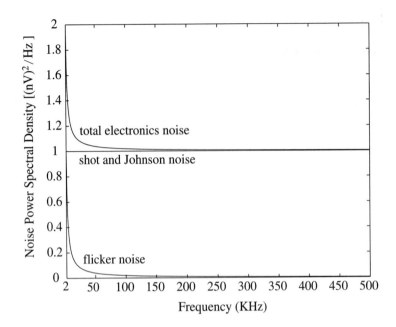

FIGURE 9.7.   Electronics noise (white noise and flicker noise) voltage vs. frequency.

distortion, and interference sources in magnetic recording that the list in Table 9.1 is not yet exhaustive.

As mentioned previously, there are many definitions of SNR in the literature. The most commonly used one is defined as

$$ SNR \equiv \frac{\text{zero-to-peak signal power}}{\text{noise power}} = \frac{S}{N} = \frac{V_{0-p}^2}{V_{rms,n}^2}, \tag{9.72} $$

where the rms noise voltage $V_{rms,n}$ could broadly include any disturbances such as medium noises, interferences, or distortions, and the peak signal voltage can be defined either as the isolated signal voltage peak or as the square wave amplitude recorded at the signal frequency. SNR is most often quoted in the units of dB:

$$ SNR \text{ (dB)} \equiv 10 \log \frac{S}{N} = 20 \log \frac{V_{0-p}}{V_{rms,n}}. \tag{9.73} $$

TABLE 9.1. Noises, Distortions, and Interferences in Magnetic Recording

Noises	Medium noises	Transition noise Particulate noise Modulation noise
	Head noises	Johnson noise Barkhausen noise
	Electronics noises	Shot noise Johnson noise Flicker noise
Distortions	Write distortions	Nonlinear transition shift Hard transition shift Partial erasure
	Read distortions	AMR head nonlinearity GMR head nonlinearity
Inferences	On-track	Linear transition shift Residual old information
	Off-track	Side read, track edge effect Head position misregistration[a]

[a]Misregistration refers to the mechanical fluctuation of the radial head position relative to disk from revolution to revolution. See Chapter 14.

If there are many noise sources, $V_{n,j}$, $j = 1, 2, 3, 4, \ldots$, in a magnetic recording system, then the total average noise power is

$$N = \left\langle \left(\sum_j V_{n,j} \right)^2 \right\rangle = \sum_j \langle (V_{n,j})^2 \rangle + \sum_{j,\, l \neq j} \langle V_{n,j} V_{n,l}^* \rangle.$$

If the noise sources are *uncorrelated,* then the total average noise power is

$$N = \sum_j V_{rms,n,j}^2 \tag{9.74}$$

where the rms noise voltage is defined as

$$V_{rms,n,j} \equiv \sqrt{\langle (V_{n,j})^2 \rangle}. \tag{9.75}$$

Assume that the noises from medium, head and, electronics are *uncorrelated,* then the total noise power spectral density is the sum of each noise power spectral density:

$$PSD_{tot} = PSD_{medium} + PSD_{head} + PSD_{electronics}. \tag{9.76}$$

Depending on the error correction code and recording system error rate requirement, the minimum required SNR in magnetic recording is typically 20–25 dB. The relation between error rate and SNR will be addressed in Chapters 11 and 12, where we will learn that SNRs in different points of the data detection channel are different! For example, white electronics noise becomes frequency dependent after going through a filter.

In designing a magnetic recording system, we want to design the system such that each noise sources are about equal. It is not efficient to reduce one noise source without doing the same to the other noise sources.

9.5 EXPERIMENTAL MEASUREMENTS OF SNR

Measurements of SNR are essential in the characterization of magnetic recording systems. This section will discuss several approaches to measure signal and noise that provide different insights into the performance of recording system components.

According to a simplified *additive* model, the on-track readback voltage signal (including noise) in time domain is given by

$$s(t) = V(t) + n_e(t) + n_m(t, V(t)), \qquad (9.77)$$

where $V(t)$ is the on-track readback signal, $n_e(t)$ is the electronics noise, and $n_m(t, V(t))$ is the signal-dependent medium noise. As shown in Fig. 9.8, different noise sources are most conveniently captured with a spectrum analyzer. The electronics noise is measured by unloading the read head from the magnetic disk (trace 1). When a reading head is to be characterized, the medium noise is usually eliminating by DC-erasing the magnetic medium. This results in a spectrum (trace 2) containing electronics noise, head noise, and DC-erased medium noise (which is signal-independent). The main contribution to thin film medium noise is the transition jitter, which is signal dependent. As seen from Fig. 9.8, square waves with different frequencies (traces 3 and 4) result in different noise spectra. The medium noise rises as the recording frequency increases. If only the medium transition noise is to be evaluated, the standard deviation of transition jitter can be measured with a time interval analyzer.

There are many methods of SNR measurements in the literature, which can be distinguished according to the following four aspects[13]:

1. The recording data pattern used for the signal measurement, which can be the low frequency square wave pattern (isolated transitions), the high-frequency square wave pattern, or the pseudo-random data sequence

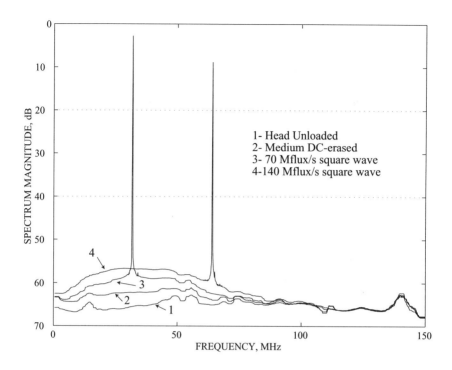

FIGURE 9.8. Noise and signal spectra in magnetic recording channel.

2. The definition of the signal, which can be the peak-to-peak voltage amplitude, the zero-to-peak voltage amplitude, the root-mean-square (rms) signal power, the fundamental harmonic spectrum power, or the slope of the voltage signal
3. The recording data pattern used for the noise measurement, which can be the DC-erased medium, the AC-erased medium, the square wave pattern, or the pseudo-random data sequence
4. The definition of noise, which can be the noise floor of the signal spectrum, the integrated noise power, the rms noise value, or the autocorrelation noise power

Here we will discuss three representative methods of SNR measurements.

9.5.1 Peak-signal-to-erased-medium-noise measurement

The signal used for this measurement is usually the low-frequency square wave pattern (isolated transitions) in order to avoid signal amplitude reduction caused by intersymbol interference (ISI). At first, the zero-to-peak amplitude V_{0-p} of the isolated readback pulses is measured. Then, the medium is either DC- or AC-erased. In the former case, the medium is erased by applying a high DC current to the write head. In the later case, the medium is erased by applying an AC field whose amplitude gradually decreases from far above the medium coercivity to zero. The AC-erase process can be approximated by writing at very high frequencies. If the readback voltage from the erased medium $n(t)$ is captured for a period of T, the rms noise voltage is then

$$V_{rms,n} = \sqrt{\frac{1}{T} \int_0^T n^2(t)dt}, \qquad (9.75)$$

which includes both electronics noise, head noise, and part of medium noise (either DC-erased or AC-erased medium noise). Note that the erased medium noise reflects the particular medium noise, and is different from the transition noise or modulation noise. The measured SNR (in dB) is calculated as the logarithmic ratio of the peak signal amplitude to the rms noise voltage, multiplied by 20:

$$SNR \text{ (dB)} = 20 \log \frac{V_{0-p}}{V_{rms,n}}. \qquad (9.73)$$

9.5.2 Spectrum analyzer measurement of SNR

There are several versions of spectrum analyzer measurements of SNR which vary by the recording patterns used and by the spectrum analyzer devices used. The most commonly used recording pattern is a single-frequency square wave pattern, resulting in a spectrum shown in Figure 9.8. Commercial spectrum analyzer devices use different measurement approaches, so device-dependent corrections of SNR values may be required when performing SNR measurements.

The typical architecture of a spectrum analyzer is shown in Fig. 9.9. The input broadband signal is input to the mixer where it is multiplied by the carrier frequency of an oscillator. The mixer's output contains both the difference and sum frequencies of the oscillator and the signal, which is then passed through a band-pass filter having a resolution bandwidth BW (typically 10–100 KHz). The band-pass filter is often called the intermediate frequency (IF) filter. The frequency response of the band-pass filter is approximated by a rectangular function centered at the intermediate frequency. The resolution bandwidth BW is defined by the two −3 dB

Mixer

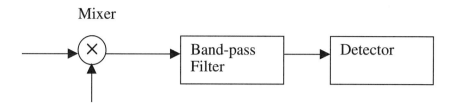

FIGURE 9.9. Block diagram of spectrum analyzer.

points at which the voltage gain drop to $1/\sqrt{2}$ of the maximum value. Only the difference or the sum signal passes through the filter, while the other frequency is filtered out.

The quasi-sine wave signal from the output of the band-pass filter is eventually passed to the detector. The measured spectrum value depends on the particular implementation of the detector. There are at least three common approaches of signal spectrum detection:

1. *Peak envelope amplitude.* The signal is rectified, and then passed to a peak detector whose output signal is proportional to the envelope peak amplitude.[14] For a sine wave with a zero-to-peak amplitude of 1 V, this will result in a signal equal to 1 V.
2. *rms voltage.* To calculate the rms voltage, the signal is squared, integrated and then averaged. The square root of this value is the rms voltage. For a sine wave this will result in a signal equal to $1/\sqrt{2}$ of the zero-to-peak amplitude. The rms value can also be obtained from the peak detector with a scaling factor of 0.707.
3. *Effective value.* The signal is rectified, resulting in a unipolar voltage signal. Then it is low-pass filtered. For a sine wave signal, the DC output of the low-pass filter is $2/\pi$ ($= 0.636$) times the zero-to-peak amplitude of the input. This is a standard scheme to measure the effective voltage value of a sine wave signal.

We will assume that the correction factors are applied to both signal and noise measurements.

After a signal is written, there are usually several spectral harmonic peaks. If each harmonic peak has a rms power of V_i, $i = 1, 2, 3, \ldots, N$, then the total rms signal power is equal to

$$S = \sum_{i=1}^{N} V_i^2. \tag{9.78}$$

To calculate the noise power, the noise spectrum must be integrated. Usually noise power is measured by integrating the noise spectrum (after removing signal harmonic peaks) from DC to the highest channel frequency using a specified number of steps M. This determines the frequency step Δf and results in the series of rms noise voltages $n(1), n(2), \ldots,$ $n(M)$. The minimum step between the frequencies for noise measurement equals to the resolution bandwidth BW. Since each of the rms noise voltages is measured with the resolution bandwidth, the total rms noise power is calculated as:

$$N = \sum_{i=1}^{M} \frac{n^2(i)}{BW} \Delta f. \qquad (9.79)$$

If the noise spectrum is displayed as the noise power spectral density $PSD_n(f)$, then the noise power can be calculated as

$$N = 2 \int_{f_1}^{f_2} PSD_n(f) df. \qquad (9.26)$$

where $[f_1, f_2]$ is the interested frequency range.

Finally the spectrum SNR (in dB) is calculated as the logarithmic ratio of the rms signal power over the rms noise power, multiplied by 10:

$$SNR_{rms} \text{ (dB)} = 10 \log \frac{S}{N}. \qquad (9.80)$$

If the zero-to-peak amplitude of isolated pulses or high frequency square-wave pattern is known, then the peak-signal-to-rms-noise ratio can be calculated as:

$$SNR \text{ (dB)} = 10 \log \frac{V_{0-p}^2}{N}. \qquad (9.81)$$

9.5.3 Cross-correlation measurement of SNR

This method is based on the property of *cross-correlation coefficient* of two different random variables.[15] The cross-correlation coefficient of two DC-free periodic signal $s_1(t)$ and $s_2(t)$ with a period of T is defined as

$$
r(\tau) = \frac{\dfrac{1}{T} \displaystyle\int_{-T/2}^{T/2} [s_1(t) - \langle s_1(t) \rangle][s_2(t) - \langle s_2(t) \rangle] dt}{\sqrt{\dfrac{1}{T} \displaystyle\int_{-T/2}^{T/2} [s_1(t) - \langle s_1(t) \rangle]^2 dt} \sqrt{\dfrac{1}{T} \displaystyle\int_{-T/2}^{T/2} [s_2(t) - \langle s_2(t) \rangle]^2 dt}}
$$

$$
= \frac{\dfrac{1}{T} \displaystyle\int_{-T/2}^{T/2} s_1(t) s_2(t) dt}{\sqrt{\dfrac{1}{T} \displaystyle\int_{-T/2}^{T/2} [s_1(t)]^2 dt} \sqrt{\dfrac{1}{T} \displaystyle\int_{-T/2}^{T/2} [s_2(t)]^2 dt}}, \qquad (9.82)
$$

where $\langle s_1(t) \rangle = \langle s_2(t) \rangle = 0$, $s_1(t) = V(t) + n_1(t)$, $s_2(t) = V(t + \tau) + n_2(t)$, $V(t)$ is the readback signal from the pseudo-random data sequence with a period of T, and s_1 and s_2 are written such that $\tau = mT$ (m = an integer) is much longer than the correlation time of the two noise sources n_1 and n_2. Therefore, n_1 and n_2 are uncorrelated, n_1 is uncorrelated to $V(t + mT)$, and n_2 is uncorrelated to $V(t)$. Consequently,

$$\frac{1}{T}\int_{-T/2}^{T/2} [V(t) + n_1(t)][V(t + mT) + n_2(t)]dt = \frac{1}{T}\int_{-T/2}^{T/2} [V(t)][V(t + mT)]dt$$

$$= \frac{1}{T}\int_{-T/2}^{T/2} [V(t)]^2 dt = S,$$

where S is the rms signal power. Assume that the noise is statistically uniform in the period T, i.e.,

$$\frac{1}{T}\int_{-T/2}^{T/2} \{2V(t)n_1(t)\}dt = 0,$$

then,

$$\sqrt{\frac{1}{T}\int_{-T/2}^{T/2} [s_1(t)]^2 dt} \sqrt{\frac{1}{T}\int_{-T/2}^{T/2} [s_2(t)]^2 dt} \approx \frac{1}{T}\int_{-T/2}^{T/2} [s_1(t)]^2 dt$$

$$= \frac{1}{T}\int_{-T/2}^{T/2} \{[V(t)]^2 + [n_1(t)]^2 + 2V(t)n_1(t)\}dt$$

$$= \frac{1}{T}\int_{-T/2}^{T/2} \{[V(t)]^2 + [n_1(t)]^2\}dt = S + N,$$

where N is the rms noise power. Therefore,

$$r(mT) = \frac{S}{S + N} = \frac{1}{1 + N/S}, \tag{9.83}$$

$$SNR_{rms} \text{ (dB)} = 10 \log \frac{S}{N} = 10 \log \frac{r(mT)}{1 - r(mT)}.$$

The cross-correlation measurement of SNR should give similar results as the spectrum SNR [Equation (9.80)] because both are the ratio of rms signal power to rms noise power. However, the peak-signal-to-rms-noise-ratio is typically higher because the zero-to-peak signal amplitude is $\sim\sqrt{2}$ times the rms signal amplitude. For example, for a reasonably good combination of recording head and medium, the peak-signal-to-rms-noise-ratio should be better than 23–25 dB, and the spectrum or cross-correlation SNR should be better than 20–22 dB.

References

1. For example, B. Sklar, *Digital Communication*, (Englewood Cliffs, NJ: Prentice Hall, 1988). R. A. Witte, *Spectrum and Network Measurements* (Englewood Cliffs, NJ: Prentice Hall, 1991).

2. H. N. Bertram, *Theory of Magnetic Recording*, (Cambridge, UK: Cambridge University Press, 1994).

3. T. C. Arnoldussen and L. L. Nunnelley (eds), *Noise in Digital Magnetic Recording*, World Scientific, New Jersey, 1992.

4. N. R. Belk, P. K. George, and G. S. Mowry, "Noise in high performance thin-film longitudinal magnetic recording media," *IEEE Trans. Magn.*, **21**, 1350, 1985.

5. A. M. Barany and H. N. Bertram, "Transition noise model for longitudinal thin-film media," *IEEE Trans. Magn.*, **23**, 1776, 1987.

6. J. G. Zhu and H. N. Bertram, "Recording and transition noise simulations in thin film media," *IEEE Trans. Magn.*, **24**, 2739, 1988.

7. J. G. Zhu, "Micromagnetics of thin film media," in *Magnetic Recording Technology*, edited by C. D. Mee and E. D. Daniel, McGraw-Hill, New York, 1995.

8. G. J. Tarnopolsky and P. R. Pitts, "Media noise and SNR estimates for high areal density recording," *J. Appl. Phys.*, **81**, 4837, 1997.

9. H. N. Bertram, H. Zhou, and R. Gustafson, "Signal to noise ratio scaling and density limit estimates in longitudinal magnetic recording," *IEEE Trans. Magn.*, **34**, 1845, 1998.

10. J. Li, *Preparation and Characterization of Sputtered Barium Ferrite Magnetic Thin Films*, Ph.D. Dissertation, Department of Materials Science and Engineering, Standord Univ., 1995.

11. H. N. Bertram, *Theory of Magnetic Recording*, p. 273, Cambridge University Press, 1994.

12. G. H. Lin, X. Xing, K. E. Johnson, and H. N. Bertram, "Texture induced noise and its impact on system performance," *IEEE Trans. Magn.*, **33**, 950, 1997.

13. Y. S. Tang, private communication, 1997.

14. E. J. Kennedy, *Operational Amplifier Circuits*, p. 397, (New York, NY: Holt, Rinehart and Winston, 1988).

15. G. Mian and T. D. Howell, "Determining signal to noise for an arbitrary data sequence by a time domain analysis," *IEEE Trans. Magn.*, **29**, 3999, 1993.

CHAPTER 10

Nonlinear Distortions

In this chapter we will consider various nonlinear distortions which occur during write and read processes, such as overwrite, transition shifts, partial erasure, and MR read nonlinearity. These distortions depend on both the write/read head and media, and should be minimized when a magnetic recording system is designed.

10.1 HARD TRANSITION SHIFT AND OVERWRITE

10.1.1 Hard transition shift

In digital magnetic recording the erasure of old information is accomplished by direct overwrite of new data pattern. When a new data pattern is written on top of the old magnetization, the head field can either coincide or oppose the incoming magnetization. When the head field is in the direction of the incoming magnetization, the "easy" transition is written. Easy transitions are not distorted. Transitions, which are written against the medium magnetization are called "hard" transition. Hard transitions require more head field to saturate the magnetization in the gap region due to increased demagnetizing field from the leading edge of the write bubble. Hard transitions are delayed and in general their shape may become different from the shape of easy transitions—they may have smaller amplitude and larger width. The resulting distortion of the readback waveform is schematically illustrated in Fig. 10.1 for a case of writing a square wave pattern on the uniformly magnetized (DC-erased) medium. For this particular case an alternating sequence of hard and easy transitions is written.

To explain the difference between easy and hard transitions, the write process and the influence of the demagnetization fields have to be consid-

267

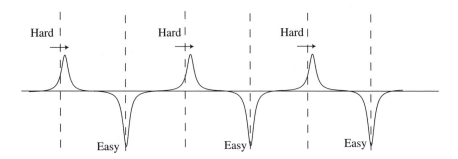

FIGURE 10.1.   Hard and easy transitions written over DC-erased medium.

ered in more detail. The process of forming a hard transition is shown in
Fig. 10.2. The magnetic medium is initially magnetized in one direction
(to the right), and the head moves to the right direction. When a hard
transition is written, the head magnetic field in the gap region H_h is
directed opposite to the medium magnetization. The medium is to be
remagnetized inside the initial write bubble, shown in Fig. 10.2 by the

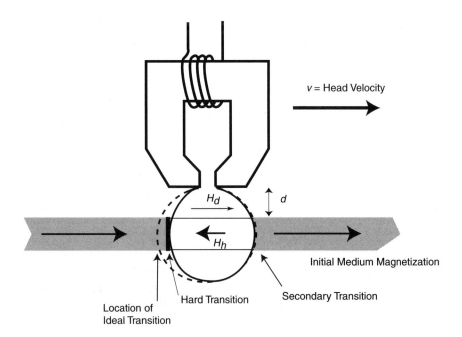

FIGURE 10.2.   Writing of a hard transition.

dashed line. Inside this initial write bubble, $H_h > H_c$, where H_c is the coercivity of the medium.

Once a head field is applied, a secondary transition is instantaneously formed at the leading edge of the head gap (to the right of the write bubble). This happens because the magnetic medium in front of the head is already magnetized (DC-erased). While the actual hard transition written by the trailing edge of the write bubble will remain on the medium, the secondary transition written by the leading edge of the write bubble is dynamically overwritten by the write bubble as the head is moving to the right. However, immediately after the secondary transition is formed, a demagnetization field H_d appears inside the medium. This field has a direction opposite to the head field. Now to remagnetize the medium, we need $H_h - H_d > H_c$. This means that the initial write bubble of the head will be reduced in size. As a result, the hard transition will be written later than what is determined from the initial write bubble.

Figure 10.3 provides an illustration of how demagnetizing field interacts with the magnetic field of the head. To estimate the amount of hard transition shift (HTS), we must determine the location of the actual written

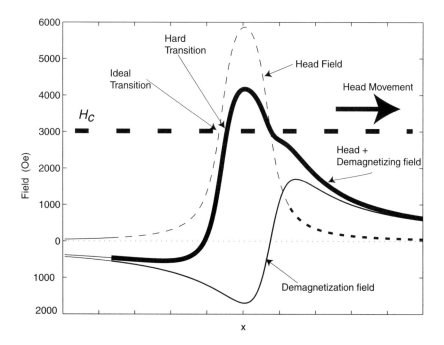

FIGURE 10.3.   Interaction of the head field with the demagnetization field.

transition.[1] The transition will be written at the location where the sum of the head field H_h and of the opposing demagnetization field H_d equals the coercivity:

$$H_h(x) + H_d(x) = H_c. \tag{10.1}$$

The unperturbed transition location x_0 is given by

$$H_h(x_0) = H_c.$$

Assuming that HTS is small, we obtain:

$$\Delta = x - x_0 = -\frac{H_d(x_0)}{dH_h/dx|_{x=x_0}}. \tag{10.2}$$

Therefore, the transition shift is directly proportional to the magnitude of the demagnetization field at the transition location and is inversely proportional to the head field gradient.

The longitudinal demagnetizing field at the center of the written hard transition due to the secondary transition at the leading edge of the write bubble is

$$H_d = -\frac{2M_r}{\pi}\left[\tan^{-1}\left(\frac{a+\delta/2}{R}\right) - \tan^{-1}\left(\frac{a}{R}\right)\right] \approx -\frac{M_r\delta}{\pi R}, \text{(if } a, \delta << R) \tag{10.3}$$

where R is the horizontal size of the write bubble, δ is the thickness of magnetic medium, and the transition shape is assumed to be arctangent with a transition parameter a. The head field gradient at the center of the hard transition is

$$\frac{dH_h}{dx} = \frac{QH_c}{d+\delta/2}. \tag{10.4}$$

Therefore, HTS is

$$\Delta = \frac{M_r\delta(d+\delta/2)}{\pi QH_cR}. \tag{10.5}$$

Note that HTS is positive, i.e., the hard transition is written *later* than desired. Note the sign of transition shift is dependent on the choice of coordinates. The medium is chosen to be stationary in this chapter, while the head is stationary in Chapters 3 and 4. Therefore, the directions of x-axes in the two coordinates are physically opposite. However, coordinate changes will not alter the fact that HTS is always written later than intended.

Taking the imaging effect into account,

$$\Delta = \frac{M_r\delta(d+\delta/2)}{\pi QH_cR}F(d, \delta, R), \tag{10.6}$$

where function $F(d,\delta,R)$ describes the head imaging effect which tends to reduce the demagnetizing field and the HTS. The function is no greater than 1 and equals $(d + \delta/2)^2/R^2$ in the case of ideal imaging.

The shape of the magnetic transition is also affected by the demagnetization field from the secondary transition, which causes a change of transition slope and increases the transition parameter. The shape of the isolated pulse is directly dependent on the shape of the magnetic transition $M(x)$ because the isolated pulse of voltage is the convolution of dM/dx with the head field. As a result, the isolated pulse of voltage for the hard transition is not only delayed, but also may have a reduced voltage amplitude and a larger pulse width. This effect is often referred to as *pulse broadening*. While shape distortions are usually not significant, they may become important for specific head/medium combinations.

Now let us examine what happens to an easy transition shown in Fig. 10.4. First of all, the easy transition can be written only if the medium magnetization at the moment of writing is opposite to the initial magnetization of the medium, meaning that a hard transition was previously written. At the moment when the easy transition is to be written, the write head field is reversed and its direction now coincides with the

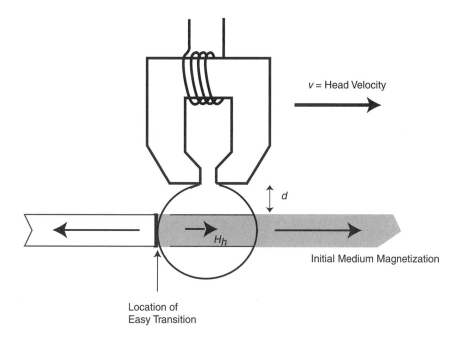

FIGURE 10.4. Writing of an easy transition.

medium magnetization to the right of the magnetic head. In this case, the secondary transition does not appear. Therefore, the demagnetization field from the leading edge of the write bubble will not interfere with the magnetic field of the head and the write process will be determined only by the trailing edge of the write bubble. The situation shown in Fig. 10.4 corresponds to writing an isolated transition, which is written at the trailing edge of the gap without nonlinear distortions.

Note that when the medium is AC-erased, all transitions are "50% hard" and are delayed from their nominal positions by one half of the hard transition shift. This happens because the AC-erased medium creates approximately 50% of the equivalent magnetic charge at the leading edge of the write bubble, for *both* directions of magnetization. Effectively, there is no hard transition shift in the AC-erased medium.

A simple way of measuring the HTS is the asymmetry test. This test is based on measuring the timing difference between positive and negative peaks of a relatively low-frequency signal written on the magnetic medium after the medium was DC-erased (Fig. 10.5). The measured asymmetry value is calculated as:

$$\text{Asymmetry} = \text{Average}[T(p - n) - T(n - p)],$$

where $T(p - n)$ is the distance between positive and negative peaks and $T(n - p)$ is the distance between negative and positive peaks. While the asymmetry value is not equal to hard/easy (H/E) transition shift Δ, it is related to it. If the positive transition is easy and negative transition is hard, then $T(p - n) = T + \Delta$ and $T(n - p) = T - \Delta$. Therefore,

$$|\text{Asymmetry}| = 2\Delta. \qquad (10.7)$$

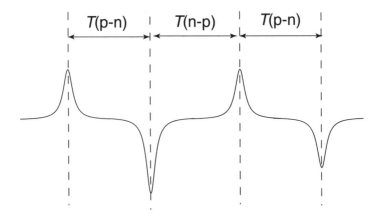

FIGURE 10.5.   Measurement of hard transition shift.

10.1.2 Overwrite

When a new data pattern is written on top of the old information, the write field must be sufficient to rewrite the old information. However, a simple magnetic recording experiment demonstrates that the residual signal from the old data pattern will always be present in the spectrum of the new data pattern. Therefore, the requirement is to reduce any residual signal of the old pattern to the levels that are low enough not to cause errors while reading the new data.

Consider the following experiment, which is the most commonly cited method of measuring the overwrite ratio:

1. A pattern with a low frequency of f_1 is written first and its average amplitude A1 is measured.
2. A pattern with a higher frequency of f_2 is then written on the same track over the old pattern.
3. A residual signal at frequency f_1 is measured using a band-pass filter or a spectrum analyzer. This pattern has an amplitude of A2.
4. The *overwrite ratio* is calculated as 20 log(A2/A1), which reflects the ability of a new data pattern to suppress the old data previously written on the magnetic medium. Typically the overwrite values should be kept below −30 dB.

When a pattern with frequency f_1 is overwritten with a new pattern having a frequency f_2, the new pattern is modulated at a frequency of the old pattern. Figure 10.6 presents the channel spectra of a typical overwrite experiment. First, a 20 Mflux/s square wave is written on the magnetic medium. Then, an 80 Mflux/s square wave is written on top of the 20 Mflux/s pattern. The overwrite spectrum from the new pattern contains a number of residual features related to 20 Mflux/s recording. The most evident is the residual harmonic at 10 MHz, which is the fundamental harmonic of the 20 Mflux/s square wave. There are also significant side-band signals around the fundamental and the second harmonic of the 80 Mflux/s square wave (at 30, 50, 70, and 90 MHz). Note that these residual overwrite harmonics disappear from the spectrum after a second writing of the 80 Mflux/s signal. Some details of the spectra will be analyzed in Sections 10.1.4 and 10.1.5.

Overwrite is a complicated phenomenon. At least several factors explain the residual signal observed in an overwrite experiment:

1. If the write current is not sufficient, the residual signal is due to incomplete erasure of the f_1 signal. This is typically not common in normal magnetic recording environment.

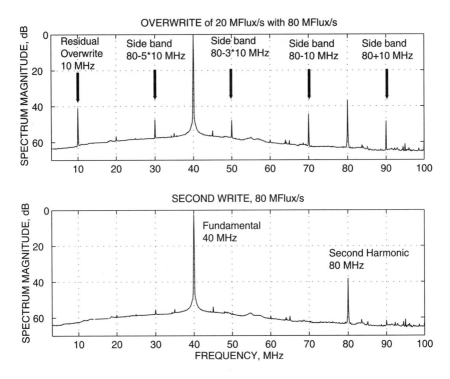

FIGURE 10.6. Spectra of overwrite signals: first overwriting of 20 Mflux/s square wave with 80 Mflux/s square wave (top), and second consecutive writing with 80 Mflux/s square wave (bottom).

2. *Track-edge effects.* When a new pattern is written, the track edges of the old information are erased incompletely. The track-edge effect is frequency-dependent, i.e., the larger the ratio between the frequencies of the old and the new data patterns, the stronger the residual signal from the track edges. This is explained by the dependence of gap loss on the wavelength of the recorded pattern. Figure 10.7 depicts the spatial distribution of the residual signal at cross-track position for two different overwrite ratios. The cross-track amplitude and residual signals are asymmetrical due to read/write offset of the MR head and skew angle between the head and the track. Note that the track-edge effect can be reduced by a "write wide, read narrow" scheme (Chapter 14).

3. HTS of f_2 signal at f_1 rate due to magnetostatic fields from the incoming f_1 pattern. These shifts appear to be the main contribution

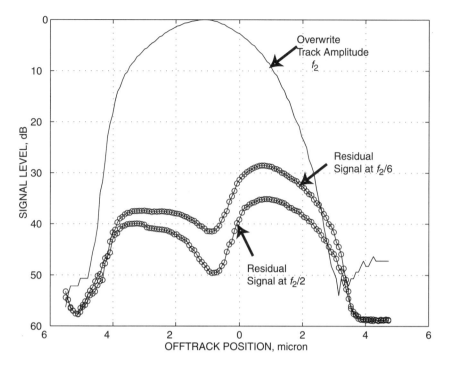

FIGURE 10.7. Cross-track distribution of residual overwrite signals relative to the fundamental of the f_2 pattern for $f_1 = f_2/6$ and $f_1 = f_2/2$. Note the increased residual signal at $f_1 = f_2/6$.

to the overwrite signal and they are most difficult to overcome. We will analyze this source of residual signal in more detail.

10.1.3 Overwrite at a frequency ratio of 2

To explain the origin of the overwrite signal, we need to examine what actually happens when a new data pattern is written over the old data. For simplicity we consider the case when a square-wave pattern is over-written by another square-wave pattern at twice the density of the original pattern, as shown in Fig. 10.8, in which the head is moving to the right. We will assume a simple rule that the transition is hard when its trailing edge is written against the corresponding magnetization of the old pattern. This is valid because distance between the leading and trailing edge (the

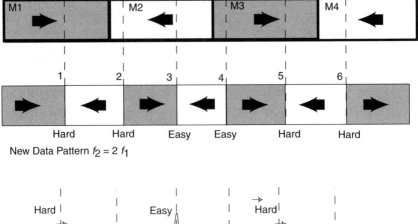

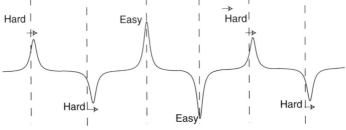

FIGURE 10.8. Explanation of overwrite signal generation: transition patterns (top) and readback waveform (bottom).

size of the write bubble) causes only a phase offset in the sequence of easy and hard transitions.

Transition 1 of the new data is written when the trailing (left) edge of the write bubble is at position "1" and the head field is directed to the left. The old data are in the region M1 and have a magnetization opposite to the direction of the head magnetic field, so this transition is a hard transition. When transition 2 is written, the head field is directed to the right and the initial medium magnetization is within region M2 of the old data and is opposite to the head field. Therefore, transition 2 will also be a hard transition. However. when transition 3 is written, the head field is directed to the left and the initial medium magnetization is still within the M2 region. Therefore, this transition will be an easy transition. Similarly, transition 4 will also be an easy one, and transitions 5 and 6 will again become hard transitions. Therefore. the overwrite signal for frequency f_2

$= 2f_1$ will consist of a repeated pattern of two hard transitions followed by two easy transitions: ... HHEEHHEE. . . .

Note that the write pattern is modulated with the period of the old pattern. Indeed, whatever the frequency of the old pattern, it will periodically change the sequence of hard and easy transitions in the overwrite pattern. In our example this pattern is two hard and two easy transitions. However, the sequence of hard and easy transitions will become more complicated for higher overwrite ratios, as to be shown in Sections 10.1.4 and 10.1.5.

For the particular case of $f_2 = 2f_1$, we can obtain an analytical expression which relates the value of overwrite as $20 \log(A1/A2)$ in dB to the value of the HTS.[1,2] To do that we will calculate the Fourier transform of the old pattern and the overwrite pattern with hard/easy transitions.

Assume that the hard transition generates a pulse $h(t)$ and the easy transition generates $e(t)$. The transitions of the old pattern are written with a period of $4T$ (the distance between positive and negative transitions is $2T$) and the transitions of the overwrite pattern are written with a period of $2T$. All the transitions in the old pattern are easy transitions and, therefore, a single period of the old signal is given by:

$$f_{\text{old}} = e(t) - e(t - 2T). \tag{10.8}$$

The corresponding period of the overwrite pattern is given by:

$$f_{\text{ow}} = h(t - \Delta) - h(t - T - \Delta) + e(t - 2T) - e(t - 3T), \tag{10.9}$$

where Δ is the hard transition shift in the overwrite pattern. When the Fourier transform is calculated, each timing shift changes the phase of the signal. Therefore,

$$FT[f_{\text{old}}] = E(\omega) - E(\omega)e^{-iw2T},$$
$$FT[f_{\text{ow}}] = H(\omega)e^{-iw\Delta} - H(\omega)e^{-iw(T+\Delta)} + E(\omega)e^{-iw2T} - E(\omega)e^{-iw3T}, \tag{10.10}$$

where ω is the angular frequency, $E(\omega)$ is the spectrum of the easy transition pulse, and $H(\omega)$ is the spectrum of the hard transition pulse. For the old information pattern, $\omega_1 = 2\pi/4T$, at which many of the complex exponents are integer multiples of $\pi/2$, therefore they result in one or i. The Fourier components at ω_1 in the old pattern and the overwrite pattern are:

$$A1 = |\, E(\omega_1) + E(\omega_1)\, |= 2|E(\omega_1)|,$$
$$A2 = |\, (1 + i)[H(\omega_1)e^{-iw_1\Delta} - E(\omega_1)]|. \tag{10.11}$$

If we assume that the shapes of the easy and hard transition pulses are approximately the same, i.e., $E(\omega) = H(\omega)$, then the overwrite ratio (in dB) is

$$OW = 20 \log \left(\sqrt{1 - \cos\left(\frac{\pi}{2T}\Delta\right)} \right)$$

$$= 20 \log \left(\sqrt{2} \sin \frac{\pi}{4T}\Delta \right) \qquad (10.12)$$

If the value of Δ is small, we can use approximation: $\sin(x) \approx x$ for $x \ll 1$. Therefore,

$$OW \approx 20 \log \left(\frac{\sqrt{2}\pi}{4T}\Delta \right). \qquad (10.13)$$

This means that overwrite ratio is proportional to the logarithm of the ratio of the HTS to the bit period of the overwrite pattern. Therefore, a larger HTS and a shorter bit length will result in a larger residual signal and a smaller absolute overwrite ratio.

For example, if $T = 10$ ns, $\Delta = 1$ ns, overwrite ratio will be equal to -20 dB. However, if $T = 5$ ns, $\Delta = 1$ ns, the overwrite ratio will be equal to only -13 dB. In practice, the desirable levels of overwrite ratio should be below -30 dB.

10.1.4 Overwrite at different frequency ratios

The overwrite ratio measured at a frequency ratio $f_2/f_1 = 2$ actually gives an optimistic estimate of overwrite. It must be cautioned that the overwrite ratio is frequency dependent, which is one of the most evident manifestations of the complicated interactions between the old pattern and new pattern. Figure 10.9 presents typical experimental results, where each of the curves gives the overwrite of different f_1 patterns with the fixed f_2 frequency.

Several distinct effects are evident from Fig. 10.9. First, if we fix the low-frequency (LF) signal and write higher frequency overwrite patterns, the absolute overwrite values will decrease. We may end up with a drop of more than 12–14 dB in overwrite value! The second effect is that if we write LF patterns with a higher frequency and overwrite with the high-frequency (HF) pattern of a fixed frequency, then the overwrite will improve. An improvement as much as 12–15 dB can be obtained. The third

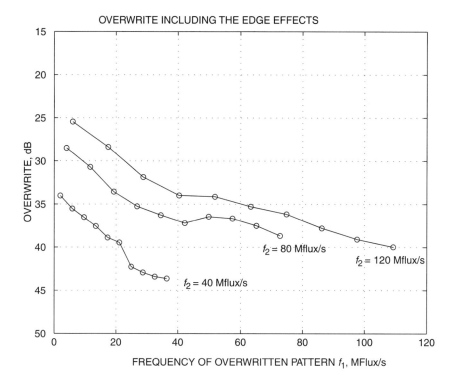

FIGURE 10.9. Dependence of overwrite on the frequency of LF (f_1) pattern. Frequencies f_2 of HF patterns are 40, 80, and 120 Mflux/s, respectively.

effect is that the magnitude of the improvement depends on the particular low and high frequencies and not only on their ratio.

The strong frequency dependence of overwrite is in part explained by the track edge effects. As we mentioned previously, the amplitude of the residual signal at the track edges depends on the frequency ratio between the old and the new patterns. However, even when the track edges are erased, the overwrite still demonstrates significant frequency dependence as shown in Fig. 10.10. To explain the frequency dependence of overwrite, we must first find how much the overwrite signals are at different overwrite frequencies.

When the frequency of the overwrite pattern is not exactly two times higher than the frequency of the old pattern, the sequence of easy and hard transitions becomes complicated.[3,4] For example, let us consider the magnetization of the old pattern and of the new pattern. The frequency of the overwrite pattern in Fig. 10.11 is five times higher than the frequency

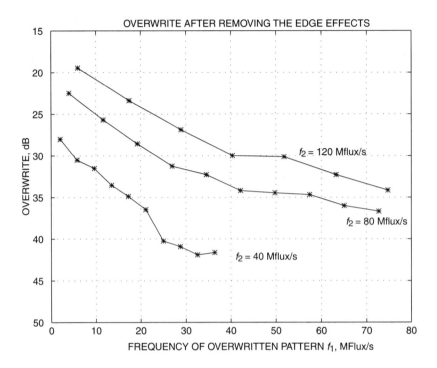

FIGURE 10.10.   Frequency dependence of overwrite after erasure of track edges.

of the old pattern. Therefore, the sequence of hard and easy transitions is quite complicated: ... HEHEEHEHEE. ...

Define the average pulse $a(t)$ and the error pulse $d(t)$ as follows:

$$a(t) = [e(t) + h(t)]/2, \quad d(t) = [e(t) - h(t)]/2$$
$$e(t) = a(t) + d(t), \qquad h(t) = a(t) - d(t). \tag{10.14}$$

If we now write the sequence of easy and hard pulses in Fig. 10.11 using the sum and difference of $a(t)$ and $d(t)$ and alternate polarities, a regular modulation of new pattern by the old pattern emerges. The average pulses, as expected, behave like the undistorted sequence of a square recording waveform:

$$\ldots + a - a + a - a \ldots$$

The sequence of the difference pulses $d(t)$ are periodic with exactly the period of the old pattern:

$$\ldots - d - d - d - d - d + d + d + d + d + d - d - d - d - d - d \ldots$$

Old Data Pattern f_1

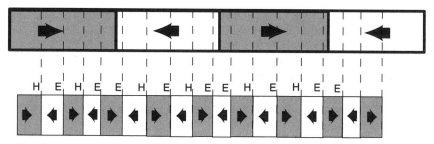

New Data Pattern $f_2 = 5 f_1$

Readback Waveform of New Data Pattern

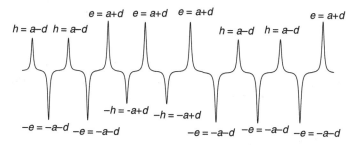

FIGURE 10.11. Overwrite signal generation at $f_2/f_1 = 5$.

The periodicity of the $d(t)$ sequence is general and holds for any integer ratio of overwrite frequencies. This property allows us to analyze the overwrite signal analytically.

Let the ratio of the high frequency to the low frequency be N, i.e., the low angular frequency is

$$\omega_1 = 2\pi/(2NT) = \pi/NT.$$

Assume that the transitions of the low-frequency pattern are all easy transitions, then one period of the old signal is given by the function:

$$f_{\text{old}} = e(t) - e(t - NT). \tag{10.15}$$

As a general rule, the corresponding period of the overwrite pattern is modulated by the period of the old pattern. Obviously, the sequence of the average pulses is not modulated by the old pattern and has no harmonic peak at the frequency of the old pattern. The sequence of the error pulses is given by:

$$f_{ow} = \sum_{m=0}^{N-1} d(t - mT) - \sum_{m=N}^{2N-1} d(t - mT). \tag{10.16}$$

Only this signal is of interest because the alternating average signal will not create a harmonic at the low frequency. The Fourier transforms of Equations (10.15) and (10.16) at the low frequency of π/NT are

$$\left| FT(f_{old}) \right| = \left| E\!\left(\frac{\pi}{NT}\right) - E\!\left(\frac{\pi}{NT}\right) e^{-i\frac{\pi}{NT}NT} \right| = 2\left| E\!\left(\frac{\pi}{NT}\right) \right|, \tag{10.17}$$

$$\left| FT(f_{ow}) \right| = \left| D\!\left(\frac{\pi}{NT}\right) \right| \cdot \left| \sum_{m=0}^{N-1} \left(e^{-im\frac{\pi}{N}} - e^{-i(m+N)\frac{\pi}{N}} \right) \right| = \frac{2\left| D\!\left(\dfrac{\pi}{NT}\right) \right|}{\sin(\pi/2N)}.$$

If we assume that the shapes of the easy and hard transitions are approximately the same, the difference between the easy and hard transition is only in the timing shift Δ. Therefore,

$$\left| D\!\left(\frac{\pi}{NT}\right) \right| = \left| 0.5E\!\left(\frac{\pi}{NT}\right)\left[\exp\!\left(-i\frac{\pi}{NT}\Delta\right) - 1 \right] \right|$$

$$= \left| E\!\left(\frac{\pi}{NT}\right) \right| \sin\!\left(\frac{\pi}{2NT}\Delta\right). \tag{10.18}$$

The overwrite ratio (in dB) is

$$OW = 20 \log \left[\frac{\sin\!\left(\dfrac{\pi}{2NT}\Delta\right)}{\sin\!\left(\dfrac{\pi}{2N}\right)} \right]. \tag{10.19}$$

According to Equation (10.19), the value of the overwrite should be almost independent of the frequency ratio N because it is a very slow function of N. When Δ is small, $\sin(\pi\Delta/2NT) \approx \pi\Delta/2NT$, and $\sin(\pi/2N) \approx \pi/2N$ for $N \geq 2$, then

$$OW \approx 20 \log\!\left(\frac{\Delta}{T}\right). \tag{10.20}$$

Equations (10.19–20) indicate that the overwrite should not change dramatically when the frequency of the high frequency pattern (or T) is fixed. In other words, all the three curves shown in Fig. 10.10 should be almost horizontal. Something must be missing in the above model, or the experiment is wrong.

To explain the dependence of the overwrite signal on the low-frequency pattern, let us examine what will happen to a transition when the edge of the write bubble is close to the previously written transition of the old information pattern as shown in Fig. 10.12. When the current transition is a hard transition, the leading edge of the write bubble may be close to a transition of the old pattern. This old transition creates an additional demagnetizing field that is opposite to the demagnetizing field of the secondary transition and, therefore, reduces the HTS.[4,5] This observation is an important clue to solving the puzzle of Fig. 10.10.

If we consider that the HTS Δ is frequency-dependent, then we can explain the dependence of overwrite on the frequency ratio. If the old

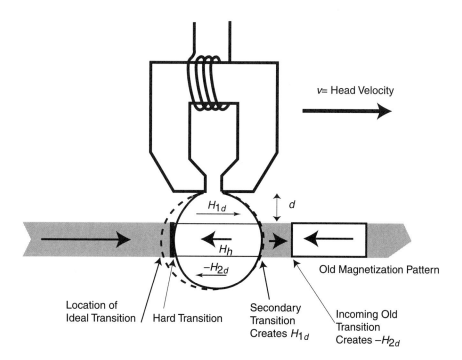

FIGURE 10.12. Explanation of the proximity effect on HTS.

pattern frequency f_1 is low, the value of Δ is large because it is not affected by the old pattern demagnetizing field. This is reasonable because very few transitions of the f_2 pattern will fall within the proximity of the transitions of the f_1 pattern. As the frequency of the old pattern is increased, more transitions of the f_2 pattern will fall within the proximity of the old transitions, so the HTS becomes smaller. This is referred to as the *proximity effect*. Note that this attenuation of HTS has a statistical character and, in principle, may depend on the relative phase between the f_1 and the f_2 patterns. However, the relative phase shift is usually randomized due to the spindle instability of the disk drive motor and thus can be neglected. These considerations lead to us to modify the expression of the overwrite ratio as follows[4]:

$$ OW = 20 \log \left[\frac{\Delta}{T} \, \Psi(f_1) \right], \tag{10.21} $$

where function $\Psi(f_1)$ describes the attenuation of HTS caused by the proximity effect, and Δ is the HTS value at large frequency ratios. Note that $\Psi(f_1)$ is a function of the old pattern frequency f_1.

It is very difficult to derive the shape of $\Psi(f_1)$ analytically, but it can be easily obtained from the experimental data. From the overwrite measurements we can calculate an equivalent HTS:

$$ \Delta \cdot \Psi(f_1) = T \cdot 10^{OW/20}. \tag{10.22} $$

The results of this calculation based on Fig. 10.10 are shown in Fig. 10.13. The equivalent HTS decreases from about 0.5 ns to approximately 0.15 ns as the frequency of the old pattern is increased. Note that the equivalent HTS follows a single curve for different densities of the f_2 pattern as it is expected to be a function of f_1 only.

Additional degradation of overwrite may be caused by density-related effects such as the interactions between HTS and nonlinear transition shift,[6] which will be considered in Section 10.2.

10.1.5 Spectrum of the overwrite signal

In this section we will give an explanation of the spectra shown in Fig. 10.6. If a periodic pattern consisting of pulses of alternating polarity with a bit period of T is written, the spectrum of the pulse sequence consists of the odd harmonics of fundamental frequency $\omega_2 = \pi/T$. The envelope of the spectrum repeats $P(\omega)$, which is the continuous spectrum of an

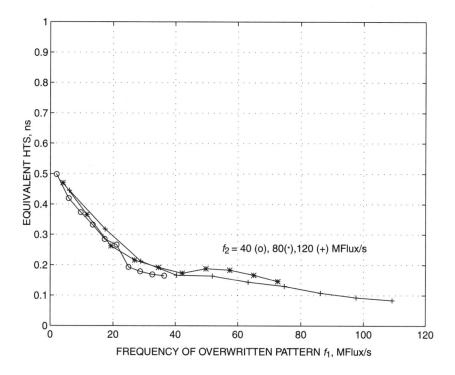

FIGURE 10.13.   Calculation of the equivalent HTS from the overwrite data shown in Fig. 10.10.

isolated transition $p(t)$. This can be easily seen after performing the Fourier transform of the pulse sequence.[7] An equivalent result in the spatial frequency domain has been derived previously in Section 3.3.

If the shape of the isolated pulse is given by $p(t)$, the pulse sequence is described by the convolution of the isolated pulse with an infinite sequence of alternating delta-functions. The period of the pattern is $2T$, and the readback signal can be expressed as

$$r(t) = \sum_{n=-\infty}^{\infty} p(t-2nT) - \sum_{n=-\infty}^{\infty} p(t-2nT-T) \qquad (10.23)$$

$$= p(t)* \left[\sum_{n=-\infty}^{\infty} \delta(t-2nT) - \sum_{n=-\infty}^{\infty} \delta(t-2nT-T) \right].$$

where n is any integer. The Fourier transform of the readback signal is given by:

$$R(\omega) = 2\omega_2 P(\omega) \sum_{m=\text{odd}} \delta(\omega - m\omega_2). \qquad (10.24)$$

When a periodic pattern is overwritten on some old information pattern, it is generally impossible to predict which of the transitions will become distorted. In general, HTS and shape changes will cause random or periodic modulation of overwrite pattern. When the old information pattern is periodic with a lower angular frequency of ω_1, the spectrum of the overwrite signal may be predicted. As seen in the previous section, a sequence of easy and hard transitions in the new pattern has a period of the old pattern.

Any distorted overwrite pattern can be represented as a sequence of $a(t) \pm d(t)$ with appropriate spacing and polarity, as shown in Fig. 10.11. The $a(t)$ sequence will just produce a spectrum containing only the odd harmonics of ω_2. In contrast, the $d(t)$ sequence is modulated by the frequency of the old pattern ω_1, as expressed by Equation (10.16), which can be written as the following convolution[7]:

$$o(t) = d(t) * \left\{ \left[\sum_{n=-\infty}^{\infty} \delta(t - nT) \right] S(t) \right\}. \qquad (10.25)$$

where $S(t)$ is a square wave of alternating ± 1 amplitude with an angular frequency of ω_1. The Fourier transform of Equation (10.25) is given by:

$$O(\omega) = D(\omega) \left\{ \pm i \frac{4\omega_2}{\pi} \sum_{n=-\infty}^{\infty} \sum_{m=\text{odd}} \frac{1}{m} \delta[\omega - (2n\omega_2 \pm m\omega_1)] \right\}. \qquad (10.26)$$

which represents the spectrum of the residual signal in the overwrite pattern. It contains side bands which are odd harmonics of the old frequency around the even harmonics of the overwrite pattern. Note that the amplitude of the side-band peaks can be used to quantify the value of the HTS. This is called the *frequency domain* method, as opposed to the *time domain* method shown in Fig. 10.5.

Now we can indeed explain Fig. 10.6. The residual overwrite spectral harmonics are side bands of the second harmonic of the overwrite pattern. The residual signal peaks are located at 80 MHz $- m \times 10$ MHz, where m is an odd integer. However, the presence of the 80-MHz signal (the second harmonic of the overwrite pattern) in the spectrum is contrary to the prediction of Equation (10.26). This peak is not due to HTS but rather due to the nonlinear response of the MR head, which causes peak asymmetry and even harmonic distortions as described in Sections 6.4 and 10.5.

10.2 NONLINEAR TRANSITION SHIFT (NLTS)

10.2.1 NLTS in dibit

Consider two adjacent transitions, which are often called a *dibit* transition. It is usually assumed that the dibit transition is separated from the rest of the data, i.e., there are no transitions written in the vicinity of the dibit. Readback signal from the dibit transition is called a *dipulse*. As the distance between the two transitions is decreased, or equivalently, the ratio of PW_{50} to the transition separation is increased, the amplitude of the dipulse signal decreases and its peaks are moved further apart than the transition separation. This effect is due the linear superposition of two pulses with opposite polarity, and is therefore called the *linear peak shift*, or the *intersymbol interference* (ISI).

Unfortunately, magnetic transitions are nonlinearly distorted at high recording densities. When two transitions are written close enough, the demagnetizing field from the previous transition affects the writing of the next transition in a manner similar to HTS, as shown in Fig. 10.14.

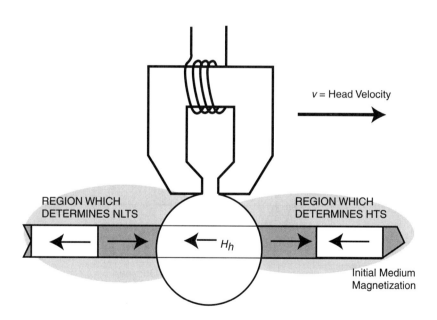

FIGURE 10.14. Generation of NLTS and hard transition shift (HTS).

This results in a shift of the transition location and is referred to as the *nonlinear transition shift* (NLTS).

It is important to distinguish the origins of HTS and NLTS. HTS is related to the initial magnetization of the magnetic medium. When the head moves over the magnetic medium as shown in Fig. 10.14, HTS is produced by the magnetization of the magnetic medium to the right of the head, which is in the region not yet written. In contrast, NLTS depends on the transitions that are already written, i.e., the magnetization of the medium to the left of the magnetic head.

In this section, we will ignore the HTS by assuming that the medium is AC-erased and consider only the influence of the previous transition on the transition being written. This influence is produced by the demagnetizing fields of the previous transition as shown in Fig. 10.15. When a transition (the first transition in a dibit) is written, it creates a static demagnetizing field H_d. When the subsequent transition (the second transition in the dibit) is written, the magnetic field of the head H_h switches

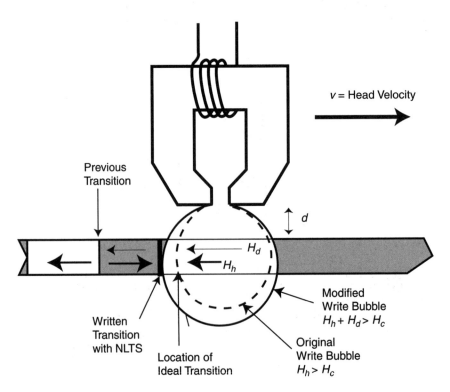

FIGURE 10.15.   Explanation of nonlinear transition shift.

its direction. However, the demagnetizing field of the previous transition has a direction that coincides with the head field. Therefore, the write bubble is expanded and the second transition is written earlier than intended.[1,8,9] NLTS will always occur when two transitions are written, since the direction of the demagnetizing field will always be the same as the direction of the head field. Note also that the effect of NLTS is to "attract" the second transition closer to the first one. In some sense it is opposite to the linear peak shift caused by ISI and to the HTS which delays the hard transition.

Let us now find an analytical expression of NLTS. As shown in Fig. 10.16, the demagnetizing field, H_d, from the previous transition is now added to the head field H_h. The expression which allows us to estimate the value of NLTS is similar to Equation (10.2), which was used to calculate the value of the HTS. The transition is written at the location $x_0 - \Delta$ *defined by the trailing edge of the write bubble:*

$$H_d(x_0 - \Delta) + H_h(x_0 - \Delta) = H_c. \tag{10.27}$$

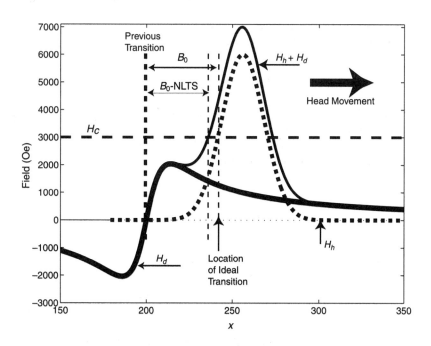

FIGURE 10.16. Interaction of head field with the demagnetizing field.

where x_0 is the location of the ideal transition (without NLTS), Δ is the amount of NLTS. Assuming that the value of NLTS is small, we can use the Taylor series expansion around x_0 and obtain that the value of Δ is proportional to the value of the demagnetizing field:

$$\Delta = +\frac{H_d(x_0)}{\left(\frac{dH_h}{dx}\Big|_{x=x_0}\right)}. \tag{10.28}$$

Let the nominal separation between two transitions be B_0. The longitudinal demagnetizing field at the center of the second transition due to the first transition (assumed to be arctangent) of a dibit is

$$H_d = \frac{2M_r}{\pi}\left[\tan^{-1}\left(\frac{a+\delta/2}{B_0}\right) - \tan^{-1}\left(\frac{a}{B_0}\right)\right] \approx \frac{M_r\delta}{\pi B_0}, \text{ (if } a, \delta << B_0) \tag{10.29}$$

The head field gradient at the center of the second transition x_0 is

$$\frac{dH_h}{dx} = \frac{QH_c}{d + \delta/2}, \tag{10.30}$$

where Q is the numerical factor determined by the head geometry (Chap. 4). Therefore, the nonlinear transition shift is

$$\Delta = +\frac{M_r\delta(d + \delta/2)}{\pi QH_cB_0}. \tag{10.31}$$

Note that this shift is *positive*, therefore the transition is indeed written *earlier* than desired. Note that the positive direction of NLTS in Fig. 10.16 is opposite to that of HTS in Fig. 10.3. They are indeed opposite with both being of positive values. Again, note the coordinates in Fig. 10.16 are the same as that in Fig. 10.3, but different from that in Chapters 3 and 4.

Taking the imaging effect into account,

$$\Delta = +\frac{M_r\delta(d + \delta/2)}{\pi QH_cB_0}F(d, \delta, B_0). \tag{10.32}$$

where function $F(d,\delta,B_0)$ depends on head imaging. This function is no greater than 1 and is $(d + \delta/2)^2/B_0^2$ in the case of ideal imaging.

Generally speaking, *NLTS tends to pull the dibit closer*. It is smaller if the transition parameter and head-medium spacing are smaller and bit length is larger. A medium with a smaller magnetic moment and a higher coercivity will have a smaller value of NLTS. Similarly, the imaging effect will reduce NLTS.

It has been experimentally observed that the dependence of NLTS on the distance between transitions is usually well described by some power of distance:

$$\Delta \approx K/B^{\gamma},$$

where the power γ is experimentally measured, typically in the range of 1.5–3.

In addition to a shift in location, the second transition in a dibit tends to produce a larger pulse width and a smaller readback amplitude. This is due to the influence of the demagnetizing field as described by the Williams-Comstock model. This effect is very similar to the shape distortion of a hard transition pulse. The *pulse broadening* effect can be a significant source of distortion.

Depending on head and medium designs, NLTS generally becomes serious at recording densities above 100–150 Kbits/in. NLTS values are usually measured in percents of the distance between transitions B, or in absolute units like nanoseconds or nanometers. Unfortunately, the values of NLTS in high-density recording may reach as much as 50%. For example, if the distance between transitions (bit period T) is 10 ns, the values of $\Delta = 2–4$ ns are not unusual. However, typical NLTS values are kept within 10–20% of the bit period to be precompensated effectively.

The distortions of a dipulse readback signal caused by NLTS are shown in Fig. 10.17. The dipulse signal is directly measured from a disk. By capturing a pair of isolated positive and negative pulses, a linear dibit (without NLTS) can be constructed. The amplitude of the signal with NLTS is reduced and the position of the second transition is shifted closer to the first transition. Even if the shape distortions of the second pulse by NLTS are negligible, the fact that the transitions are actually written closer to each other will lead to a smaller dipulse amplitude because of stronger linear ISI.

NLTS is especially critical for detection channels such as the partial-response maximum likelihood (PRML) channels, which will be described in Chapter 12. PRML detection is capable of handling large amounts of linear ISI, but it is based on the assumption that the recording channel is linear. Even moderate amounts of NLTS can cause high error rates in PRML systems.

10.2.2 Precompensation of NLTS

Since NLTS is a nonlinear distortion that occurs during the write process, it cannot be eliminated with any linear filters. The practical way to reduce

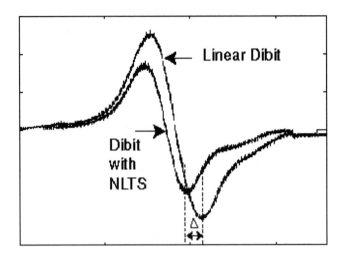

FIGURE 10.17. Comparison of a dipulse without NLTS with the dipulse distorted by 25% NLTS.

the influence of NLTS and to achieve better linearity of the recording channel is to use *write precompensation*. The principle of write precompensation is illustrated in Fig. 10.18. If the second transition in a dibit is being written at the correct location, it will be pulled closer to the previous transition due to NLTS. If the writing of this transition is delayed by some precompensation parameter p, it will also be pulled closer to the previous transition. The main idea of precompensation is that the resulting shift of the delayed transition should place this transition to the correct location. Hopefully, shape distortions caused by NLTS will also become smaller and the resulting readback signal will become approximately linear.

To find the appropriate precompensation parameter p, the dependence of NLTS on the distance between transitions, $\Delta = N(B)$, must be known. To achieve an ideal precompensation, we need to satisfy the following equality:

$$B_0 + p - N(B_0 + p) = B_0, \tag{10.33}$$

where B_0 is the required distance between transitions, $N(B_0 + p)$ is the "attraction" of the precompensated transition to the previous transition.

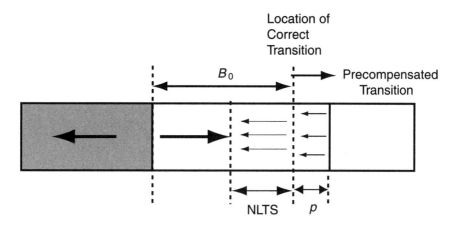

FIGURE 10.18. Write precompensation principle.

Equation (10.33) means that *the NLTS of the precompensated transition should exactly compensate the precompensation delay:*

$$p = N(B_0 + p). \tag{10.34}$$

This gives the optimum value of precompensation p.

Figure 10.19 illustrates how the optimal precompensation parameter can be found graphically. In this example, the value of B_0 is 160 nm and $N(B_0) = 25$ nm. To find the value of $p = N(B_0 + p)$, plot a diagonal line at 45° starting from the point of B_0. The intersection of this line with the curve $N(B)$ gives the desired precompensation parameter of $p = 17$ nm. Note that the value of NLTS for the targeted transition separation is 25 nm, which is larger than the value of the precompensation parameter. This reflects the nonlinearity of function $N(B)$.

10.2.3 Interactions of NLTS with HTS

NLTS is caused by the demagnetizing fields of the previously written transitions, However, the influence of the medium magnetization in front of the head will also affect the transition location. In other words, *the actual transition shift will be modulated by both previous pattern transitions and old information.*

What happens with NLTS when the medium in front of the head write bubble has initial magnetization? Consider a simple case when the medium is DC-erased as shown in Fig. 10.20. The situation corresponds

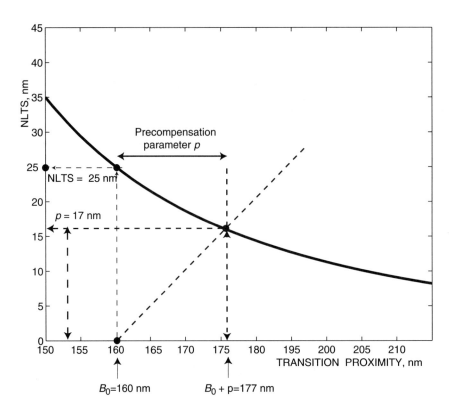

FIGURE 10.19. Determination of optimal precompensation parameter.

to the writing of an easy/hard (E/H) dibit, i.e., the first transition in the dibit is an easy transition and the second a hard transition.

Since the first transition in the dibit is written in the direction of the initial medium magnetization, its shape and location are not modified. However, when a second transition is written, the head field switches direction and becomes opposite to the initial medium magnetization. Now the head field is influenced by two demagnetization fields. One of them, H_{d1}, is caused by the first transition in the dibit. It has a direction which coincides with the head field H_h and expands the write bubble. However, at the right edge of the write bubble a secondary transition is formed, which instantaneously creates another demagnetizing field, H_{d2}, opposite to the head field.[10]

The actual transition will be written at the location where

$$H_h\,(x_0 + \Delta_{E/H}) + H_{d_1}(x_0) + H_{d_2}(x_0) = H_c = H_h(x_0). \qquad (10.35)$$

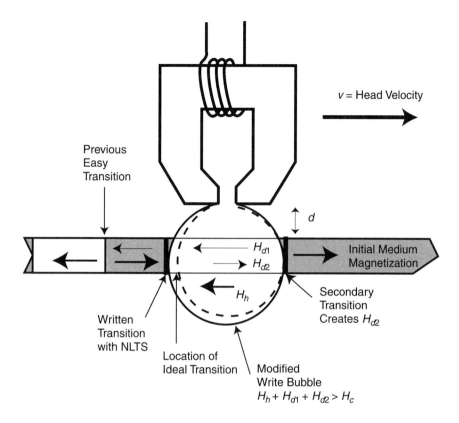

FIGURE 10.20. Formation of easy/hard dibit.

The resulting shift of the transition is

$$\Delta_{E/H} = -\frac{H_{d1} + H_{d2}}{\left(\frac{dH_h}{dx}\bigg|_{x=x_0}\right)},\tag{10.36}$$

or

$$\Delta_{E/H} = \Delta_1 + \Delta_2,\tag{10.37}$$

where Δ_1 is the value of NLTS without the HTS, and Δ_2 is the value of the HTS. Note that their signs are *opposite*. Therefore, for an easy/hard (E/H) dibit, NLTS and the HTS have opposite directions, so the resulting transition shift will be smaller than that for an AC-erased medium. Usually the magnitude of NLTS is greater than that of HTS, so the second transition ends up slightly *earlier* than intended.

Let us now consider the writing of an H/E dibit, as shown in Fig. 10.21. When the first transition in the dibit is written, its direction is opposite to the initial medium magnetization. Therefore, the first transition is distorted by the hard transition shift and is written with a delay. When the second transition in the dibit is written, its direction coincides with the initial medium magnetization. Therefore, the second transition is an easy transition and is affected only by the demagnetizing fields of the previous transition.

The previous transition in this case is written *later* by the amount of hard transition shift Δ_2. Therefore, the actual value of NLTS for the second transition will become larger because of the density dependence of NLTS:

$$\Delta_{H/E} = N(B - \Delta_2). \tag{10.38}$$

In other words, the second transition in the H/E dibit will be written earlier than the second transition in the E/H dibit.

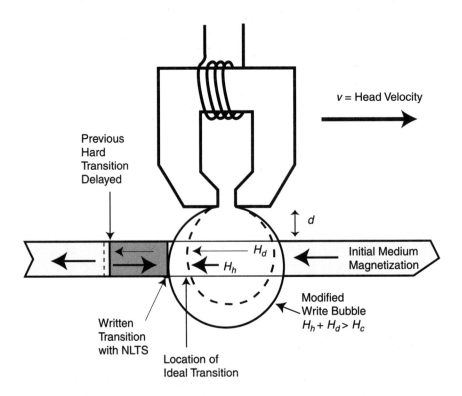

FIGURE 10.21. Formation of hard/easy dibit.

Figure 10.22 summarizes the difference between an E/H dibit and an H/E dibit. Although the shifts of transition edges are somewhat exaggerated, it is certain that the modulation of NLTS caused by the HTS can be significant. For example, if the NLTS is 25%, $\Delta = N(B) \approx K/B^3$, and the HTS is 6% of the bit period, then the value of the H/E shift is 30%, while the E/H shift is only about 18%.

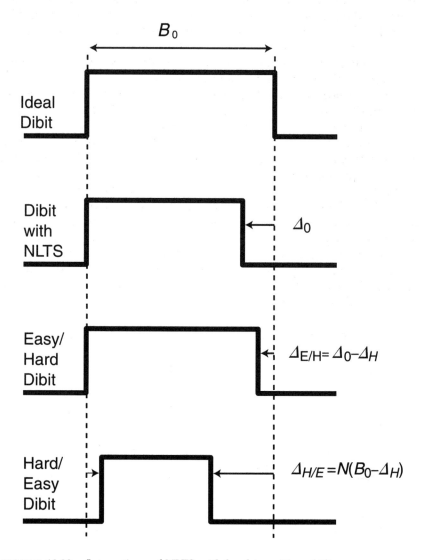

FIGURE 10.22. Interactions of NLTS with hard transition shift.

Interactions of NLTS with HTS can be easily observed using a pattern of data consisting of a dibit, an isolated transition, and another dibit, as shown in Figure 10.23. In this case, the two dibits have opposite polarity with respect to the initial magnetization because they are separated by an isolated transition. In the left picture the first dibit is an H/E dibit and the other is an E/H dibit. When the initial medium is in the opposite direction, the E/H and H/E dibits are reversed. It can be seen from Fig. 10.23 that the H/E dibit has a much smaller readback amplitude than the E/H dibit. When the medium is effectively AC-erased, both dibit pulses become symmetrical, with their voltage amplitude being approximately the mean amplitude of the E/H and H/E dibits.

The precompensation of the data pattern affected by both NLTS and hard transition shift is possible only if the medium is DC-erased before writing the pattern. In this case different precompensation should be applied to hard/easy and easy/hard dibits. However, when an existing information pattern is written on the medium, it is impossible to predict the correct precompensation of each bit in the new data pattern. NLTS for each of the transitions in the new pattern will be modulated approximately within the values of $\Delta_{H/E}$ to $\Delta_{E/H}$. The degree of this modulation depends on the value of HTS and the dependence of $N(B)$, which are in turn determined by parameters of the magnetic medium and the gradient of the head field. The steeper the dependence $N(B)$, the larger the expected variation of NLTS. Optimal head and medium designs should be based on careful consideration of both NLTS and HTS in order to achieve a linear recording channel.

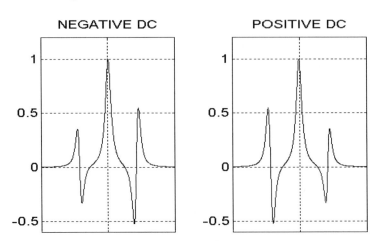

FIGURE 10.23.   Readback voltages for different medium magnetization.

10.2.4 NLTS in a series of transitions

Nonlinear transition shift considered in the previous sections is based on the assumption that a dibit is relatively isolated from the rest of the data. In reality, a pattern of data written on the magnetic medium may have long sequences of transitions following one another. This is called *a series of transitions*. The demagnetizing field of the transitions in series may interact with one another and produce a complex pattern of transition shifts.

A simple illustration of multiple-transitions interaction is presented in Fig. 10.24. let us consider qualitatively what will happen with NLTS of each transition in the series. Transition 1 is not affected by NLTS since no transitions have been written before it. Transition 2 will be distorted by a standard dibit NLTS value equal to Δ_2. However, when transition 3 is written, we should take into account that transition 2 has already been shifted in the direction of transition 1. This means that the actual separation between transition 2 and transition 3 is larger than the nominal distance between transitions B_0. Since NLTS is determined by the demagnetizing fields from the previous transitions, we can expect that the value Δ_3 will be smaller than the value Δ_2. When transition 4 is written, the actual distance between transition 4 and transition 3 is smaller than the distance between transition 3 and transition 2. It means that the value of NLTS Δ_4 will become larger than the value of Δ_3. However, it should still be slightly smaller than the value of Δ_2.

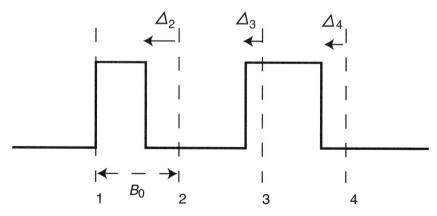

FIGURE 10.24. Series of four consecutive transitions with different values of NLTS.

It is clear from this example, that when several consecutive transitions are written, each transition will have different value of NLTS. In fact, NLTS will *oscillate* between larger and smaller values.

The main contribution to the transition shift of the current transition comes from the demagnetizing fields of the previous transitions. If only the demagnetizing field from the nearest previously transition is significant, the value of NLTS depends on the distance between transitions B as: $\Delta = N(B) \approx K/B^\gamma$, $\gamma = 1.5 \sim 3$. If the nominal separation of transitions equals B_0, then NLTS for the second transition $\Delta_2 = K/B_0^\gamma$. However, the distance between the second transition and the third transition equals $B_0 + \Delta_2$, so $\Delta_3 = K/(B_0 + \Delta_2)^\gamma$. Similarly, for the fourth transition, $\Delta_4 = K/(B_0 + \Delta_3)^\gamma$. In general we obtain:[11]

$$\Delta_k = N(B_0 + \Delta_{k-1}), \qquad k = 3, 4, 5, \ldots \qquad (10.39)$$

The process of finding a sequence of Δ_k can be visualized using the simple plot shown in Fig. 10.25. A sequence of Δ_k values gradually converges to

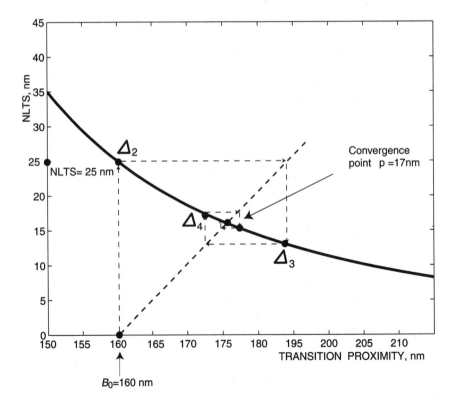

FIGURE 10.25. Finding consecutive values of NLTS for a series of transitions.

a stable value of p. Comparing Fig. 10.25 to Fig. 10.19, we can immediately see that the value p equals the optimal precompensation parameter for a dibit pattern.

This convergence is a general property of Equation (10.39) for any monotonously decreasing function $N(B)$. In fact, the mathematical requirement for the convergence of NLTS for a series of transitions is given by the contraction property of function $N(B)$, which reflects the fact that NLTS changes more slowly than the distance between transitions. More precisely, NLTS for a series of transitions will converge if $|N(B_1) - N(B_2)| < |B_1 - B_2|$ for any two values of the transition separation B_1 and B_2. This property will hold for all realistic forms of $N(B)$.

A more general model of transition interactions includes the influence of demagnetization fields from all previous transitions. For example, in the case of ideal imaging, when $\Delta = N(B) = K/B^3$, the influence of the second adjacent transition is given by: $N(2B) = K/8B^3$. While the shift caused by the second adjacent transition is about eight times smaller than the shift of the closest adjacent transition, it is still significant and will affect the current NLTS value. Note that due to the fact that the second adjacent transition has the opposite direction to that of the first adjacent transition, the NLTS from the second adjacent transition will have the opposite sign.

If a periodic pattern is written, we have to consider the impact of *all* previous transitions in the series on the current kth transition. The total NLTS of the kth transition is given by the following expression[11]:

$$\Delta_k = \sum_{m=1}^{k} (-1)^{m+1} N(mB_0 + \Delta_{k-m}), \tag{10.40}$$

which describes the total influence of all previous transitions separated by mB_0 from the current transition. Note that the sum has alternating signs, i.e., the first transition ($m = 1$) results in positive value, which means that the current transition is attracted toward the previous one. The NLTS from the second adjacent transition has the opposite sign, meaning that it pushes the current transition away from the previous transition. The NLTS from the third adjacent transition again attracts the current transition. The signs continue to alternate.

Figure 10.26 illustrates the result calculated from Equation (10.40) for a series of 15 consecutive transitions. It is assumed that the nominal time between transitions is 16 ns, the NLTS value for a dibit is 4 ns (25% of the bit period), and $N(B) = K/B^2$. The variations of NLTS are clearly seen. For example, the NLTS for the second transition in the series is only 1.5 ns (vs 4 ns for a dibit!). After approximately 10 transitions, the value of NLTS settles at approximately 2.4 ns.

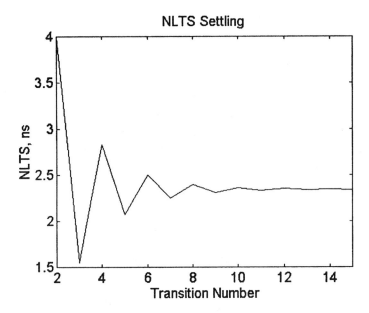

FIGURE 10.26.   NLTS settling for a series of 15 transitions.

Gradually decreasing oscillations of NLTS shown in Fig. 10.26 are only found for AC-erased media. If the magnetic medium has magnetization, which is typical for a realistic overwrite environment, the NLTS and HTS will interact with each other.

If the current kth transition is hard, we will have to subtract the value of the hard transition shift Δ_H from the total shift of the current transition as given by Equation (10.40). Therefore,

$$\Delta_{\text{hard}}(k) = \sum_{m=1}^{k} (-1)^{m+1} N(mB_0 + \Delta_{k-m}) - \Delta_H, \qquad (10.41)$$

$$\Delta_{\text{easy}}(k) = \sum_{m=1}^{k} (-1)^{m+1} N(mB_0 + \Delta_{k-m}).$$

Once a hard transition is met, the value of the HTS is added to the total transition shift caused by all previous transitions. It means that an HTS will take part in modulating the location of current and consecutive transition in a series.

Consider the case of a DC-erased medium. Every second transition in the series written on the DC-erased medium will be a hard transition. The HTS will modulate the location of the next transition. The interaction between the HTS and NLTS will result in a complicated transient process for the initial transitions in the pattern, depending upon whether the first transition in a series is easy or hard. Figure 10.27 demonstrates the results of NLTS measurements and modeling based on $\Delta = N(B) \approx K/B^3$ with the following parameters: hard transition shift is 5%, dibit NLTS without hard transition shift equals 25% of the bit period. The two cases plotted correspond to the first transition in series is either easy or hard.

Experimental values of NLTS in Fig. 10.27 were measured using a spectral elimination method, which will be discussed in Section 10.3. The measured data has an excellent match to the modeling result. When the first transition in series is hard, the HTS reinforces NLTS: the second transition has high value of NLTS, the third transition will have small transition shift. However, when the first transition in series is easy, the HTS of the second transition will slightly compensate NLTS and will damp the oscillations of transition shift for the first several transitions. This transient effect affects only several first transitions in series, as seen from Figure 10.27, the amplitude of the transition shift oscillations be-

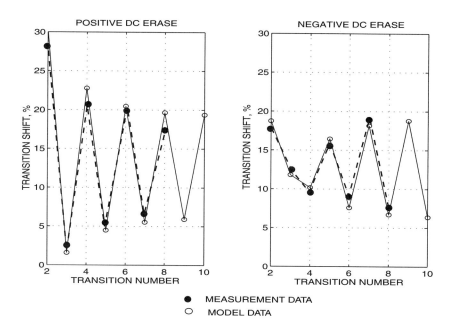

FIGURE 10.27. Modulation of NLTS for series of transitions. Left: the first transition is hard (positive DC-erase); right: the first transition is easy (negative DC-erase).

comes the same for both cases after eight to nine transitions. Another interesting observation is that the amplitude of transition shift oscillations in a series of transitions is larger than the value of the HTS. This is explained by interactions between NLTS and HTS.

Most of the encoded data pattern will contain relatively short sequences of consecutive transitions. Therefore, the problem of interactions between NLTS and HTS becomes severe in high-density recording due to the following two reasons: First, complicated settling processes occur at the edges of the repetitive patterns (the first several transitions), therefore bursts of data bits separated by zeros will be severely affected. Second, when a data pattern is written over old information, the NLTS values for individual transitions will be modulated randomly, resulting in a complicated pattern of timing shifts for individual transitions in the pattern.

One manifestation of the interactions between HTS and NLTS in the overwrite environment is a degradation of overwrite. In the density regions with high NLTS, overwrite values may degrade as much as 6 dB. Consider the modeling results shown in Fig. 10.28, which illustrates the transition shifts for HTS = 0.5 ns and different levels of NLTS. When NLTS is taken to be 30% of the bit period and the overwrite frequency ratio is 2, the transition shifts in the new pattern oscillate, but with a moderate amplitude of ~0.5 ns. The fact that the amplitude of these oscillations is similar to the nominal value of HTS is explained by the

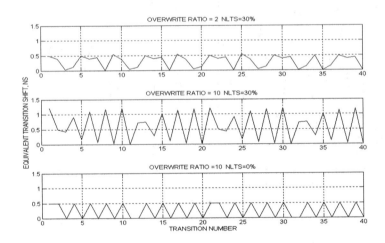

FIGURE 10.28. Oscillations of transition shifts for overwrite frequency ratios of 2 and 10.

small ratio between the frequencies of the old and the new square waves. However, when the overwrite frequency ratio is 10, each period of the old square wave becomes similar to a DC-erased medium. In this case we have a situation similar to data presented in Fig. 10.27, and the oscillations of transition shift in a new pattern are amplified, exceeding 1 ns.[6]

Figure 10.29 presents an experimental measurement of equivalent HTS, similar to the measurement presented in Figure 10.13.[6] While the density of the f_2 pattern is kept below 100 KFCI (curves 1–4), the equivalent HTS behaves similarly to Figure 10.13. However, at higher densities (curves 5–7), the equivalent HTS increases. This increase (and corresponding degradation of overwrite by up to 6 dB) is caused by interactions between HTS and NLTS. The amplification of the equivalent HTS corresponds to the density region in which NLTS exceeds 10% of the bit period (above 100 KFCI) as shown in Figure 10.30.

Finally, the precompensation for a series of transitions becomes complicated. Theoretically, standard precompensation strategy, discussed in

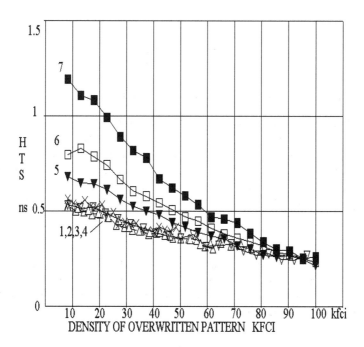

FIGURE 10.29. Equivalent HTS (ns) vs. the density of the LF pattern. HF densities (curves 1 to 7): 50, 66, 83, 100, 133, 150, 180 KFCI.

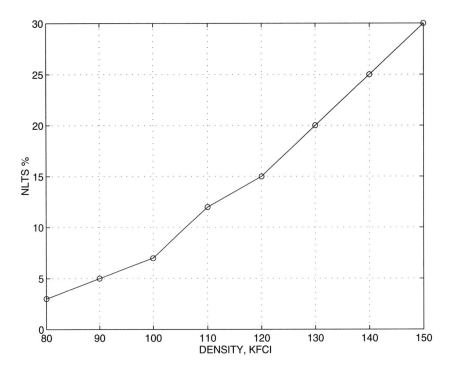

FIGURE 10.30.   Density dependence of NLTS for data shown in Fig. 10.29.

Section 10.2.2, can be successful only when the interactions between transitions in series can be ignored. Consider that the kth transition in series is written with some precompensation delay p_k. This transition is "attracted" by the demagnetization field of all previous transitions by some amount Δ. The correct precompensation is obtained when the total shift Δ at precompensated position $T + p_k$ exactly equals the value of precompensation p_k:

$$T + p_k - \Delta = T. \tag{10.42}$$

The shift Δ is determined by the distances to all previous transitions. If the previous transitions are correctly precompensated, these distances are exactly $T + p_k$, $2T + p_k$, etc. Therefore, the correct value of precompensation for the current transition is found from the following equation:

$$p_k = \sum_{m=1}^{k} (-1)^{m+1} N\,(mT + p_k). \tag{10.43}$$

In principle, every transition in series must have different values of precompensation. In practical applications, realization of Equation (10.43) is complicated, so only a few terms are considered, because the demagnetizing fields many bit periods away are small. Interactions between transitions in series and HTS set a practical limit for a good-quality recording system. The values of NLTS should be kept below 15–20% of the bit period in order to avoid oscillations of transition shifts.

10.2.5 Data rate effects and timing NLTS

The transition shifts considered in the previous sections are caused by demagnetizing fields inside the magnetic medium and appear at high recording densities. The second source of transition shifts is caused by a finite head flux (or field) rise time, which becomes critical at high recording frequencies.

Figure 10.31 illustrates the origin of the rise time problems. Ideally, a rectangular pulse of write current is supplied to a write head, and the

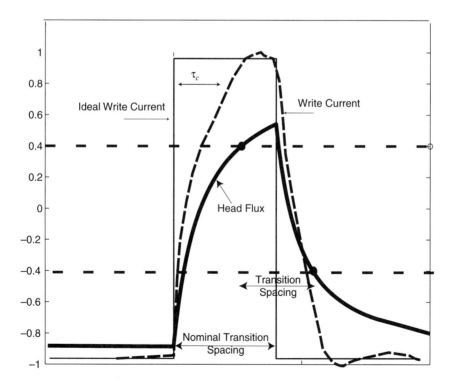

FIGURE 10.31. Dibit write current and head flux responses. The two dashed horizontal lines represent $\pm H_c$.

gap field instantaneously follows the current. However, the write current shape is not ideal and has a certain rise time τ_c, caused by the limited bandwidth of the write driver. Furthermore, the head is a transformer that cannot switch instantaneously. When the magnetization of the soft core changes, the domain structures in the yoke take certain time to move and change their geometrical configuration. The presence of eddy current in the magnetic yokes causes the so-called "skin effect,"[1,12] which makes the magnetization near the center of the yoke change either very slowly or not at all. The total effect of these mechanisms is that the head flux response is characterized by some rise time τ_f, as defined in Section 5.4.2. The head flux response can be modeled using a step response function of the type:

$$H(t) = \frac{t - \tau}{t + \tau}. \tag{10.44}$$

where τ is a time constant related to τ_f. This equation is slightly different from Equation (5.14).

Another effect which has to be taken into account is the head core saturation. When the write current of the same polarity is supplied to the write head for a sufficient period of time, the core becomes saturated.[12] However, when a short current pulse is supplied to the write head, the head flux swings from the saturated state and may not have enough time to swing to another saturated state (Fig. 10.31). When the second edge of the dibit current pulse is applied to the head, it pushes magnetization back, the head field swings to the initial state. Due to the flux rise time limitation, it takes longer to reach the coercivity level $+H_c$ starting from the saturation state, compared to the time required to reach an opposite level $-H_c$. As a result, the first transition in the dibit is written "late," while the second transition is written "less late." The absolute shift of the first transition is not important, since the rise time introduces a constant phase delay to all written transitions. However, the fact that the second transition is written *relatively early* has serious consequences. This timing shift pulls the second transition in the dibit closer to the first one, which is equivalent to the density-related NLTS and is often called the *timing NLTS*.

The above timing shift is determined by the rise time of the write current driver circuit and by the writing head. If the time interval between adjacent transitions is much larger than the corresponding time constant, the timing shifts become negligible. The timing constants of the state-of-the-art write drivers and heads are on the order of nanoseconds. If this time constant equals 5 ns, then the data rate of magnetic recording is limited to ~200 Mbit/s (~25 MB/s). Increasing recording rates to 40–80

MB/s requires fast write driver circuits and writing heads, with a total flux rise time below 1–2 ns.

Let us now analyze how these timing shifts interact with the density-related transition shifts.[13] Consider writing a dibit transition and assume that an additional timing shift τ of the second transition is introduced due to the finite flux rise time. This timing shift is translated into a spatial shift of $\Psi = v\tau$, where v is the medium velocity, so the transition separation is reduced from B_0 to $B_0 - \Psi$. If the recording density is so low that the demagnetizing field from the previous transition is negligible, then the second transition in the dibit is actually written with a proximity shift Ψ. However, if the demagnetizing field at the proximity of $B_0 - \Psi$ is not negligible, the timing shift will be reinforced by the demagnetizing field. Note that the timing shift τ is itself a function of the time between the adjacent transitions: $\tau = \tau(B_0/v)$. Therefore, the total NLTS of the second transition is given by:

$$\Delta = -v\tau - \frac{H_d(B_0 - v\tau)}{dH_h/dx|_{x=x_0}}. \tag{10.45}$$

To verify the validity of Equation (10.45), an experiment is carried out and shown in Fig. 10.32. Three data rates are used: 15, 30, and 45 MB/s, which are kept constant by simultaneously changing spindle RPM

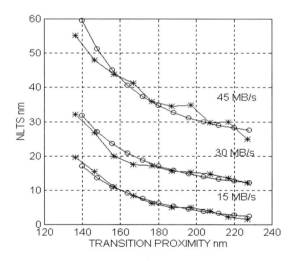

FIGURE 10.32. NLTS for a dibit transition at 15, 30, and 45 MB/s data rates. (*) = Measurements, (○) = Modeling based on Equation (10.45).

and linear density. Apparently, NLTS increases with both the linear density and the data rate. To estimate the density-related fraction of NLTS, we use the NLTS dependence on transition proximity at a data rate of 15 MB/s. At this data rate the minimum bit period equals 7.4 ns, and we may neglect the timing shift. The density dependence of NLTS at 15 MB/s is best fit by K/B^3.

At the low density with a transition separation of 220 nm, the density effects on NLTS are negligible. Therefore, the transition shifts at this density are caused entirely by rise time effects, as shown in Fig. 10.33. For a range of bit periods between 3.5 and 7 ns, the rise-time-induced NLTS is larger at smaller bit periods. Based on the fitted curve in Fig. 10.33, along with the K/B^3 dependence of NLTS, the experimental data in Fig. 10.32 can be matched by Equation (10.45) very nicely.

The problems introduced by timing-related NLTS are compounded when a series of several transitions is written.[13] The writing of the first transition starts from the saturated state, therefore this transition is written late. The second transition has a smaller delay than the first one. The timing shift of the third transition may be different from that of the second transition since the writing of this transition starts from yet another state. These are visualized in Fig. 10.34, where a simple linear head model is used. In this model, the head field is described by a linear step response given by Equation (10.44). The head field swings from saturated state and crosses the coercivity level at points shown by small solid circles. The

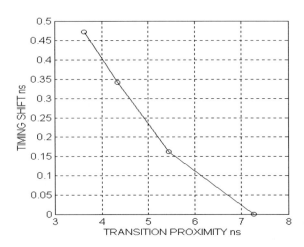

FIGURE 10.33.   Estimated timing shift dependence on the bit period.

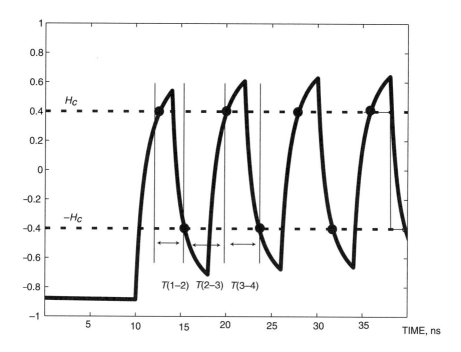

FIGURE 10.34. The effect of field rise time on a series of transitions.

relative timing shift for each of the first few transitions is different. When the head flux swing becomes symmetrical, the timing shifts settle to a steady value.

The variable timing shifts of the kth transition in series, add to the total transition shifts as given by Equation (10.45) and modulate the demagnetizing fields. Therefore, oscillations of transition shifts caused by the density effects are reinforced by the timing shifts caused by the field rise time.

To study the oscillating behavior of transition shifts at high data rates, a sequence of transition bursts (two, three, and more consecutive transitions) is written at different data rates. An example of such a pattern recorded at data rates of 15 and 35 MB/s data rates while keeping a constant linear density of 150 KFCI is shown in Fig. 10.35. Both waveforms are read back (with averaging) at a data rate of 15 MB/s, so the readback signals can be easily aligned. While the waveform recorded at 15 MB/s is symmetrical, that at 35 MB/s has certain periodic features. The latter has more significant NLTS because its dibit has a smaller amplitude. Note

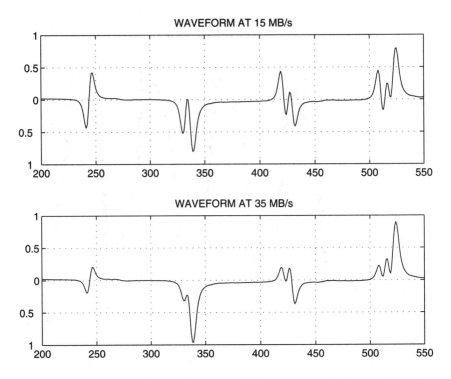

FIGURE 10.35. Averaged waveforms recorded at constant densities of 15 and 35 MB/s, respectively. The data pattern consists of isolated bursts of dibit, tribit, four transitions, and five transitions. The horizontal axis indicates the sample numbers.

also that the amplitude of the third transition is higher, that of the fourth transition is lower, and that of the fifth transition is again higher. This periodic amplitude variation of even and odd transitions suggests that transition shifts are oscillating,

The readback waveform can be fitted using the isolated pulse waveforms, which provides the best waveform approximation and gives the values of transition shifts. The results of this calculation are shown in Fig. 10.36.

In another experiment, the transition shifts for second, third, and fourth transitions in series were measured at 15 and 30 MB/s data rates using the spectral elimination method (refer to the next sections) for a range of densities. The results are shown in Fig. 10.37. At the data rate of 15 MB/s, the dibit has the largest values of NLTS. The third transition in series (tribit) has the smallest NLTS, while the NLTS for the fourth

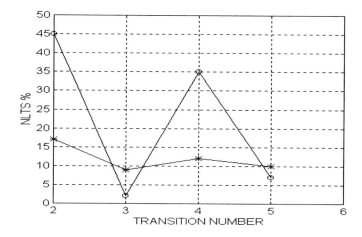

FIGURE 10.36. Transition shifts of individual transitions written at 35 MB/s and 15 MB/s, estimated from readback waveform in Fig. 10.35. Nominal dibit spacing is 170 nm. (∗ = 15 MB/s, ○ = 35 MB/s).

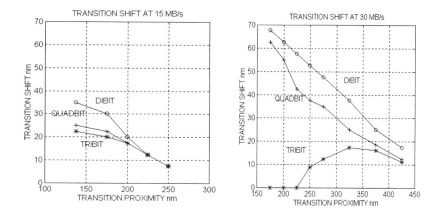

FIGURE 10.37. Transition shifts for dibit, tribit, and quadbit recorded at 15 MB/s (left) and 30 MB/s recording (right).

transition (quadbit) is in between the dibit and the tribit values. At the data rate of 30 MB/s, the shift of the third transition in series becomes negligibly small at transition proximity less than 250 nm, while the shift of the fourth transition is comparable with the shift of the second transition. The small shift of the third transition is caused by the timing oscillations of transition shift due to the finite rise time.

The oscillations of transition shift caused by the rise time present a serious problem in magnetic recording. First of all, unlike the density-related NLTS, the timing NLTS is more difficult to precompensate. The timing shift is usually not a simple function of transition separation. Second, the timing NLTS modifies the recording density and interacts with the density-related NLTS. These interactions result in complicated variations of total transition shift during writing process. Therefore, the performance of magnetic recording channel at high recording frequencies is limited by the field rise time. In practice, the total head field rise time should be kept at least 1.5–2 times smaller than the channel bit period.

10.3 MEASUREMENT OF NONLINEAR TRANSITION SHIFT

The amount of NLTS and optimal precompensation are important parameters in the characterization of magnetic media and heads. However, measurements of NLTS are complicated. Since NLTS becomes important at high recording densities, it is usually very difficult to separate transition shift effects from strong linear ISI, pulse broadening, and partial erasure.

Several methods for measuring NLTS are available. The first group of methods is usually referred to as *spectral (harmonic) elimination*.[14–16] This method is based on writing a special pattern on the magnetic medium and measuring spectral components (harmonics) of the signal using a spectrum analyzer or digital Fourier signal processing. The second group of methods is based on *pseudo-random sequences*, such as the *dipulse extraction* and the *time correlation* methods.[17–21] These methods take advantage of the special properties of pseudo-random sequences, and allow independent measurement of different nonlinear distortions (e.g., NLTS and hard transition shift).

10.3.1 Method of spectral elimination.

This method is also referred to as the *"fifth harmonic method."* The first implementation of spectral elimination consists of the following basic steps:

1. Write a special pattern on the disk and measure the amplitude of a particular (e.g., the fifth) harmonic of this pattern, $|D(k\omega_0)|$. The

pattern is constructed in such a way that if the superposition of all pulses is linear, the corresponding harmonic is equal to zero, or *eliminated*.

2. Write a square wave pattern which has the same period as the special pattern above, and measure the amplitude of the same harmonics of this pattern $|X(k\omega_0)|$.

3. Compute the value of NLTS as $|D(k\omega_0)|/[k\omega_0|X(k\omega_0)|]$. The result will be derived next.

Consider a data pattern consisting of a sum of delayed pulses:

$$y(t) = p(t) - p(t - T_1) + p(t - T_2) - \ldots - (-1)^N p(t - T_N). \quad (10.46)$$

where $p(t)$ is an isolated pulse. If $P(\omega)$ is the Fourier transform of the isolated pulse, then the spectrum of the pattern $y(t)$ is

$$Y(\omega) = P(\omega)[1 - e^{-i\omega T_1} + e^{-i\omega T_2} - \ldots + (-1)^N e^{-i\omega T_N}]. \quad (10.47)$$

Elimination of a specific harmonic ω is achieved when $Y(\omega) = 0$.

When an arbitrary pattern of some length NT is repeated periodically, its spectrum will consist of discrete spectrum harmonics at $k\omega_0 = k2\pi/NT$. For a square wave pattern with a fundamental angular frequency of ω_0, it contains only odd harmonics:

$$R(\omega) = 2\omega_0 P(\omega) \sum_{k=odd} \delta(\omega - k\omega_0). \quad (10.24)$$

Now consider a special pattern shown in Fig. 10.38, which consists of dipulses followed by isolated pulses. Using NRZI notation, this pattern is written as:

$$1100\ldots(m \text{ } 0s)100\ldots(n \text{ } 0s)1100\ldots(m \text{ } 0s)100(n \text{ } 0s)$$

The period of this pattern equals $2m + 2n + 6$ bit periods.

Using Equation (10.47), we find that the spectrum of a single period of the special pattern is given by

$$Y(k\omega_0) = P(k\omega_0)[1 - (-1)^k]\left[1 - e^{-i\frac{k\pi}{m+n+3}} + e^{-i\frac{(m+2)k\pi}{m+n+3}}\right], \quad (10.48)$$

where $\omega_0 = 2\pi/(2m + 2n + 6)T$. If m and n are multiples of 6, i.e., $m = 6p$ and $n = 6q$, then the $(2p + 2q + 1)$-th harmonic is zero.[15] It is common to choose $p = q = 1$, then the fifth harmonic is eliminated. Note that the

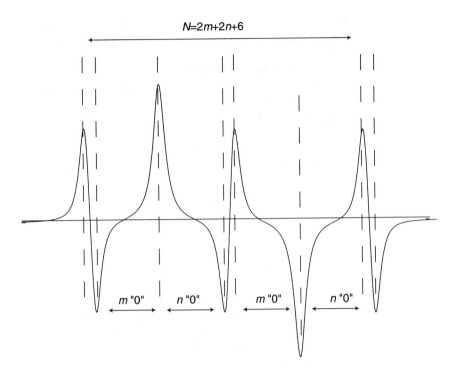

FIGURE 10.38.   Example of special pattern used for spectral elimination.

spectrum of the whole waveform shown in Fig. 10.38 can be expressed as

$$Y(k\omega_0) = R(k\omega_0)\left[1 - e^{-i\frac{k\pi}{m+n+3}} + e^{-i\frac{(m+2)k\pi}{m+n+3}}\right], \qquad (10.49)$$

where $R(k\omega_0)$ is the spectrum of a square wave pattern, in which the transitions are equally spaced with a distance of $MT = (m + n + 3)T$, as shown in Fig. 10.39. The square wave pattern is the same as the special pattern in Fig. 10.38 if the dibits are removed.

   What will happen with the readback signal if the second pulse in each dipulse is distorted by NLTS? In this case, a dipulse readback signal is written as $p(t) - p(t - T + \Delta)$. It means that the second pulse will be shifted closer to the first pulse by some value Δ. If the shift is small, the error signal caused by this shift is

$$d(t) = p(t) - p(t - \Delta) \approx \Delta\frac{dp}{dt}. \qquad (10.50)$$

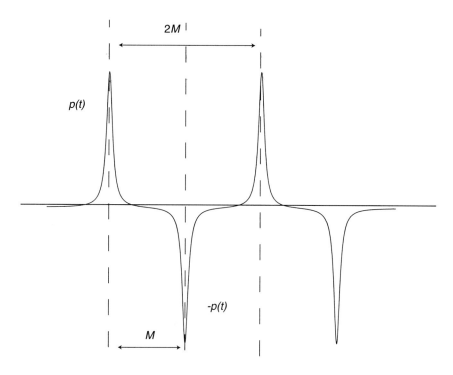

FIGURE 10.39.   Reference square wave pattern for spectral elimination method.

Therefore, the spectrum of the error signal at the kth harmonic of the pattern is

$$D(k\,\omega_0) = ik\,\omega_0 P(k\,\omega_0)\cdot\varDelta. \tag{10.51}$$

which is proportional to the value of the NLTS. For the special pattern shown in Fig. 10.38, the actual spectral amplitude of the error signal is

$$|D(k\omega_0)| = k\omega_0|R(k\omega_0)|\cdot\varDelta. \tag{10.52}$$

Therefore, the value of NLTS for the pattern shown in Fig. 10.38 is

$$\varDelta = \frac{1}{k\omega_0}\left|\frac{D(k\omega_0)}{R(k\omega_0)}\right|. \tag{10.53}$$

The second implementation of spectral elimination is very similar to the first except that the special pattern used is now more general.[14,16]

Consider first a pattern consisting of a pair of positive pulses, as shown in Fig. 10.40, because any bipolar pattern could be obtained from a unipolar sequence of pulses by subtraction. Let the repetition period of the whole waveform be NT, where T is the bit period, then the fundamental angular frequency is $\omega_0 = 2\pi/NT$.

The spectrum of two unipolar pulses separated by an integer number of periods, pT, is

$$Y(\omega) = P(\omega)(1 + e^{-i\omega pT}),\qquad(10.54)$$

The kth harmonic is

$$Y(k\omega_0) = P(k\omega_0)(1 + e^{-i2\pi\frac{kp}{N}}).\qquad(10.55)$$

Therefore, the kth harmonic is eliminated when

$$1 + e^{-i2\pi\frac{kp}{N}} = 0,$$

or

$$kp = \frac{2m + 1}{2}N, \; m \text{ is an interger.}\qquad(10.56)$$

Equation (10.56) dictates that N must be an even number and that the zero harmonic will always be odd.

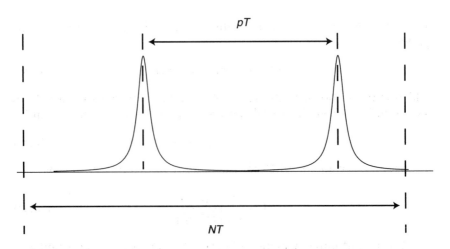

FIGURE 10.40.   Two pulses separated by pT in a waveform with a period of NT.

A simple pattern for $k = 5$ that satisfies the above requirement is shown in Fig. 10.41. This pattern has a repetition period $30T$. The distance between the two positive pulses (P1 and P3) is $9T$, while the distance between two negative pulses (P2 and P4) is $15T$. If the superposition of the pulses is linear, both the positive and negative pulses will give zero spectral amplitude at the fifth harmonic of the pattern.

The distance between the first positive and first negative pulse is d, which can be arbitrary and very small, (e.g., $d = T$). If the negative pulse P2 is distorted by NLTS, it will introduce an error signal, which has a fifth harmonic as given by Equation (10.51). However, note that there is only *one* distorted dibit per period in this case, while there are two distorted dibits per period in Fig. 10.38. Therefore, for the special pattern shown in Fig. 10.41, the actual spectral amplitude of the error signal is

$$|D(k\omega_0)| = \frac{1}{2}k\omega_0|R(k\omega_0)|\cdot\Delta,$$

or

$$\Delta = \frac{2}{k\omega_0}\left|\frac{D(k\omega_0)}{R(k\omega_0)}\right|. \tag{10.57}$$

Comparing Equation (10.57) with Equation (10.53), we see that the factor

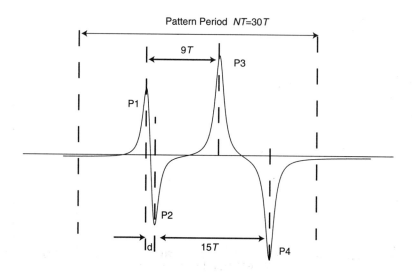

FIGURE 10.41. Spectral elimination pattern combined of pairs of unipolar pulses.[13]

of 2 appears in (10.57). This is because the error spectral signal $D(k\omega_0)$ for the pattern in Figure 10.41 is generated by only one dibit and, therefore, is two times smaller than for the pattern in Fig. 10.38 if the NLTS is the same in both cases.

Applying spectral elimination algorithms at various bit periods allows us to map the density dependence of NLTS. Figure 10.42 illustrates a typical density dependence of NLTS measured using this technique.

Spectral elimination can also be useful for finding the optimal precompensation parameters for NLTS. Take the pattern in Figure 10.41—we can change the precompensation, i.e., delay pulse P2 during the write process, and then measure the fifth harmonic. This procedure gives the dependence of NLTS on the precompensation parameter, as shown in Fig. 10.43. At some precompensation value, which is the optimal precompensation value, the NLTS is at a minimum. If the precompensation is larger than needed (overcompensation), the nonlinear distortion spectral peak will increase again.

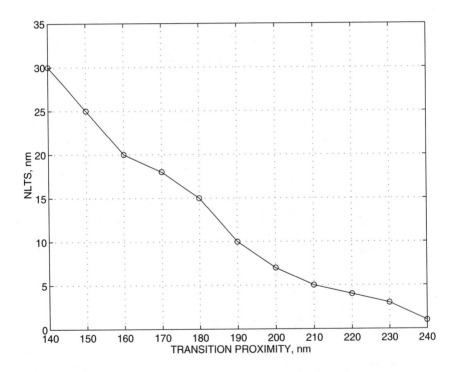

FIGURE 10.42. Typical dependence of NLTS on the channel bit period.

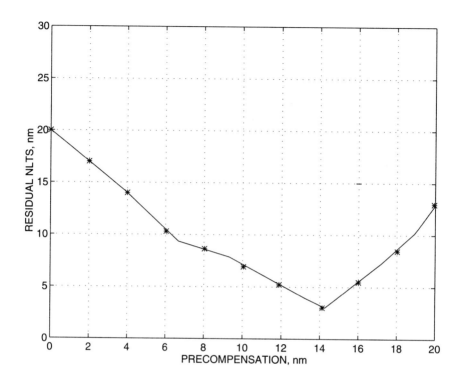

FIGURE 10.43.   Dependence of NLTS on precompensation.

Note that the original NLTS value is 20 nm, but the optimal precompensation is ~14 nm. Note also that the fifth harmonic content is never zero because of the noise level in the system, partial erasure, residual nonlinearity, etc. This demonstrates that the best possible precompensation will not eliminate all nonlinearities.

The method of spectral elimination is subject to a number of distortions. First, let us consider what happens with the readback signal for a pattern shown in Figure 10.41 when it contains amplitude loss α due to partial erasure (PE). The dipulse signal distorted by both NLTS and PE is given by:

$$f(t) = (1 - \alpha)p(t) - (1 - \alpha)p(t - T + \Delta). \tag{10.58}$$

The difference (error signal) between the ideal dipulse and the distorted signal given by Equation (10.58) is

$$d(t) \approx \alpha p(t) - \alpha p(t - T) + (1 - \alpha)\Delta\frac{dp}{dt}. \tag{10.59}$$

The spectrum of the error signal is

$$D(\omega) = P(\omega)[\alpha - \alpha e^{-i\omega T} + i\Delta(1 - \alpha)\omega]. \qquad (10.60)$$

At the kth harmonic,

$$D(k\omega_0) = P(k\omega_0)[\alpha - \alpha e^{-ik\omega_0 T} + i\Delta(1 - \alpha)k\omega_0].$$

Equation (10.60) is used to calculate the plots shown in Fig. 10.44, which gives the dependence of the measured value of NLTS using the fifth harmonic on the actual value of NLTS at various amounts of partial erasure (0–50% with 5% steps). Both NLTS and PE affect the spectral elimination results. For example, if 20% NLTS and 20% amplitude loss are present in the signal, the spectral method will measure approximately 35% of NLTS. On the other hand, if NLTS is absent, the measured signal will be completely due to partial erasure.

Let us now consider the impact of additive noise on NLTS measurement. As we mentioned previously, the voltage amplitude of the measured harmonic is proportional to rms value of signal and noise if they are uncorrelated:

$$P^*(k\omega_0) = \sqrt{|P(k\omega_0)|^2 + |N(k\omega_0)|^2}. \qquad (10.61)$$

Obviously, the larger the bandwidth of the band-pass filter, the more noise

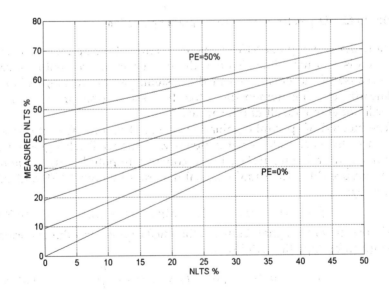

FIGURE 10.44. Dependence of measured NLTS on partial ensure.

will be in the spectral signal. Figure 10.45 demonstrates the dependence of the measured NLTS value on original NLTS for different levels of SNR (10–40 dB in steps of 5 dB).

Finally, the last factor which affects NLTS measurements using the spectral elimination technique, is the MR head nonlinearity. Transfer curve of the MR head may cause saturation of isolated pulses. The spectral elimination method is based on the principle of linear superposition of isolated transition and the MR nonlinearity will affect the NLTS measurements made using the spectral elimination method. One way of dealing with this type of distortion is to apply digital correction of the MR transfer curve. This procedure will be discussed in Section 10.5.

10.3.2 Methods based on pseudo-random sequences

These methods are based on special pseudo-random patterns with a sufficient period (127 bits or more). They include the following basic steps:

1. Write a pseudo-random sequence.

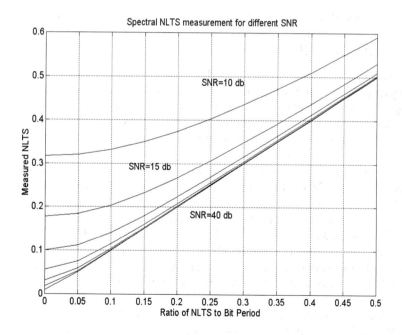

FIGURE 10.45. Measured NLTS for different SNR levels.

2. Sample a full period of readback signal with a high-resolution analog-to-digital converter (ADC).
3. Process the resulting digital signal using either discrete Fourier transform or time correlation analysis.

The *pseudo-random sequence* of data is a sequence of bits that are random within a long period. It has a regular structure and a number of important properties. A pseudo-random sequence is generated by a generating polynomial.[17] For example, let the generating polynomial be $x^7 + x^3 + 1$. This means that each new bit the sequence $y(i)$ is generated by the following rule:

$$y(i + 7) = y(i) \oplus y(i + 3),$$

where $\oplus$ stands for exclusive OR operation (XOR). It is guaranteed that this sequence will contain all possible sequences of 7 bits and the period of the sequence will be $2^7 - 1 = 127$ bits. Obviously, the sequence can not start from all zeros. However if we initially choose 7 bits of the sequence to be {1000000}, an appropriate pseudo-random sequence will be generated. Here is the standard $x^7 + x^3 + 1$ pseudo-random sequence that is used for most NLTS tests:

```
10000001  00010011  00010111  01011011  00000110  01101010
01110011  11011010  00010101  01111101  00101000  11011100
01111111  00001110  11110010  1100100
```

What is so special about pseudo-random sequences except for the simple rule for their generation? They possess a "shift and add" property: $y_k \oplus y_{k+p} = y_{k+M}$, where M is an integer. This means that the exclusive OR of two pseudo-random sequences taken with some shift p produces the initial sequence that is also shifted by some number of bits M. For the $x^7 + x^3 + 1$ sequence, $y_k \oplus y_{k+1} = y_{k+31}$.

Polynomial $x^7 + x^3 + 1$ is one of the most popularly used. However, other polynomials are also used. For example, the sequence generated by $x^6 + x^4 + x^3 + 1$ has a period of $2^6 - 1 = 63$ bits. For each generating polynomial, we have a different "shift and add" property.[17]

The detailed mathematical derivation of the *dipulse extraction* method was given in a seminal paper by D. Palmer, P. Ziperovich and R. Wood.[17] This derivation contains several assumptions, one of which is that PW_{50} of pulse is much larger than the bit period T as encountered in PRML channels.

Values of input pattern $a(k)$ are defined in NRZ format: $a(k) = 1$ corresponds to one polarity of the medium magnetization, and $a(k) = -1$ corresponds to an opposite polarity. This definition is very important because the product of 2 bits becomes similar to XOR operation, as shown below:

a1	a2	Product a1×a2	x1	x2	XOR x1 ⊕ x2
+1	+1	+1	0	0	0
−1	−1	+1	1	1	0
+1	−1	−1	0	1	1
−1	+1	−1	1	0	1

A readback voltage from the head can be written as the following sum:

$$V(t) = \frac{1}{2}\sum_k [a(k + 1) - a(k)]p(t + kT), \qquad (10.62)$$

where $p(t)$ is an isolated pulse. When the pattern is defined in NRZ format, a transition corresponds to a change from +1 to −1 (negative readback pulse) or to a change from −1 to +1 (positive readback pulse). If NLTS equal to Δ is present, the readback waveform is distorted:

$$V(t) = \frac{1}{2}\sum_k [a(k + 1) - \qquad (10.63)$$

$$a(k)]p\left\{t + kT - \frac{\Delta}{4}[a(k + 1) - a(k)][a(k) - a(k - 1)]\right\}.$$

If at least one of the differences $a(k + 1) - a(k)$ or $a(k) - a(k - 1)$ is equal to zero, the corresponding isolated pulse in Equation (10.63) is unshifted. However, if the current transition is preceded by a previous one, the term Δ is added to the delay of the $p(t + kT)$.

A similar trick is used to include different nonlinearities into Equation (10.62). For example, an H/E transition shift equal to ε is described by the following term:

$$-\frac{\varepsilon}{2} + \frac{\varepsilon}{4}[a(k + 1) - a(k)].$$

For a particular polarity of the transition, $a(k + 1) - a(k) = 2$, and the resulting shift is zero. For an opposite polarity of transition, the shift equals to ε.

The next step in analyzing Equation (10.63) is to expand the function $p(t)$ into the Taylor series around $t + kT$, discarding the high order terms and approximating the derivative of $p(t)$ with a finite difference at a bit period interval. The Taylor series expansion results in quadratic terms of pseudo-random sequence coefficients. The main contributions to the playback signal from nonlinear distortions end up in the cross-terms of the pseudo-random sequence. For example, the dominant contribution for NLTS is given by the term $a(k-1)a(k)a(k+1)$. For polynomial $x^7 + x^3 +1$, this term causes the shift of $M = -25$ bit periods. The resulting waveform is approximately written as:

$$V(t) \approx V\left(t+\frac{\Delta}{2}\right) + \frac{\Delta}{2T}V\left(t+\frac{\Delta}{2}-\frac{3T}{2}\right) - \frac{\Delta}{2T}V\left[t+\frac{\Delta}{2}+\left(M-\frac{1}{2}\right)T\right]. \quad (10.64)$$

The distorted readback signal consists of several shifted versions of the original signal, including a main peak $V(t + \Delta/2)$ and two *echoes* shifted by $1.5T$ and $25.5T$, respectively.

Similarly, when an HTS is present, its main contribution to the signal is determined by the product $a(k)a(k+1)$. This results in an echo at the $-30.5T$ location. For the second adjacent transition NLTS, i.e., NLTS caused by a transition which is written two periods earlier than the current transition, the term $a(k-2)a(k)a(k+1)$ gives a different echo at the $45.5T$ location.

Note that the amount of shifts caused by nonlinearities depend on the pseudo-random polynomial used. When several nonlinearities are present, different echoes may be overlapped. The study of each new nonlinear effect using pseudo-random sequences requires a lot of care.

The method of pseudo-random sequences to measure NLTS can be summarized as follows:

1. Write a pseudo-random sequence and process at least one full period of the sequence, assuming that the pattern is repeated infinitely before and after this period.
2. Digitize more than one full period of pseudo-random with oversampling, i.e., acquire at least several samples per bit period.
3. Extract exactly one full period of the readback waveform.
4. Resample data to obtain an integer number of samples per bit period. This step is required for subsequent calculation of discrete Fourier transform.
5. Create an oversampled pseudo-random sequence. This is required to match the sizes of the digitized readback data and of the binary pseudo-random sequence. Nor example, if M samples per bit period is obtained, insert $M-1$ zeros between all ones and zeros of the data pattern. If $M = 4$ and the data pattern is (111010), the oversampled data would be {1000100010000010000000}.

6. Calculate the discrete Fourier transforms of the oversampled read-back waveform $D(k)$ and the oversampled pseudorandom sequence $P(k)$. Take *inverse* Fourier transform of the $D(k)/P(k)$. The result gives the *dipulse extraction plot* shown in Figure 10.46, where the echoes characterize nonlinearities.

7. Measure the amplitudes of the main pulse (A at the zero location) and the echoes (B at $+25.5$ bit period, C at -30.5 bit period, and D at $+45.5$ bit period). The NLTS is $2B/A$ fraction of the bit period, HTS is $2C/A$ fraction of the bit period, and the second adjacent transition NLTS is $2D/A$ fraction of the bit period.

A typical dipulse extraction plot is shown in Fig. 10.46. Three significant echoes are present to the right and to the left of the main peak. The right echo is at a location of $25.5T$ and corresponds to the NLTS. The second small echo to the right is at $45.5T$ and reflects the second adjacent NLTS. The left echo at $-30.5T$ corresponds to the HTS. These echoes disappear when the magnetic recording channel is linear. In this case,

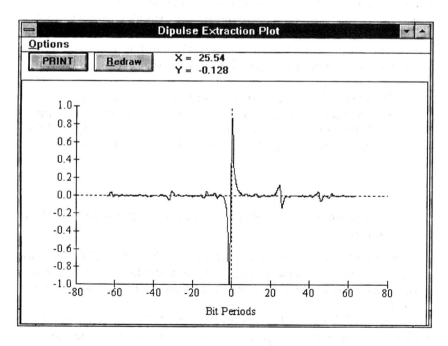

FIGURE 10.46.   A typical dipulse plot with approximately 20% NLTS, 7% second adjacent NLTS, and 10% HTS.

the dipulse extraction plot describes the ideal response of the magnetic recording channel to an isolated dibit.

The dipulse extraction technique is generally a robust method for NLTS estimation. It is reasonably accurate when PW_{50}/T is over 2. for smaller ratios, the measured NLTS is exaggerated and the real NLTS values are 20–50% smaller than the measured ones.

Note that any read nonlinearities may result in increased noise, oscillations, or asymmetrical peak distortions, making the amplitudes of the echoes maybe different from the correct ones. For example, the main peak of the dipulse is often asymmetrical due to the echoes close to the zero location. Partial erasure creates an echo that is only $0.5T$ from the NLTS echo, so both effects are mixed in the dipulse extraction plot. The effects of PE and NLTS are approximately additive: if 20% NLTS and 20% PE are present in the signal, the total NLTS meaured by the dipulse extraction method is close to 40%.

Another method for NLTS measurement called the *time domain correlation* method is based on calculating the cross-correlation function of two periods of the pseudo-random sequence. The basic idea is to follow the same procedure in the dipulse extraction method, but to calculate the time correlation function instead of the dipulse extraction plot. The correlation of the original data with its copy is the maximum, while the correlation of the signal with noise is close to zero. The nonlinear distortions give additional correlation peaks. The algorithm for the correlation analysis is described in the literature.[19-21]

A typical correlation plot is shown in Fig. 10.47. Since the correlation is an even function, all echoes are "wrapped" around zero. For example, the peak of the correlation function at $25.5T$, corresponding to NLTS, is now at the same side of the peak at $30.5T$, which corresponds to the HTS.

10.3.3 Comparison of spectral elimination and pseudo-random methods

It is important to remember that different methods for NLTS measurements may have different advantages and drawbacks. Therefore, we now compare the spectral elimination methods with pseudo-random sequence methods.

First of all, both methods report wrong NLTS values when other nonlinear distortions are present. Neither spectral elimination nor pseudo-random methods can distinguish NLTS from partial erasure (amplitude loss). However, it appears that pseudo-random sequence

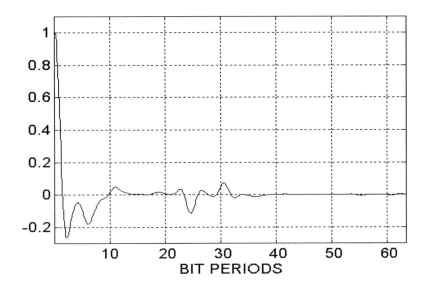

FIGURE 10.47. Example of time domain correlation plot. NLTS echo is at $25.5T$ and HTS echo is at $30.5T$. Their values are $\sim$20% and $\sim$10% of bit period, respectively.

method is less sensitive to MR head readback nonlinearity than spectral elimination method. This is due to the fact that MR head saturation creates an additional echo that is separated from the main NLTS echo in the dipulse extraction plot. On the contrary, the spectral elimination method is based on the principle of linear superposition, and any readback nonlinearity distorts the signal spectrum and the corresponding NLTS measurements.

An important difference between the described methods for NLTS measurement comes from the fact that spectral elimination method measures NLTS of a particular transition in series, e.g., a dibit. The method of pseudo-random sequences is based on writing the random pattern of data and reports an "average" NLTS value for a random pattern.[11] For each of the polynomials, the measurement result will be determined by one period of the pseudo-random sequence. For example, consider the $x^7 + x^3 + 1$ polynomial, which produces one period of 127 bits. A careful analysis reveals that it contains a total of:

> eight dibit transitions,
> four tribit transitions (three adjacent transitions),
> two series of four adjacent transitions,

one sequence of five adjacent transitions,
one sequence of seven adjacent transitions.

If we denote the dibit transition shift as $\Delta(2)$, then a total of 16 transitions will have this value of NLTS. In one period of the pseudo-random sequence, we have six distinct values of transition shifts:

16 transitions with value of $\Delta(2)$,
8 transitions with value of $\Delta(3)$,
4 transitions with value of $\Delta(4)$,
2 transitions with value of $\Delta(5)$,
1 transition with value of $\Delta(6)$,
1 transition with value of $\Delta(7)$,

where $\Delta(n)$ denotes the NLTS of the nth transition in a series of adjacent transitions. When an echo is measured using either dipulse extraction method or time correlation method, an average value of all the NLTS values is obtained:

$$\Delta = \frac{16\Delta(2) + 8\Delta(3) + 4\Delta(4) + 2\Delta(5) + \Delta(6) + \Delta(7)}{32}. \qquad (10.65)$$

This equation explains why the NLTS measured using spectral elimination tends to be larger than that using pseudo-random sequence method. For example, for an inductive head at approximately 85 KFCI, the value of NLTS measured at a 16-ns bit period equals 3.75 ns using the dipulse extraction technique, but it is 4.6 ns using the harmonic elimination method. This is because the former measures the average NLTS while the latter measures $\Delta(2)$. Take an experimentally measured dependence of NLTS on transition separation, Equation (10.65) predicts a value of 3.8 ns, which is very close to the experimentally measured NLTS value using the former method.

Generally speaking, the average NLTS could be of more interest to drive manufacturers, but head/medium characterization requires measurement of NLTS for individual transitions in series such as dibit and tribit. Spectral elimination method is more appropriate to get this type of information. For example, the pattern for measuring the NLTS of the third transition in the tribit, $\Delta(3)$, is shown in Fig. 10.48. The pattern consists of a tribit {A,B,C}, followed by an isolated transition D, a dibit {E,F}, and two isolated transitions G and H. The pairs of transitions (A–E), (B–F), (C–G), and (D–H) compensate each other in spectral domain, so there is no fifth harmonic if the channel is linear. Since the NLTS of transition B is the same as for transition F, only the transition shift of C

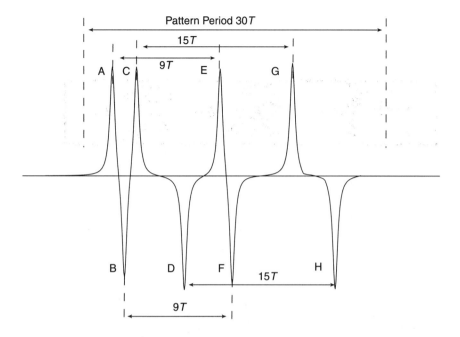

FIGURE 10.48. Special elimination pattern for measuring transition shift of the tribit.

contributes to the fifth harmonic. Therefore, the value of $\Delta(3)$ is measured. Similar patterns for measuring $\Delta(4)$, $\Delta(5)$, $\Delta(6)$, $\Delta(7)$, etc., can be constructed.

10.4 PARTIAL ERASURE

Partial erasure (PE), also referred to as *nonlinear amplitude loss* or *percolation*, occurs at high recording densities when two transitions are written at a small separation. Partial erasure results in a reduction of the readback voltage amplitude.

The main cause of PE is "local annihilation" of magnetic transitions, as shown in Fig. 10.49. When isolated transition 1 is written, it has a typical zigzag structure and is isolated from the rest of the transitions. Transitions 2 and 3 form a closely spaced dibit. When transition 3 is being written, it creates a demagnetizing field opposite to the direction of medium magnetization. This field is the largest at the zigzag tips, which

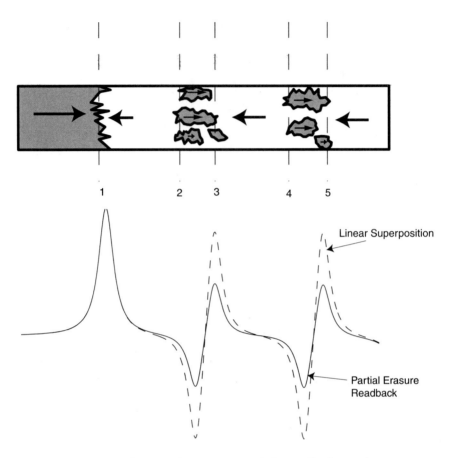

FIGURE 10.49.   Partially erased transitions and the readback signal.

are closest to the previously written transition.[1,14] It may happen that two zigzag tips from transition 2 and 3 are so close to each other that the demagnetizing field reverse the magnetization between the tips. In other words, transition 2 and 3 are *partially erased* by breaking into isolated strips of magnetization. The erased areas between the isolated strips are called the *percolation* regions.

Note that this effect is density-dependent or pattern-dependent: only the high-density transitions (dibits or series of transitions) will be affected by PE. Since partial erasure depends on the magnitude of the demagnetizing field, the reduction of $M_r\delta/H_c$ and the head-medium separation is important. Numerical micromagnetic simulations and experimental magnetic force microscope (MFM) measurements indicate that large exchange-coupling between grains produces large, but few percolation regions, while low exchange-coupling produces many narrow percolation regions. Partial erasure can *not* be precompensated, so it is a serious limitation in magnetic recording.

The main manifestation of the partial erasure in the readback signal waveform is the nonlinear amplitude loss. The readback signal from transition with partial erasure is only from the magnetized region, so the effective data track width is reduced. An amplitude loss factor α is often introduced so that the readback signal with PE can be written as

$$V_{PE}(x) = (1 - \alpha)V(x). \tag{10.66}$$

In practical applications it is not easy to separate the nonlinear amplitude loss from NLTS. Both types of nonlinear distortions cause amplitude degradation of dibits and high-frequency transition bursts. The main difference between these two types of distortions is that PE does not cause transition shifts.

The amplitude loss resulting from PE can be measured using a simple overwrite-type technique. If we write a long square wave pattern consisting of adjacent transitions, each transition in this pattern will be shifted by the same amount of NLTS toward the previous transition as illustrated in Fig. 10.50. Therefore, a square wave pattern does not contain NLTS, and its amplitude is only affected by PE. Note that the magnetic medium should be effectively AC-erased, or demagnetized prior to writing this square wave in order to avoid NLTS oscillations, caused by the HTS.

A simple way to measure PE is to write two square wave patterns: low-frequency (LF) pattern and high-frequency (HF) pattern. The bit period of the HF pattern is T and that of the LF pattern is kT. The ratio is chosen such that only the HF pattern is affected by the PE.

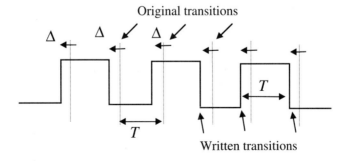

FIGURE 10.50. Square wave pattern with a constant NLTS.

In the absence of PE, a square wave pattern contains only the odd harmonics of the fundamental angular frequency, which is $\omega_0 = 2\pi/2T$ for the HF pattern and $\omega_0/k = 2\pi/2kT$ for the LF pattern, respectively. If $P(\omega)$ is the spectrum of the isolated pulse $p(t)$, the spectra of the HF and LF pattern are given by

$$R_{HF}(\omega) = 2\omega_0 P(\omega) \sum_{m=odd} \delta(\omega - m\omega_0), \qquad (10.67)$$

$$R_{LF}(\omega) = 2\frac{\omega_0}{k}P(\omega) \sum_{m=odd} \delta(\omega - m\frac{\omega_0}{k}). \qquad (10.68)$$

Now it is easy to see that the kth harmonic (k must be odd since the pattern does not contain even harmonics) of the LF pattern is[14]

$$R_{LF}(k\omega_{LF}) = \frac{1}{k}R_{HF}(\omega_{HF}), \text{ where } \omega_{LF} = \frac{\omega_{HF}}{k}. \qquad (10.69)$$

Choosing $k = 3$ and $\omega_{HF} = \omega_0$, the first harmonic of the HF pattern is

$$R_{HF}(\omega_0) = \omega_0 P(\omega_0). \qquad (10.69)$$

The third harmonic of the LF pattern is indeed 1/3 of the HF spectral peak at the same frequency:

$$R_{LF}(\omega_0) = \frac{\omega_0}{3}P(\omega_0). \qquad (10.70)$$

In other words, the *third harmonic ratio* $R_{HF}(\omega_0)/3R_{LF}(\omega_0)$ is 1 if the channel is linear.

When the HF pattern is partially erased, its spectral peak is reduced while that of the LF pattern is unchanged. Therefore, we can easily estimate the amplitude loss factor α:

$$3R_{LF}(\omega_0)(1 - 2\alpha) = R_{HF}(\omega_0). \qquad (10.71)$$

The factor 2 is necessary since each transition has PE from its two neighbors. Therefore,

$$\alpha = \frac{1}{2} - \frac{R_{HF}(\omega_0)}{6R_{LF}(\omega_0)}. \qquad (10.72)$$

Figure 10.50 shows the experimentally measured third harmonic ratio and the corresponding PE factor α at various recording densities. The values of α may reach 20% at a high linear density of 200 KFCI. Note that the described method of PE measurement is based upon the principle of linear superposition. The non-linearity of MR heads, which will be discussed in the next section, will degrade the measurement results and must be compensated for in order to measure PE.

10.5 MAGNETORESISTIVE READ HEAD NONLINEARITY

In the previous sections various sources of nonlinear distortions in the write process were discussed. If a MR head is used in the recording system, it introduces nonlinear distortions into the read process.

The response of an anisotropic MR element is intrinsically nonlinear, varying as the squared cosine of the angle between the magnetization of the MR element and the applied read bias current. For small applied fields, the variations in the output voltage are approximately linear. However, as the applied field becomes stronger, the readback voltage becomes a nonlinear function of the field.

Typical MR read nonlinearity manifests itself in the saturation of the voltage pulses of isolated transitions. Consider the readback voltages shown in Fig. 10.52. The response of the ideal, linear MR head is shown by a dashed line, while the actual readback signal has smaller amplitudes at the peaks of the isolated transitions. Note also that the response of the MR element is asymmetrical, i.e., the distortion is stronger for the negative voltage pulses than for the positive pulses.

The nonlinear response of the MR head is not desirable. It makes the signal asymmetrical, as shown in Fig. 10.52. The amplitude asymmetry between the positive and the negative voltage pulses can be usually minimized with a proper choice of MR bias current and a careful design of MR head. However, even if the amplitude of the readback signal is symmetrical, it does not mean that the MR element is linear.

Consider Fig. 10.53 which shows two signals: the readback of a high-frequency square wave signal, and the signal obtained using linear super-position of isolated readback pulses. The actual readback signal has larger amplitude than the linear superposition of isolated voltage pulses. This means that the isolated voltage pulses are saturated, similarly to those shown in Fig. 10.52. These deviations from the linear superposition principle degrade error rates of partial-response maximum likelihood (PRML) channels and distort measurements of transition shifts and PE.

The readback distortions can be characterized by a nonlinear transfer function $V_{out} = F(V_{in})$, describing the transformation of an "ideal" voltage level into the MR readback output voltage level. The transfer function is very similar to a direct transfer function, which is usually measured by

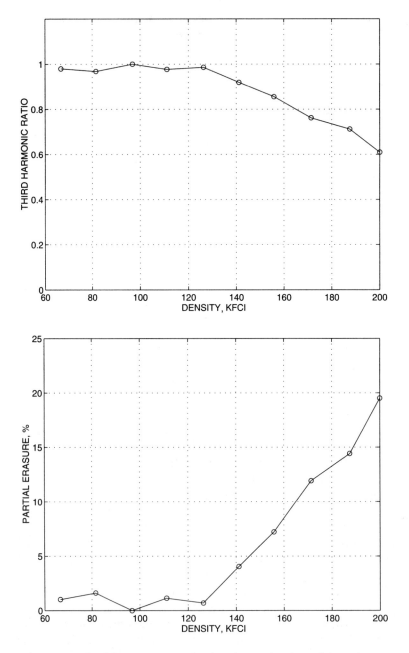

FIGURE 10.51.   Third-harmonic ratio (top) and partial erasure (bottom) vs. square wave recording density.

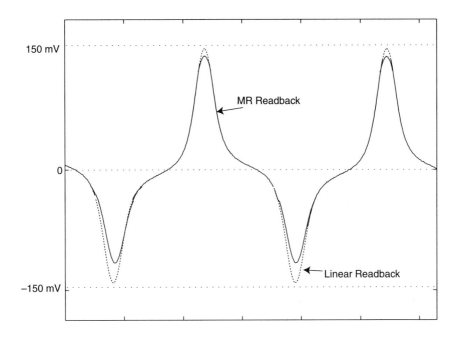

FIGURE 10.52.   Readback signal of MR head.

placing the MR head into a homogeneous magnetic field and measuring the voltage on the MR element as a function of the applied magnetic field.

The transfer function of the MR head can be measured by digitizing and comparing readback signals of two square waves having different frequencies. Figure 10.54 illustrates the principle of linear superposition which is used to approximate readback signal at a frequency three times higher than the frequency of the original pattern: $f_1 = 3f$.

If the linear superposition property holds, the readback signal of a square wave with the high-frequency $f_1 = 3f$ can be obtained as the sum of three shifted low-frequency patterns.[22,23]

$$V_{3f}(t) = V_f(t) + V_f(t + 2T) + V_f(t - 2T), \qquad (10.74)$$

where T is the period of the low-frequency pattern: $T = 1/f$. When the readback voltages are distorted by the MR transfer function, Equation (10.74) is no longer true. However, assuming that we know the inverse MR transfer function F^{-1}, we can rewrite Equation (10.74) as:

$$F^{-1}[V_{3f}(t)] = F^{-1}[V_f(t)] + F^{-1}[V_f(t + 2T)] \qquad (10.75)$$
$$+ F^{-1}[V_f(t - 2T)].$$

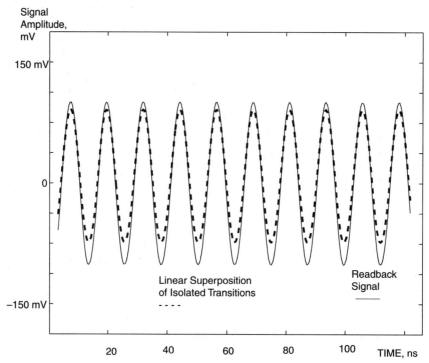

FIGURE 10.53.   Readback signal vs. linear superposition for high frequency square wave signal.

This equality means that the linear superposition principle holds for the *linearized* readback voltages.

For simplicity we can assume that the transfer function of MR head obeys:

$$F(V = 0) = 0; \; dF/dV|_{V=0} = 1.$$

To find an inverse transfer function, we can use a very simple model given by:

$$F^{-1}(V) \approx V + \alpha V^2 + \beta V^3. \tag{10.76}$$

Now substituting Equation (10.76) into Equation (10.75) yields

$$
\begin{aligned}
V_{3f}(t) + \alpha V_{3f}^2(t) + \beta V_{3f}^3(t) = \; & V_f(t) + \alpha V_f^2(t) + \beta V_f^3(t) \\
& + V_f(t + 2T) + \alpha V_f^2(t + 2T) + \beta V_f^3(t + 2T) \\
& + V_f(t - 2T) + \alpha V_f^2(t - 2T) + \beta V_f^3(t - 2T) + \varepsilon
\end{aligned}
\tag{10.77}
$$

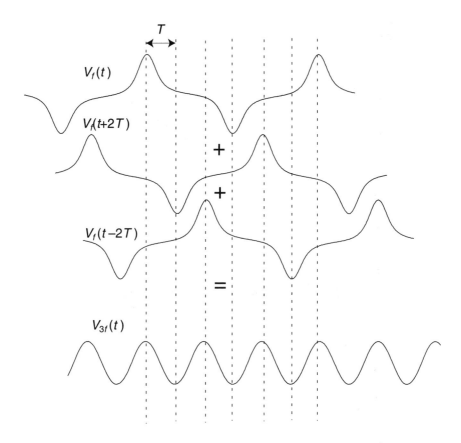

FIGURE 10.54.   Illustration of linear superposition property (Equation (10.74).

where ε represents the model fitting error. Applying Equation (10.77) repeatedly for different values of t yields an overdetermined system of linear equations. This system of equation is then written in matrix form and solved using the pseudo-inverse to find values of α and β that minimize the mean squared error $\langle \varepsilon^2 \rangle$. Fig. 10.55 shows a typical MR transfer function and its inverse.

Calculating the inverse transfer function of the MR head allows linearization of the readback signal and improvement of measurement accuracy of NLTS and PE.[23] Consider the measurement of PE (third harmonic ratio), shown in Fig. 10.56. The peak of the uncorrected third harmonic ratio is clearly seen at densities of 120–180 KFCI. This peak is a manifestation of the MR transfer function: the amplitude of the high-frequency square wave falls within a linear region of the transfer function, while the low-

Output
Voltage, mV

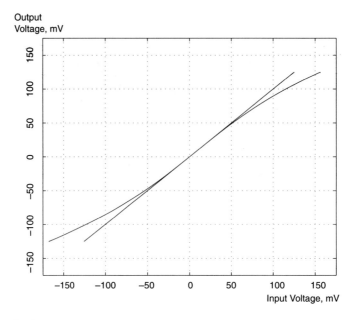

Input Voltage, mV

Input
Voltage, mV

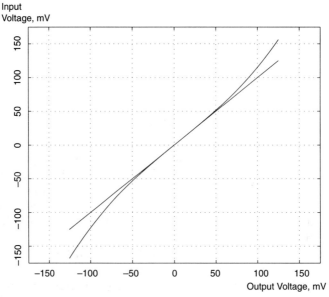

Output Voltage, mV

FIGURE 10.55.   Direct (top) and inverse (bottom) transfer functions of MR head, calculated using Equation (10.77).

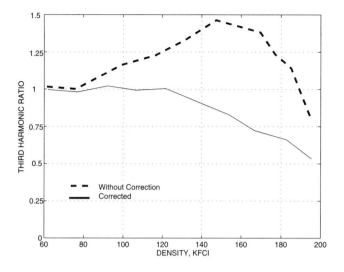

FIGURE 10.56. Measurement of third harmonic ratio with (solid) and without (dashed) correction of MR transfer function.

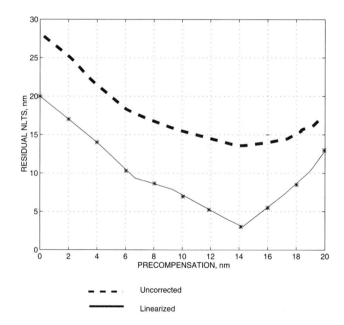

FIGURE 10.57. Measured values of NLTS vs. dibit precompensation with and without correction of MR transfer function.

frequency square wave is saturated and therefore has smaller amplitudes than would be expected for linear superposition. As a result, the calculated third harmonic ratio becomes larger than unity, which in turn distorts the calculated value of PE. Indeed, if the value of amplitude loss is calculated based on the uncorrected curve in Fig. 10.56, it will result in negative PE, which cannot be true.

The effect of linearization of the readback signal on the measurements of precompensation using spectral elimination is shown in Fig. 10.57. The method of spectral elimination without correcting the MR head nonlinearity overestimates the value of NLTS and makes it difficult to find the optimal precompensation parameter.

References

1. H. N. Bertram, *Theory of Magnetic Recording*, (New York, NY: Cambridge University Press, 1994).

2. G. Lin, Y. Zhao, and H. N. Bertram, "Overwrite in thin film disk recording system," *IEEE Trans. Magn.*, **29**, 4215, 1993.

3. Y. S. Tang and C. Tsang, "Non-linear transition shifts in magnetic recording due to interpattern proximity effects," *J. Appl. Phys.*, **74**(5), 3546, 1993.

4. J. Fitzpatrick and X. Che, "The dependence of overwrite on proximity effects," *IEEE Trans. Magn.*, **32**, 3869, 1996.

5. Y. S. Tang and C. Tsang, "Theoretical study of the overwrite spectra due to hard-transitions effects," *IEEE Trans. Magn.*, **25**, 698, 1989.

6. A. Taratorin, B. Wilson, S. X. Wang, "The dependence of overwrite on non-linear transition shift," *IEEE Trans. Magn.*, **33**, 2689, 1997.

7. Y. S. Tang and C. Tsang, "Theoretical study of the overwrite spectra due to hard transitions effects," *IEEE Trans. Magn.*, **25**, 698, 1989.

8. A. Armstrong, H. N. Bertram and J. K. Wolf, "Nonlinear effects in high density magnetic recording," *IEEE Trans. Magn.*, **27**, 4366, 1992.

9. H. N. Bertram, A. Armstrong and J. K. Wolf, "Theory of nonlinearities and pulse asymmetry in high density magnetic recording," *IEEE Trans. Magn.*, **28**, 2701, 1992.

10. C. Tsang and Y. S. Tang, "Time domain study of proximity effect induced transition shifts," *IEEE Trans. Magn.*, **27**, 795, 1991.

11. A. Taratorin, "Characterization of Magnetic Recording Systems" Guzik Technical Enterprises, San Jose, CA 1996.

12. K. Klaassen, R. Hirko, and J. van Peppen, "High speed magnetic recording," *IEEE Trans. Magn.*, **34**, 1822, 1998.

13. A. Taratorin, D. Cheng, P. Arnett, R. Olson, T. Diola, J. Fitzpatrick, S. X. Wang, and B. Wilson, "Intra- and inter-pattern nonlinearities in high density magnetic recording," *IEEE Trans. Magn.*, **34**, 45, 1998.

14. X. Che, "Nonlinearity measurements and write precompensation studies for a PRML recording channel," *IEEE Trans. Magn.*, **31**, 3021, 1995.

15. Y. Tang and C. Tsang, "A technique for measuring non-linear bit shift," *IEEE Trans. Magn.*, **27**, 5316, 1991.

16. X. Che, M. J. Peek and J. Fitzpatrick, "A generalized frequency domain nonlinearity measurement method," *IEEE Trans. Magn.*, **30**, 4236, 1994.

17. D. Palmer *et al.*, "Identification of nonlinear write effects using pseudo-random sequences," *IEEE Trans. Magn.*, **23**, 2377, 1987.

18. P. Newby and R. Wood, "The effects of nonlinear distortion on class IV partial response," *IEEE Trans. Magn.*, **22**, 1203, 1986.

19. X. Che and P. Ziperovich, "A time correlation method of calculating nonlinearities utilizing pseudo-random sequences," *IEEE Trans. Magn.*, **30**, 4239, 1994.

20. G. Mian and T. Howell, "Determining a signal to noise ratio for an arbitrary data sequence by a time domain analysis," *IEEE Trans. Magn.*, **29**, 3999, 1993.

21. G. Mian, "An algorithm for a real time measurement of nonlinear transition shift by a time domain correlation analysis," *IEEE Trans. Magn.*, **31**, 816, 1995.

22. B. Wilson, S. X. Wang, and A. Taratorin, "A generalized method for measuring read-back nonlinearity using a spin stand," *J. Appl. Phys.*, **81**, 4828, 1997; T. C. Arnoldussen and J.-G. Zhu, "Nonlinear behavior of magnetoresistive heads," *IEEE Trans. Magn.*, **34**, 36, 1997.

23. B. Wilson, S. X. Wang, and A. Taratorin, "Linearizing the read process for write nonlinearity measurements," *IEEE Trans. Magn.*, **33**, 2692, 1997.

CHAPTER 11

Peak Detection Channel

Peak detection channel is a simple and reliable method of data detection. Peak detection was the first detection channel utilized in magnetic disk drives. It was extensively used for several decades and is still found in many disk drive products. In this chapter we will describe the main principles of peak detection channel operation and consider the error rates of this channel.

11.1 PEAK DETECTION CHANNEL MODEL

A block diagram of a typical peak detection channel is shown in Fig. 11.1. It is based on the assumption that each transition results in a relatively sharp peak of voltage. The goal of the peak detection channel is to detect each individual voltage peak.

The input analog signal is passed through two paths. One path qualifies a peak of voltage by rectification and threshold detection. When a voltage level exceeds some threshold, a comparator is turned on and a rectangular pulse appears at the output of the threshold detector.[1,2] The other path consists of a differentiator and a zero-crossing detector. A voltage peak will correspond to a zero-crossing after differentiation. The zero-crossing detector generates a short rectangular pulse for each zero-crossing. If a zero-crossing is detected and it is located within the region where signal amplitude exceeds a specific threshold, a transition is detected and a "qualified" pulse appears at the peak detector output.

The "coincidence" scheme used in the peak detection channel makes it robust to find the positions of magnetic transitions. If only the threshold detector is used, the pulse generated by the rectifier is relatively wide, which cannot give the exact location of the transition. On the other hand, if only the zero-crossing detector is used, a lot of extra zero-crossings

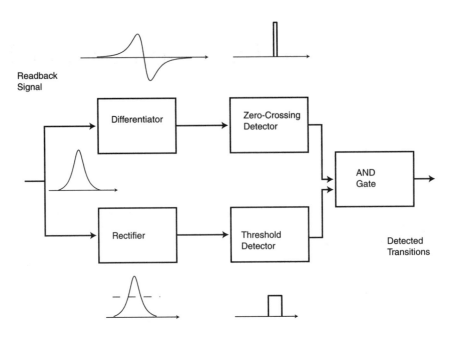

Readback Signal

FIGURE 11.1.   Block diagram of peak detection channel.

caused by noises will be mistaken as magnetic transitions. Only when both a zero-crossing and a rectified pulse are detected simultaneously, a magnetic transition is found reliably.

To distinguish between adjacent transitions and to combat instabilities of the disk rotational speed, each pulse of voltage is detected inside an appropriate *detection window,* also called a *timing window* and should be equal to the *channel bit period.* A special phase-locked loop (PLL) system is used to provide a detection window for each channel bit. The PLL updates its frequency based on detected pulses. Each incoming transition or voltage pulse is searched inside its detection window. As shown in Fig. 11.2, each pulse should be detected after the previous channel bit and before the next channel bit, so the timing window is equal to a channel bit period or *bit cell.* If a peak detection channel uses (1,7) modulation encoding, the detection window is equal to 50% of the minimum timing distance between two transitions that are written in the magnetic medium.

The performance of a detection channel is often characterized by channel bit rate as well as *bit error rate* (BER). Bit error rate P_e is the probability of mistaking a "0" as a "1", or mistaking a "1" as a "0" due to the noises, distortions, or interferences in the channel. The reciprocal of P_e means 1 error per $1/P_e$ bits transferred in the channel. Obviously,

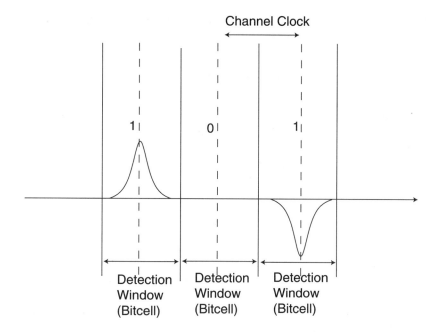

FIGURE 11.2. Clock and detection window in peak detection channel.

we would like P_e to be as small (ideally zero) as possible. The BER can be reduced through error detection and correction. At present, the *corrected* BER is usually in 10^{-12}–10^{-11} range, while the *raw* (uncorrected) BER is typically 10^{-9}–10^{-7}.

The error rate at the threshold detector is determined by the probability of drop-outs when the pulse amplitude falls below the specified threshold or the probability of strong noise outbursts when total media and electronic noise exceeds the specified signal level. The error rate at the zero-crossing detector is determined by random shifts of the zero-crossing position from the correct peak location. Random noises cause fluctuations of the zero-crossing position in the differentiated readback signal. An error occurs when the zero-crossing position falls beyond the detection window, i.e., when zero-crossing is detected earlier or later than the current bit cell. Next we will examine how to calculate BER.

11.2 BER AT THE THRESHOLD DETECTOR

The error rate of the threshold detector may be calculated from the channel SNR. If the zero-to-peak signal voltage amplitude is V_{0-p} and the rms noise voltage is $V_{rms,n}$, then the SNR at the threshold detector is defined as:

$$SNR(dB) = 20 \log \frac{V_{0-p}}{V_{rms,\,n}}. \tag{11.1}$$

To estimate the BER of the threshold detector in the peak detection channel, we assume that the noise voltage n in the recording channel is approximately Gaussian with a zero mean value and a standard deviation of $\sigma = V_{rms,n}$. This means that the probability density of the noise voltage n is given by the Gaussian distribution:

$$f(n) = \frac{1}{\sqrt{2\pi}\sigma} \exp\left(-\frac{n^2}{2\sigma^2}\right). \tag{11.2}$$

Now we will consider the probability of errors in the threshold detector with a threshold fixed at 50% of the zero-to-peak signal amplitude. If there is no peak of voltage in the channel, we may detect a false transition when the noise outburst exceeds one-half of zero-to-peak signal voltage. The probability of this error event, i.e., mistaking a "0" for "1", is given by the following integral of the Gaussian distribution function:

$$P_{0/1} = \int_{V_{0-p}/2}^{\infty} \frac{1}{\sqrt{2\pi}\sigma} \exp\left(-\frac{n^2}{2\sigma^2}\right) dn. \tag{11.3}$$

Define the following integral of the Gaussian distribution as the complementary error function:

$$\text{erfc}(x) = \frac{2}{\sqrt{\pi}} \int_{x}^{\infty} \exp\left(-y^2\right) dy. \tag{11.4}$$

Alternately, one can define a $Q(x)$ function:

$$Q(x) = \frac{1}{\sqrt{2\pi}} \int_{x}^{\infty} e^{-y^2/2} dy = \frac{1}{2} \text{erfc}(x/\sqrt{2}),$$

which represents the probability that a unit-variance zero-mean Gaussian noise exceeds x. Then we can easily obtain that the BER is.

$$P_{0/1} = \frac{1}{2} \text{erfc}\left(\frac{V_{0-p}}{2\sqrt{2}\sigma}\right) = \frac{1}{2} \text{erfc}\left(\frac{\sqrt{SNR}}{2\sqrt{2}}\right) = Q\left(\frac{\sqrt{SNR}}{2}\right). \tag{11.5}$$

Similarly, a "1" signal will be mistaken as "0" if a negative noise outburst exceeding a value of $-V_{0-p}/2$ occurs. Such a BER can be derived as:

$$P_{1/0} = \int_{-\infty}^{-V_{0-p}/2} \frac{1}{\sqrt{2\pi}\sigma} \exp\left\{-\frac{n^2}{2\sigma^2}\right\} dn$$

$$= \frac{1}{2}\text{erfc}\left(\frac{\sqrt{SNR}}{2\sqrt{2}}\right) = Q\left(\frac{\sqrt{SNR}}{2}\right). \tag{11.6}$$

Therefore, the error rate at any bit is a constant if the signal amplitude is stable and the channel noise is Gaussian. The BER is a strong function of the channel SNR. The expected BER at the threshold detector for a channel SNR of 20 dB is 3×10^{-7}, while that for a channel SNR of 24 dB at the threshold detector is $<10^{-15}$. However, the factors not accounted for by the above model, such as the signal amplitude instability and medium defects, may greatly increase the threshold errors predicted above.

11.3 BER AT THE ZERO-CROSSING DETECTOR

Zero-crossing detector locates the transition by looking at the time derivative of voltage signal. The readback signal $V(t)$ is mixed with noise $n(t)$, both of which are differentiated, so the output signal of the differentiator is equal to $V'(t) + n'(t)$. As a result, the zero-crossing locations in the signal are shifted from the transition locations, as shown in Fig. 11.3. These zero-crossing shifts are also called the *peak-shifts* or *bit-shifts*.[1,2]

An error will occur at the zero-crossing detector if the zero-crossing shifts out of the detection window. Assume that the zero-crossing shift t_s has a Gaussian distribution with a standard deviation σ_t, then the probability of a bit error due to the zero-crossing shift can be calculated as:

$$P_e = 2\int_{T_w/2}^{\infty} \frac{1}{\sqrt{2\pi}\sigma_t} \exp\left\{-\frac{t_s^2}{2\sigma_t^2}\right\} dt_s = \text{erfc}\left(\frac{T_w}{2\sqrt{2}\sigma_t}\right) = 2Q\left(\frac{T_w/2}{\sigma_t}\right). \tag{11.7}$$

The error rate at the output of the zero-crossing detector can be derived for the all "1"s NRZI pattern, which has an approximately sinusoidal readback waveform: $V(t) = V_0 \sin(\omega t)$, where $\omega = 2\pi f = \pi/T_w$ is the recording angular frequency. The signal at the output of the differentiator is:

$$V^*(t) = V'(t) = \frac{dV(t)}{dt} = \omega V_0 \cos(\omega t). \tag{11.8}$$

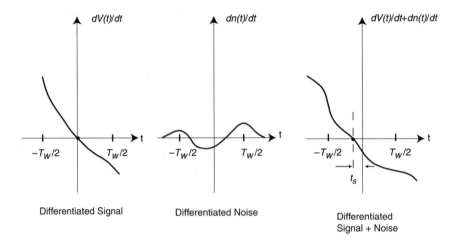

Differentiated Signal   **Differentiated Noise**   **Differentiated
Signal + Noise**

FIGURE 11.3. Zero-crossing and noise (after differentiation) in peak detection channel. T_w is the detection window or channel bit period.

Therefore, the zero-crossings occur when

$$V^*(t_n) = 0, \quad t_n = (2n + 1)T_w/2, \quad n = 0, \pm1, \pm2, \ldots$$

If the random noise after the differentiator is n^*, then the zero-crossing is shifted by t_s, and they are related by the following:

$$V^* (t_n + t_s) + n^* \approx V^*(t_n) + \frac{dV^*}{dt}\bigg|_{t_n} t_s + n^* = 0,$$

$$n^* \approx -t_s \frac{dV^*}{dt}\bigg|_{t_n} = -t_s \frac{d^2V}{dt^2}\bigg|_{t_n}.$$

Calculating the rms value of noise at the differentiator output, we obtain that

$$V^*_{rms,\, n} = \sqrt{\langle (n^*)^2 \rangle} = \left| \frac{d^2V}{dt^2} \right|_{t_n} \sigma_t. \tag{11.9}$$

Combining Equations (11.8), (11.9), and (11.7), the BER at the zero-crossing detector becomes

$$P_e = \text{erfc}\left(\frac{T_w \left| \dfrac{d^2V}{dt^2} \right|_{t_n}}{2\sqrt{2} V^*_{rms,\, n}} \right) = \text{erfc}\left(\frac{T_w \omega V^*_{0-p}}{2\sqrt{2} V^*_{rms,\, n}} \right) = \text{erfc}\left(\omega T_w \frac{\sqrt{SNR^*}}{2\sqrt{2}} \right). \tag{11.10}$$

where $V_{0-p}^* = \omega V_0$ is the zero-to-peak signal amplitude after differentiation, and $SNR^* = (V_{0-p}^*/V_{rms,n}^*)^2$ is the SNR at the output of the differentiator.

Assume that the noise power spectral density from the readback head is approximately constant (white noise):

$$\eta(\omega) \approx \eta,$$

which is valid within the system bandwidth as determined by the cutoff angular frequency ω_c. The ideal differentiator has the following frequency response:

$$H(i\omega) = i\omega, \qquad \omega < \omega_c.$$

Therefore, the noise power after the differentiation is given by the integral:

$$N^* = (V_{rms,\ n}^*)^2 = \int_0^{\omega_c} \omega^2 \eta\, d\omega = \frac{\omega_c^3}{3}\eta = \frac{\omega_c^2}{3}N, \qquad (11.11)$$

where $N = \omega_c \eta$ is the noise power before differentiation. Since the differentiated signal has an amplitude of ωV_0, the SNR after differentiation is

$$SNR^* = \frac{\omega^2 V_0^2}{\omega_c^2 N/3} = 3\frac{\omega^2}{\omega_c^2}\, SNR, \qquad (11.12)$$

where $SNR = V_0^2/N$ is the SNR before differentiation. Most of the signal energy is concentrated in the frequency range where $\omega < \omega_c$, so SNR^* (after differentiation) may be smaller than SNR (before differentiation). For example, if $\omega = \omega_c/2.25$, then the SNR is reduced by 2.3 dB due to differentiation. For the all "1"s pattern without modulation encoding, $\omega = \pi/T_w$, so the bit-shift error rate becomes

$$P_e = \mathrm{erfc}\!\left(\pi\frac{\sqrt{SNR^*}}{2\sqrt{2}}\right) = \mathrm{erfc}\!\left(2.42\frac{\sqrt{SNR}}{2\sqrt{2}}\right).$$

The argument of the complimentary function is 2.42 times that in Equation (11.6). Therefore, in this case, the bit-shift error rate at the zero-crossing detector is much smaller than the threshold error rate at the threshold detector. In contrast, for the all "1"s pattern with (d, k) encoding, $\omega = \pi/T_w(d + 1)$, i.e., the detection window is now smaller than transition period. As a result, the bit-shift error rate will dominate if $d \geq 2$. In general, the total BER in a peak detection channel can be expressed as

$$P_{e,\mathrm{tot}} = P_{e,\mathrm{bs}} + P_{e,\mathrm{th}} - P_{e,\mathrm{bs}}P_{e,\mathrm{th}} \approx P_{e,\mathrm{bs}} + P_{e,\mathrm{th}},$$

where $P_{e,\text{bs}}$ is the bit-shift error rate at the zero-crossing detector, and $P_{e,\text{th}}$ is the threshold error rate at the threshold detector, both of which must be much smaller than 1.

It must be cautioned that SNR is not the only factor that affects the error rate at the zero-crossing detector. Both liner intersymbol interference (ISI) and nonlinear transition shift (NLTS) can cause peak-shifts. In this case, the peaks and the zero-crossings in the readback signal are shifted from the *desired* transition locations, which should be at the center of the detection window. Furthermore, noise is mixed to the distorted signal, so the zero-crossings may be shifted even more from the center of the detection window, as illustrated in Fig. 11.4. Consequently, the BER at the zero-crossing detector increases. Based on Equation (11.7), the bit-shift BER taking ISI and NLTS into consideration can be expressed as follows:

$$P_{e,\text{bs}} = \frac{1}{2}\operatorname{erfc}\left(\frac{T_w/2 - \varDelta}{\sqrt{2}\sigma_t}\right) + \frac{1}{2}\operatorname{erfc}\left(\frac{T_w/2 + \varDelta}{\sqrt{2}\sigma_t}\right),$$

where $\pm\varDelta$ is the net bit-shift, and we assumed that the peak has equal probabilities to shift early or late. Note that peak detection channels are usually used with (1,7) or (2,7) RLL codes, so the transition separations are relatively long. In other words, for peak detection channels linear ISI is more significant than NLTS. The latter is a critical factor in PRML channels (Chapter 12).

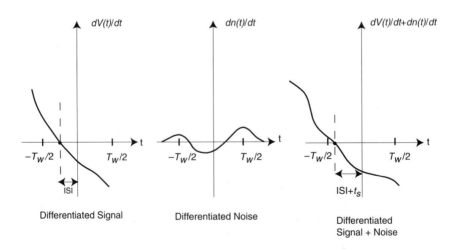

FIGURE 11.4.   Bit-shifts caused by intersymbol interference (ISI) and noise in peak detection channel.

11.4 WINDOW MARGIN AND BIT-SHIFT DISTRIBUTION

The output of the zero-crossing detector of a peak detection channel produces sharp pulses at the locations where the zero-crossings are detected, as shown in Fig. 11.5. As we have discussed in the previous section,

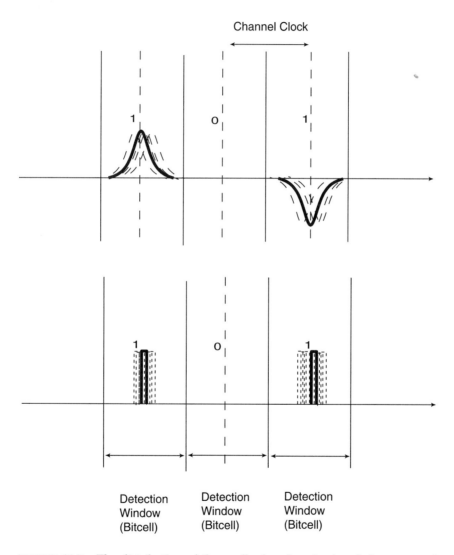

FIGURE 11.5. The distribution of the readback pulses (top) and the output of zero-crossing detector (bottom).

an error occurs when the corresponding pulse falls outside the detection window. The error rate at the output of the zero-crossing detector can be measured by counting the pulses falling outside the prescribed detection window. However, the error rate of an actual magnetic recording channel tends to be very low, so the pulse counting measurement may take a long time. In fact, the raw BER of a magnetic disk drive is typically $\sim 10^{-9}$. At such a low probability level, more than 10^{10} bits of data should be collected and analyzed to obtain reliable statistics. This would require $> 10^4$ disk revolutions and at least several minutes of measurement time for each experimental condition, just to capture the data bits.

An effective and fast method for evaluating error performance is based on the window margin analysis.[3] To understand the principle of window margin, let us imagine that we are able to measure the exact position of each pulse at the output of the zero-crossing detector and to accumulate the histogram of such positions, as shown in Fig. 11.6. The height of the histogram $H(t_k)$ at each timing position $t = t_k$ corresponds to the total number of pulses having a timing shift of t_k. The sum of $H(t_k)$ equals to the total number of detected pulses.

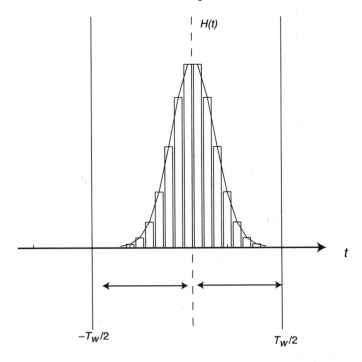

FIGURE 11.6. Histogram of peak shifts at the output of zero-crossing detector.

Once we know the histogram, the number of pulses with a bit-shift value larger than t_s, $N(t_s)$, can be expressed as the sum of $H(t_k)$ for $|t_k| > t_s$:

$$N(t_s) = \sum_{t_k > t_s} H(t_k) + \sum_{t_k < -t_s} H(t_k),$$

Obviously, the total number of pulses in the histogram is $N_{tot} \approx N(t_s = 0)$. The larger the t_s value, the smaller the $N(t_s)$ value. If we plot $N(t_s)/N_{tot}$ vs. t_s in a logarithmic scale, we get a *bit-shift plot* as shown in Fig. 11.7. The horizontal axis of this plot represents the bit-shift value (in nanoseconds), and the *half detection window* (or *half timing window*) equals to 5 ns. When the pulse from the output of the zero-crossing detector deviates more than 5 ns from its nominal position, i.e., the center of the detection window, the peak detection channel makes an error. If the half detection window is chosen to be $t_s < 5$ ns, then $N(t_s)/N_{tot}$ is the corresponding bit error rate. In this sense, the bit-shift plot shows the logarithmic BER as a function of the half timing window.

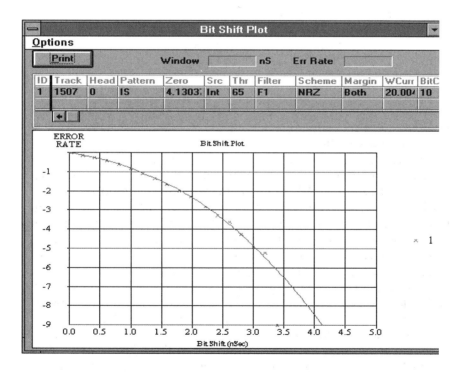

FIGURE 11.7. An example of a bit-shift plot.

The great advantage of a bit-shift plot is that it can be acquired very fast. The bit-shift plot is usually a smooth curve. Therefore, even though the actual measurement points stop at a BER of $\sim 10^{-5}$, these points can be curve-fitted and then extrapolated to a much lower BER as shown by a solid line in Fig. 11.7. It is predicted from the curve fitting that the BER is 10^{-9} if the half timing window is 4.1 ns. In other words, if we specify a BER of 10^{-9} and a half timing window of 5 ns, there is still 0.9 ns left to accommodate any additional error sources, which is often called the *timing window margin*. The timing window margin is often normalized by the half detection window, so it becomes 0.9/5 = 18% in this case. In short, the timing window margin is defined as the percentage of the half timing window (or half detection window) which is left at a specified BER level.

A parabolic curve fitting method is often used in bit-shift plots. This is based on the following mathematical approximation (for $x \gg 1$):

$$\mathrm{erfc}(x) \approx e^{-x^2}/(\sqrt{\pi}\,x),$$
$$\ln[\mathrm{erfc}(x)] \approx -x^2 - \ln(\sqrt{\pi}\,x) \approx -x^2.$$

It follows from Equation (11.7) that

$$\ln\left[\frac{N(t_s)}{N_{\mathrm{tot}}}\right] = \ln\left[\mathrm{erfc}\left(\frac{t_s}{\sqrt{2}\sigma_t}\right)\right] = -\frac{t_s^2}{2\sigma_t^2},$$

which is valid if only Gaussian noise exists at the zero-crossing detector. Obviously, the larger the standard deviation (σ_t), which measures the random peak shifts, the slower the BER drops. A small noise will result in a sharply descending bit-shift plot.

Figure 11.8 demonstrates a useful representation of a bit-shift plot: the logarithmic BER vs. the percentage of the half timing window for different SNR ratios. If SNR = 23 dB, the 50% timing window margin is reached at an error rate of 10^{-6}. In comparison, if *SNR* = 25 dB, the 50% margin is reached at an error rate of 10^{-10}. Both curves predict error rates well below 10^{-12} at the full half detection window (0% timing window margin). Generally speaking, if the same timing window margin is required, a higher SNR tends to give a much lower BER.

A simple and practical way to measure the bit-shift distribution of a peak detection channel is to vary the detection window of its zero-crossing detector and to count the actual number of pulses detected outside a given detection window. The number of error bits divided by the total number of bits counted is the BER. In a real disk drive, the bit-shift distribution reflects complicated interactions among noises, linear ISI, and nonlinear distortions.[2,3] Since the generation of bit-shift plot is fast, it is

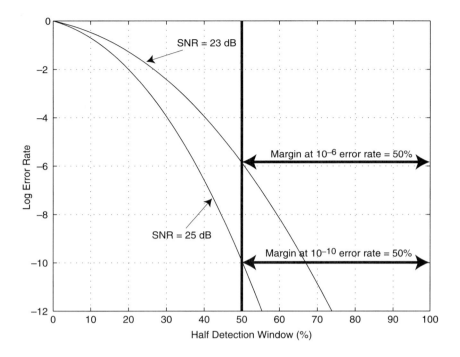

FIGURE 11.8. The dependence of the 50% window margin on system SNR.

an efficient tool for testing magnetic recording components and channel performance.

As mentioned previously, the slope of the bit-shift plot reflects the noise level. A small noise leads to a steep bit-shift plot. Conversely, a large noise level results in a relatively flat bit-shift plot. Peak-shifts due to ISI typically cause a flat part in bit-shift plot. The flat part arises from the fact the bit error rate increases as a result of peak-shifts, as shown in Fig. 11.4. In other words, a larger detection window is required to achieve the same BER, so the bit-shift plot is moved to the right with respect to that without ISI, as shown in Fig. 11.9. The leftmost curve is obtained from a data pattern consisting of isolated transitions with the read head at the nominal on-track position. In this case, the time margin at an error rate of 10^{-9} is 50%. The middle curve is generated for the same pattern but with the read head in an off-track position. When the head is shifted away from the track center, the readback signal amplitude drops, and at the same time more medium noise is read from adjacent medium regions. Therefore, the off-track signal has a lower SNR (Chapter 14), and this curve is less steep. Now the time margin at an error rate of 10^{-9} becomes

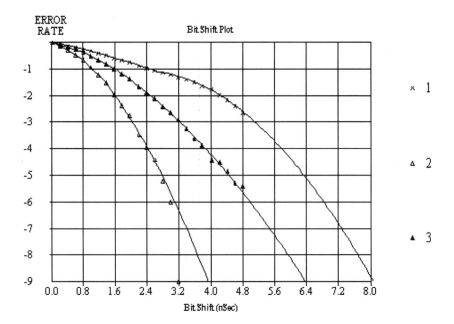

FIGURE 11.9. Examples of bit-shift plots. Left: isolated transition pattern and on-track reading; middle: same pattern in off-track position; right: data pattern with adjacent transitions.

(8.0 ns − 6.4 ns)/8.0 ns = 20%. The rightmost curve is obtained for a data pattern consisting of both isolated and adjacent transitions. A certain number of pulses are now shifted from their nominal positions, resulting in a flat part in the curve. This is the least desirable case among the three, in which there is no time margin left if the required BER is 10^{-9}, i.e., the actual error rate of the peak detector will be $\sim 10^{-9}$.

The last curve in Fig. 11.9 can be further illuminated by Fig. 11.10. The peak-shift due to ISI interacts with the random noise and the histogram of the bit-shift distributions for adjacent transitions are shifted ±2.4 ns from that of isolated transitions. The distribution of isolated transitions has an amplitude of ~ 10 times higher, meaning that approximately one out of every 10 transitions in this pattern is shifted. The initial part of curve 1 in Fig. 11.9 (until about 2.4 ns bit-shift) is slowly decreasing as the error events are dominated by those of isolated transitions. When the error rate level of $\sim 10^{-1}$ is reached at a bit-shift of about 2.4 ns, the error events are caused by both isolated and adjacent transitions. Consequently, the curve has a nearly flat part just beyond 2.4 ns as the peak shifts of adjacent transitions slow down the decrease of the BER.

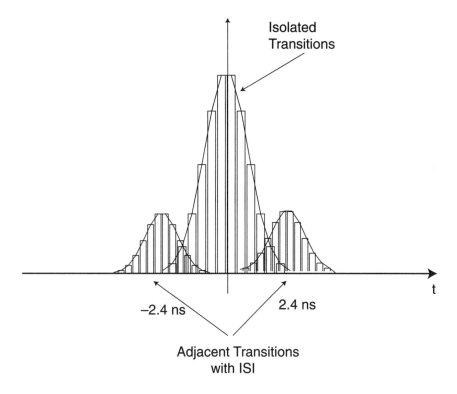

FIGURE 11.10. Bit-shift histogram for the data pattern corresponding to curve 1 (rightmost) shown in Fig. 11.9.

References
1. A. S. Hoagland and J. E. Monson, *Digital Magnetic Recording*, (New York, J. Wiley & Sons, 1991).
2. P. H. Siegel and J. K. Wolf, "Modulation and coding for information storage," *IEEE Communications Magazine*, December 1991, 12, 68–86.
3. E. Katz and T. Campbell, "Effect of bitshift distribution on error rate in magnetic recording," *IEEE Trans. Magn.*, **15**, 1050, 1979.

CHAPTER 12

PRML Channels

When recording densities are low, each transition written on a magnetic medium results in a relatively isolated voltage peak and a peak detection channel works well to recover written information. At high channel densities, however, the peak detection channel can not provide reliable data detection. Superposition of pulses (linear ISI) shifts peaks of readback signal and increases the probability of errors in the zero-crossing detector. At the same time, the signal amplitude is lowered at high densities (roll-off) and the errors in the threshold detector also increase.

The partial-response maximum likelihood (PRML) channels were proposed to overcome the problem of linear ISI and are now the dominant detection schemes in commercial magnetic disk drives.[1-7] The PRML detection method is not based on voltage peaks, but rather takes into account the fact that the signals from adjacent transitions interfere. PRML channels consist of two relatively independent parts: partial response (PR) equalization and maximum likelihood (ML) detector, which will be discussed in this chapter.

12.1 PRINCIPLE OF PARTIAL RESPONSE AND MAXIMUM LIKELIHOOD

The basic idea of *partial response* is to introduce some controlled amount of ISI into the data pattern rather than trying to eliminate it. The idea was first introduced in the field of digital communication and has proven to be a very powerful concept for magnetic information storage as well. It turns out that magnetic recording channels can be transformed into PR channels which satisfy two fundamental properties: (1) the superposition of voltage pulses from adjacent transitions is linear; (2) the shape of the readback signal from an isolated transition is exactly known and determined. A block diagram of a typical PRML channel is shown in Fig. 12.1.[7] It consists of a variable-gain amplifier (VGA), an analog equalizer, an analog-to-digital converter (ADC), a digital equalizer, an ML detector, and a clock/gain recovery circuit. The circuit blocks (except the ML detector) transform the readback signal into the partial response signal as required.

The analog readback signal from the magnetic head should have a certain and constant level of amplification. Any variation in isolated readback peaks is compensated with the VGA, which gets a control signal from the clock and gain recovery loop.

A PR channel operates within a certain bandwidth, meaning that the spectral components beyond the bandwidth have to be cut off. This is done with the continuous time filter or analog equalizer. The other function sometimes performed by the analog equalizer is to modify the frequency response of the channel. The modification of the frequency response is sometimes required to adjust the shape of the readback signal from the head. For example, it may be necessary to adjust the pulse width to make it proportional to the distance between transitions. The analog equalizer is implemented as a linear filter with a programmable frequency response including a variable cutoff frequency and boost. The analog signal at the equalizer output generally has a slightly different shape than the unmodified signal directly from the head.

The signal from the analog equalizer is sampled (or digitized) with the ADC. The sampling is initiated by a clock signal at the rate of exactly one sample per channel bit period. The frequency and phase of the clock

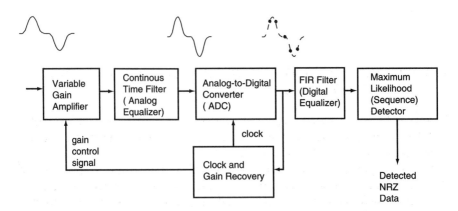

FIGURE 12.1. Block diagram of typical PRML channel.

signal is adjusted by the clock recovery loop. The signal at the ADC output constitutes a stream of digital samples or discrete-time data. Digital samples are often processed (filtered) by an additional digital filter. This digital filtering operation can improve the quality of analog equalization. The principle of the finite impulse response (FIR) filter will be discussed in Section 12.2.

The samples at the ADC output are used to detect the presence of transitions in the readback signal. If the signal quality is good, a simple threshold detector can be used to compare sample values to a threshold. However, much more reliable detection at high recording densities can be achieved with an ML detector. Unlike the peak detection channel, the PRML channels do not assume that the readback signal should contain relatively narrow peaks. The decisions of PRML channels are based on a sequence of the ADC samples of the signal, which are not necessarily taken at the signal peak level.

The partial response equalization and the ML detector are at the heart of a PRML channel. We will introduce their principles here, but defer the details to Sections 12.2 and 12.4.

12.1.1 PR4 channel

A special class of partial-response channel, Class IV partial response (PR4) system, is the first widely used PR channel. The isolated pulse shape in a PR4 system is shown in Fig. 12.2, where T is the channel bit period, and the transition is written at time instant $t = 0$. The sample values at integer number of bit periods before the transition are exactly zeroes. However at $t = 0$ and at $t = T$, the sample values of the pulse are equal to 1. The pulse of voltage reaches its peak amplitude of 1.273 at one-half of the bit period.

The samples of the isolated PR4 pulse shown in Fig. 12.2 at the output of ADC will be . . . 00011000. . . . Of course, value "1" is used for convenience and in reality it corresponds to a certain ADC level, which may be a number between 0 and 2^n-1, where n is the number of bits in ADC. The fact that the isolated transition has two nonzero samples, one at the transition location and the other at the next transition location is very important. If the next transition is written, the pulses will interfere. However, the other sample values are zero, so the interference are easily predictable.

A dibit is formed when the second transition is written immediately after the first one (i.e., one channel period later), which results in a dipulse

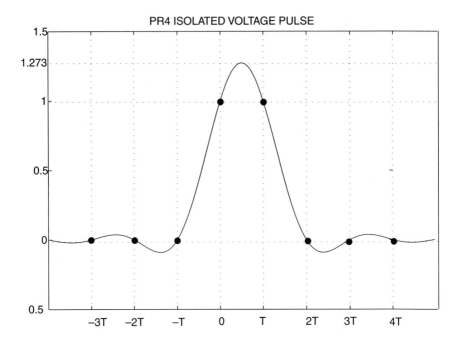

FIGURE 12.2.   Shape of isolated pulse in PR4 system.

response shown in Fig. 12.3. It is generated by the linear superposition of voltages from two isolated pulses with opposite polarity:

$$
\begin{array}{llll}
 & 0\ 0\ 0\ 1 & 1 & 0\ 0\ 0 & \text{—from the first transition} \\
+ & 0\ 0\ 0\ 0 & -1 & -1\ 0\ 0 & \text{—from the second transition} \\
\hline
= & 0\ 0\ 0\ 1 & 0 & -1\ 0\ 0 &
\end{array}
$$

The samples of a dibit are $\{\ldots,0,0,1,0,-1,0,0,\ldots\}$.

What happens if we have three consecutive transitions (tribit)? It is easy to check that the answer is $\{\ldots,0,0,1,0,0,1,0,0,\ldots\}$. Another useful pattern is the low-frequency square recording pattern consisting of transitions two channel periods apart. Obviously, the sequence of resulting samples is $\{\ldots,+1,+1,-1,-1,+1,+1,-1,-1,+1,+1,\ldots\}$.

From the sequence of samples from the ADC output we can easily reconstruct any data patterns that are written on the medium. If the data are in NRZ form (i.e., 1 and 0 represents positive and negative medium magnetization, respectively), then the current value $a(k)$ of the data pattern is the sum of the current sample $s(k)$ and the bit two channel periods earlier:

$$a(k) = s(k) + a(k-2). \tag{12.1}$$

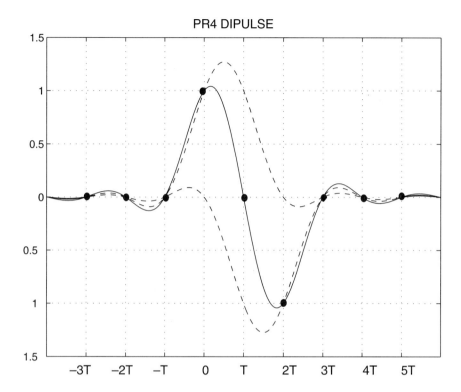

FIGURE 12.3.   A dipulse response.

Applying Equation (12.1) to an isolated transition:

Samples in PRML system:	...0000 1 1 0 0 0 0 0...
Recovered NRZ data:	...0000 1 1 1 1 1 1 1...

For a tribit:

Samples in PRML system:	...0000 1 0 0 1 0 0 0...
Recovered NRZ data:	...0000 1 0 1 1 1 1 1...

The last example is especially interesting: the second pulse in the tribit has zero samples, almost completely suppressed by the first and the third transitions due to linear superposition. However, we can still easily recover the data based on the samples. Therefore, once the pulses are reduced to a "standard" shape, the data pattern is easily recovered because the superposition of signals from adjacent transitions is known. In the last example, we know that sample "1" of a transition is canceled by "-1" of the next transition. Since a positive voltage pulse is always followed by a negative pulse, and vice versa, arbitrary linear superposition of the samples of isolated pulses leads to only three possible values: $\{-1, 0, +1\}$. Therefore, ADC output in the PR4 channel consists of only three distinct sample levels: $\{-1, 0, +1\}$. In reality, however, the sample values will deviate from the nominal values because of noises, NLTS, etc.

A practical way to characterize the quality of the PR4 channel is to analyze the statistics of samples at the ADC output. If all parts of the channel are working properly, the ADC samples should take only nominal values $\{-1, 0, +1\}$. As a result, a histogram of sample levels will consist of three distinct peaks, as shown in Fig. 12.4 (left). However, the presence of noise, NLTS, and nonideal equalization leads to the distortions of the sample values, and the three peaks of sample values may overlap one another, as shown in Fig. 12.4 (right).

Another way to characterize the quality of a PRML channel is to analyze the so-called "eye diagrams" or "eye patterns" generated by an oscilloscope. To obtain an eye pattern, a random data pattern is written on the disk. If the clock signal of the channel is input to the oscilloscope synchronization (trigger) and the analog equalizer output is taken as the oscilloscope signal, the superposition of random equalized waveforms will be observed on the oscilloscope screen. Since all these waveforms are synchronized to the channel clock, an interesting "focusing" pattern is observed: all the waveforms at clock points pass through the three points corresponding to $\{-1, 0, +1\}$, as shown in Fig. 12.5.

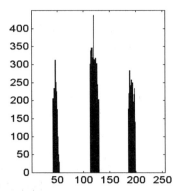 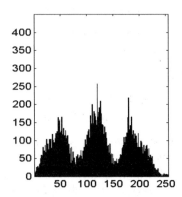

FIGURE 12.4. Sample value distributions for good (left) and poor (right) PR4 signal quality, respectively.

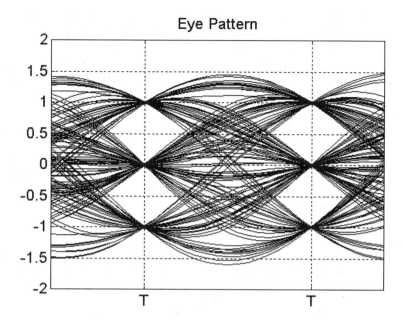

FIGURE 12.5.   An ideal eye pattern for PR4 system.

If the sample distributions do not overlap, or equivalently, the "eyes" in the eye diagram are open, signal detection can be done using a simple comparator (threshold detector). In reality, however, the samples do overlap as shown in Fig. 12.4. In this case, the ML detector is needed to achieve reliable data recovery. The gain of the ML detector over a simple comparator leads to much lower error rates, typically by about three or more orders of magnitude. While it is possible that the ML detector will still be able to decode the pattern in the case of strongly overlapping samples, the probability of errors will be much lower if the histogram is nonoverlapping. We will examine the principle of the ML detector next.

12.1.2 Maximum likelihood detector

Samples at the output of ADC ideally have a small number of levels, such as $\{-1, 0, +1\}$ for the PR4 system. A threshold detector compares the current sample value to an amplitude threshold and immediately decides what the ideal sample value should be. For example, if sample >0.5 then ideal sample $= 1$, if sample <-0.5 then ideal sample $= -1$, if $-0.5 <$ sample

< 0.5 then ideal sample = 0. For a stream of noisy samples such as {0.8 0.3 −0.7 −0.2 0.6 0.9 1.1 0.2}, the threshold detector output is {1 0 −1 0 1 1 1 0}. However, sequence "111" can not exist in the PR4 channel. Sequence "11" corresponds to an isolated voltage pulse, so the next transition should be of opposite polarity, meaning that the possible sequences are "1100", "1001 "11−1−1", etc., but not "1110". Apparently an error has occurred.

Unlike the threshold detector, the ML detector does not make immediate decisions on whether the incoming ADC sample is +1, 0, or −1. Instead, it analyzes a sequence of samples and then chooses the most probable sequence. Therefore, the ML detector is also called the sequence detector, or Viterbi detector after its original inventor. ML detector recognizes that sequence "111" is forbidden, and tries to find the most probable ideal sequence which matches the samples. In the above example, there are several allowable sequences: {1 0 −1 0 1 1 0 0}, {1 0 −1 0 0 1 1 0} or {1 0 −1 0 0 0 1 1}. These sequences can be compared with the received sequence of samples in order to choose the most probable sequence:

0.8	0.3	−0.7	−0.2	0.6	0.9	1.1	0.2	Samples
1	0	−1	0	1	1	0	0	Sequence #1
1	0	−1	0	0	1	1	0	Sequence #2
1	0	−1	0	0	0	1	1	Sequence #3

Sequence #1 takes 0.6 and 0.9 as "1", and 1.1 as "0"; sequence #2 takes 0.6 as "0", and 0.9 and 1.1 as "1"; sequence #3 takes 0.6 and 0.9 as"0", and 1.1 and 0.2 as "1". It appears that sequence #2 is the most probable. We can verify our intuition by calculating the *mean-squared distance* (MSD) between the samples $s(k)$ and the assumed sequence $b(k)$:

$$MSD = \sum_{k=1}^{N} [s(k) - b(k)]^2$$

Sequence #1: $MSD = 1.68$

Sequence #2: $MSD = 0.68$

Sequence #3: $MSD = 2.08$

Sequence #2 is the closest to the sequence of detected samples and at the same time is allowed by the PR4 channel. In other words, sequence #2 is the *most likely* among all the candidate sequences, so the ML detector takes it as the sequence that was transmitted through the channel.

The principle of ML detection can be summarized as follows:

1. Decisions are made based on a sequence of samples, instead of one current sample.

2. For each sequence of samples, a list of allowable sequences is generated.
3. Each of the allowable sequences is compared with the received sequence of samples and respective *MSD* (or another appropriate distance function) is calculated. The sequence with the minimum distance, i.e., the *maximum likelihood,* is selected to be the result of the detection.
4. Decisions of the ML detector are always made with some delay.

12.2 PARTIAL RESPONSE

Having introduced the basics of the PRML channels, now let us look at the PR channel (excluding the ML detector) in more detail.

12.2.1 PR4 channel bandwidth and frequency response

The presence or absence of a transition in a PRML channel is decided based upon the numerical values of the samples. The samples are taken only once per channel period, and the minimum separation between adjacent transitions equals to the channel period. One cannot help asking whether the sampling frequency in the PRML channel is high enough to recover the recorded transition.

The *sampling theorem* (Nyquist theorem)[3] states that any band-limited analog signal with a cutoff frequency of f_{max} can be uniquely recovered from its discrete samples taken with a sampling interval of $T \leq 1/2f_{max}$. In other words, if a band-limited analog signal is sampled with a sampling rate $\geq 2f_{max}$, the information contained in the signal is not lost. Based on the Nyquist theorem, sampling once per channel period T in the PRML channel is appropriate only if the spectrum of the analog readback signal is concentrated below the frequency $f_{max} = 1/2T$.

The spectrum of the head readback signal is discussed in Chapters 3 and 6. The frequency spectrum of a linear channel is usually defined as the Fourier transform of its *impulse response,* i.e., the dipulse response. A good approximation of the experimental spectrum of the channel is obtained if a random pattern transmitted through the channel is displayed with a spectrum analyzer. Approximately an equal number of positive and negative transitions are written on the magnetic medium, so there should be no spectrum content at zero frequency, i.e., the DC content is zero. However, the readback signal contains high frequency components.

The highest frequency in the signal spectrum corresponds to the fastest changing slope of the signal. A fast-changing narrow pulse will have a broader spectrum than a slowly changing wide pulse. Since the random pattern contains all spectral components with frequencies of $1/2nT$, where $n = 1, 2, 3, \ldots$, the experimental spectrum should resemble that of $kV_{sp}(k)$ or $fV_{sp}(f)$ [Equation (3.23)], where $V_{sp}(f)$ is the Fourier transform of the single (isolated) pulse.

If we fix the channel bit period and write a random pattern with increasing pulse widths, the spectral energy distribution will shift to a lower and lower frequency range. Figure 12.6 shows the spectral energy distribution for several different values of PW_{50}/T (channel density). For $PW_{50}/T = 0.5$, the highest spectral energy occurs at about $1/2T$, but significant spectral components extend up to the *clock frequency* ($1/T$). This means that for a system with low ISI, channel bandwidth should be close to the clock frequency. For $PW_{50}/T = 2$, however, the signal spectrum is effectively concentrated below the half of clock frequency given by $1/2T$. The tail of this spectrum is still outside the "half-bandwidth" ($1/2T$) range, but the power of these high frequency components is relatively small.[8]

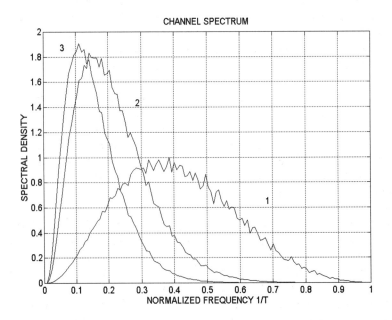

FIGURE 12.6. Magnetic recording channel spectra: (1) $PW_{50}/T = 0.5$; (2) $PW_{50}/T = 2$; $PW_{50}/T = 3$.

Therefore, the spectrum can be readily equalized into a spectrum with a cutoff frequency of 1/2T, and sampling at a rate of 1/T becomes appropriate. However, note that the PRML channel may not work properly for systems with low channel densities.

If the noise power is uniform within the bandwidth, the total noise power within $f \le 1/2T$ will be exactly half of that within $f \le 1/T$. In this case, the fact that the PRML channel bandwidth is limited to 1/2T gives the PRML channel a gain of ~3 dB in SNR over the peak detection channel.

Nyquist's *interpolation formula* states that an analog signal $g(t)$, with a cutoff frequency $\le 1/2T$ and samples of $g(nT)$, where $n = 0, 1, 2, 3, \ldots$, can be expressed as[3]:

$$g(t) = \sum_{n=0}^{\infty} g(nT)\frac{\sin[(\pi/T)(t - nT)]}{(\pi/T)(t - nT)}. \qquad (12.2)$$

In other words, each sample of "1" corresponds to a sinc(x) function (with an appropriate phase difference):

$$h(t) = \frac{\sin[(\pi/T)t]}{(\pi/T)t} = \mathrm{sinc}\left(\frac{\pi t}{T}\right),$$

which is the inverse Fourier transform of the low-pass filter. The analog signal is equal to the convolution of discrete samples and $h(t)$. The Fourier transform of Equation (12.2) is

$$G(f) = \sum_{n=0}^{\infty} g(nT)e^{-i(2\pi f)nT}, \; 0<f<1/2T. \qquad (12.3)$$

Since the PR4 channel has a bandwidth of 1/2T, the equivalent representation of PR4 channel can be given as in Fig. 12.7, where an isolated

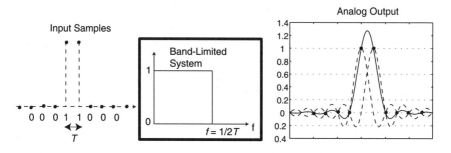

FIGURE 12.7. Nyquist interpolation formula: recovering analog waveform from discrete samples, as illustrated for the PR4 channel.

pulse represented by samples "...0110..." is passed through an ideal low-pass filter with a bandwidth of 1/2T. Therefore, the analog isolated pulse in the PR channel is given by Equation (12.2):

$$s(t) = \frac{\sin\frac{\pi t}{T}}{\frac{\pi t}{T}} + \frac{\sin\frac{\pi(t-T)}{T}}{\frac{\pi(t-T)}{T}}.$$

Now let us calculate the frequency response of the PR4 channel, which is the Fourier transform of the channel impulse response. The isolated pulse is called the "step response" of the PR4 channel, corresponding to a "step" (or transition) of magnetization. The samples due to the step response are {...0,0,0,1,1,0,0,0...}. The impulse response of the channel is the derivative of its step response, which generates samples {...0,0,1,0,−1,0,0...}, the same as the samples of a dipulse response. According to Equation (12.3), the frequency response of the PR4 channel is:

$$H(f) = 1 - \exp(-i4\pi fT), \quad |H(f)| = 2\sin(2\pi fT), \qquad 0 \le f \le \frac{1}{2T},$$

which is shown in Fig. 12.8. Note that the spectrum peaks at the midband frequency 1/4T. The PR4 spectrum is very similar to the experimental spectrum of a magnetic recording channel when the channel density is ~2, as shown in Fig. 12.6. The similarity is an important clue that magnetic recording channel could be turned into a PR4 channel with a minimum amount of equalization.

12.2.2 Partial-response polynomial and classification

The PR4 channel is only a special class of PR schemes. A convenient way of describing different partial response schemes is to use PR polynomials, which are somewhat different from those of cyclic codes in Chapter 8. The PR polynomials describe the correspondence between the NRZ input data pattern and the ideal samples. The NRZ input data pattern is generally given by a sequence of bits $\{a_k\}$. The PR polynomial P is an operator which determines the current channel sample s_k based on the current and previous NRZ bits: $s_k = P[a_k, a_{k-1}, \ldots, a_{k-n}]$. As we have already seen from Equation (12.1), the PR4 channel is described by the following operation:

$$s_k = a_k - a_{k-2}.$$

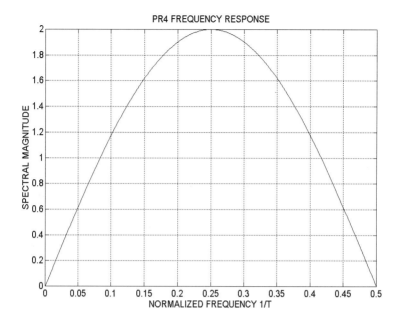

FIGURE 12.8. Frequency response of PR4 channel.

The simple operation can be presented by a PR polynomial as shown below.[4,8]

First let us define a fundamental polynomial operator D:

$$Da_k = a_{k-1},$$

which is called the "delay operator" because it delays the input bit by one bit period. The definition of delay operator allows simple arithmetical operations such as addition, subtraction, multiplication, power, etc. For example, applying the delay operator twice is defined as the second power of D:

$$D^2 a_k = D(Da_k) = Da_{k-1} = a_{k-2}.$$

When a magnetic head reads a disk, it responds only to changes of magnetization, i.e., it differentiates the NRZ bits. The differentiating function of a head is described with an operator:

$$(1 - D)a_k = a_k - a_{k-1}.$$

Since NRZ bits take values 0 and 1, operator $(1 - D)$ will result in $+1$ or -1 samples, depending on the direction of the magnetization change. Operator $(1 - D)$ is the simplest polynomial, corresponding to generating positive or negative samples at the transition locations. Taking absolute values after the $1 - D$ operation generates NRZI bits. Obviously, an isolated pulse expressed in NRZ data (or magnetization) $\{\ldots 000111\ldots\}$ is transformed into NRZI data $\{\ldots 000100\ldots\}$ after the differentiation.

In a PR4 system each voltage pulse has two samples. In other words, if a transition of magnetization occurs, it results in a sample equal to "1" at the transition location and another sample at the next sample period. This can be described with a "spreading" operator $(1 + D)$. If $a_k = 1$, $(1 + D)a_k$ will result in sequence "11". Therefore, a PR4 system can be described as a polynomial $(1 - D)(1 + D) = 1 - D^2$. In other words, the current PR4 sample is

$$s_k = (1 - D^2)a_k = a_k - a_{k-2},$$

which is the same as Equation (12.1).

Partial-response channels, as listed in the following table, were originally proposed in digital communication to combat ISI.[2] Because of the similarity of the PR4 spectrum and the frequency response of magnetic recording channels, the PR4 channel (Class IV) became the darling of the magnetic data storage industry in the 1990s.

Name	Polynomial	Impulse response samples
PR1	$1 + D$	$\ldots 0 \quad 1 \quad 1 \quad 0 \ldots$
PR2	$(1 + D)^2 = 1 + 2D + D^2$	$\ldots 0 \quad 1 \quad 2 \quad 1 \quad 0 \ldots$
PR3	$(1 + D)(2 - D) = 2 + D - D^2$	$\ldots 0 \quad 2 \quad 1 \quad -1 \quad 0 \ldots$
PR4	$(1 + D)(1 - D) = 1 - D^2$	$\ldots 0 \quad 1 \quad 0 \quad -1 \quad 0 \ldots$
PR5	$-(1 + D)^2(1 - D)^2 = -1 + 2D^2 - D^4$	$\ldots 0 \quad -1 \quad 0 \quad 2 \quad 0 \quad -1 \quad 0 \ldots$

If we ignore the differentiating operator $(1 - D)$ and look only at the coefficients of the polynomial $(1 + D)$, we get the samples of an isolated pulse: $\{1,1\}$ in the PR4 channel. The operator $(1 + D)$ determines how the transition sample is spread over the neighboring bit periods. The PR4 channel has been extended to a general family of channels as defined by the following polynomials[8]:

$$(1 - D)(1 + D)^n, \; n = 1, 2, 3, \text{ etc.} \tag{12.4}$$

The PR4 system corresponds to $n = 1$. If we set $n = 2$, the samples of the isolated pulse will be given by the operator $(1 + D)^2 = 1 + 2D + D^2$, which corresponds to $\{1,2,1\}$. This type of channel is called the "extended partial response 4" or EPR4 channel. If $n = 3$, $(1 + D)^3 = 1 + 3D + 3D^2 + D^3$, and the isolated pulse has samples $\{1,3,3,1\}$. This type of PR channel is called the E^2PR4 channel. The following table summarizes the popular PR channels used in magnetic recording:

Name	Polynomial	Isolated pulse samples	Impulse response samples
PR4	$(1 - D)(1 + D)$	$\ldots 0 \quad 1 \quad 1 \quad 0 \ldots$	$\ldots 0 \quad 1 \quad 0 -1 \quad 0 \ldots$
EPR4	$(1 - D)(1 + D)^2$	$\ldots 0 \quad 1 \quad 2 \quad 1 \quad 0 \ldots$	$\ldots 0 \quad 1 \quad 1 -1 -1 \quad 0 \ldots$
E^2PR4	$(1 - D)(1 + D)^3$	$\ldots 0 \quad 1 \quad 3 \quad 3 \quad 1 \quad 0 \ldots$	$\ldots 0 \quad 1 \quad 2 \quad 0 -2 -1 \quad 0 \ldots$

The isolated pulses in the extended PR4 channels are wider than that in PR4 channel: the EPR4 pulse extends over 3 bit periods and E^2PR4 pulse over 4 bit periods. This means that a transition in an E^2PR4 system will "interfere" with the next three transitions when they are detected. In other words, the channel density PW_{50}/T is much higher in E^2PR4 system than in PR4.

The sample values such as "2" or "3" above are somewhat arbitrary, we will normalize the maximum sample value to "1" hereafter in the text. Using this notation, an EPR4 isolated pulse has sample values $\{\ldots 0,1/2,1,1/2,0,\ldots\}$, as shown in Fig. 12.9. The EPR4 analog pulse shape is obtained with Nyquist's interpolation formula:

$$s(t) = \frac{1}{2}\frac{\sin\dfrac{\pi t}{T}}{\dfrac{\pi t}{T}} + \frac{\sin\dfrac{\pi(t - T)}{T}}{\dfrac{\pi(t - T)}{T}} + \frac{1}{2}\frac{\sin\dfrac{\pi(t - 2T)}{T}}{\dfrac{\pi(t - 2T)}{T}}. \tag{12.5}$$

The superposition of pulses in the EPR4 system results in five sampling levels: $\{-1,-1/2,0,1/2,1\}$. The current channel sample value can be found from the NRZ bit pattern using the EPR4 polynomial:

$$\begin{aligned} s_k &= (1 - D)(1 + D)^2 a_k \\ &= (1 + D -D^2 - D^3)a_k \\ &= a_k + a_{k-1} - a_{k-2} - a_{k-3}, \end{aligned} \tag{12.6}$$

which is the linear combination of the current data bit and the 3 previous data bits.

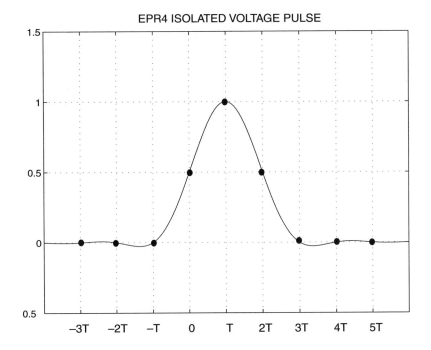

FIGURE 12.9.   Isolated pulse for EPR4 system.

The EPR4 system is often used with 8/9(0,4) or 2/3(1,7) modulation code, which results in different impulse responses. The dipulse response (impulse response) of EPR4 system with the $d = 0$ constraint can be derived as follows:

$$
\begin{array}{cccccc}
0 & \frac{1}{2} & 1 & \frac{1}{2} & 0 & \\
+ & & 0 & -\frac{1}{2} & -1 & -\frac{1}{2} & 0 \\
\hline
= & 0 & \frac{1}{2} & \frac{1}{2} & -\frac{1}{2} & -\frac{1}{2} & 0
\end{array}
$$

With the $d = 1$ constraint, the EPR4 dipulse response becomes:

$$
\begin{array}{cccccc}
0 & \frac{1}{2} & 1 & \frac{1}{2} & 0 & \\
+ & & 0 & 0 & -\frac{1}{2} & -1 & -\frac{1}{2} & 0 \\
\hline
= & 0 & \frac{1}{2} & 1 & 0 & -1 & -\frac{1}{2} & 0
\end{array}
$$

The E^2PR4 isolated pulse samples are: $\{\ldots,0,1/3,1,1,1/3,0,\ldots\}$. The shape of the analog isolated pulse is shown in Fig. 12.10. There are seven possible sample levels in an E^2PR4 system: $\{-1,-2/3,-1/3,0,1/3,2/3,1\}$ be-

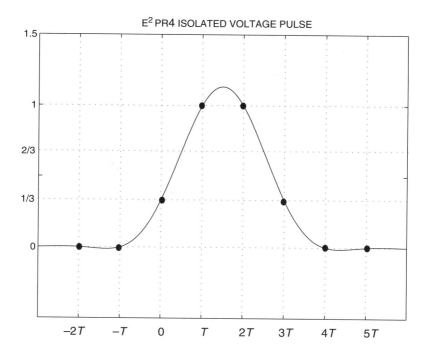

FIGURE 12.10. Isolated pulse shape for E^2PR4 system.

cause of linear superposition. An E^2PR4 system is usually used with the (1,7) encoding. The dipulse response samples for the E^2PR4 system with the (1,7) code is given by:

$$
\begin{array}{cccccccc}
0 & 1/3 & 1 & 1 & 1/3 & 0 & 0 \\
+ & & 0 & 0 & -1/3 & -1 & -1 & -1/3 & 0 \\
\hline
= & & 0 & 1/3 & 1 & 2/3 & -2/3 & -1 & -1/3 & 0
\end{array}
$$

The eye diagrams and sample value distributions of EPR4 and E^2PR4 systems are more complicated than those in PR4 channel. For example, Figure 12.11 presents the eye diagram for the EPR4 system. There are five distinct "focusing" points and the EPR4 histogram will consist of five separate peaks. Similarly, the E^2PR4 system will have seven distinct "focusing" points, as shown in Fig. 12.12, and the E^2PR4 histogram will consist of seven separate peaks.

Frequency responses of EPR4 and E^2PR4 systems are calculated similar to that of PR4 as the spectra of the dipulse responses, as shown in Fig. 12.13. Changing from PR4 to E^2PR4 with increasing order of channel

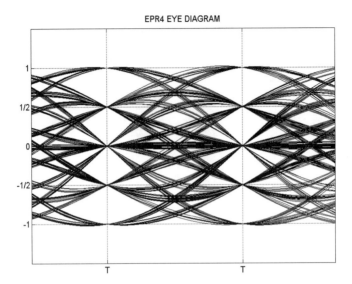

FIGURE 12.11.   EPR4 system eye diagram.

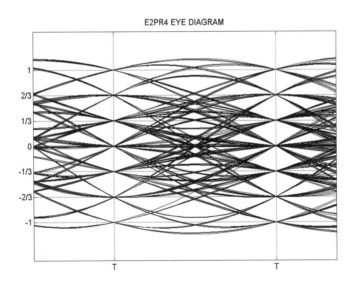

FIGURE 12.12.   Eye diagram of E^2PR4 system.

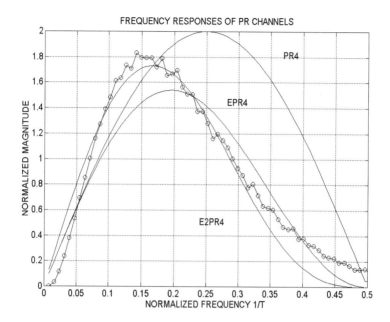

FIGURE 12.13.   Frequency responses of PR4, EPR4, and E^2PR4 systems and experimental channel frequency response for $PW_{50}/T = 2$ (circles).

polynomials, the peaks of the frequency responses shift toward the lower frequencies. This is not surprising because the isolated voltage pulses for EPR4 and E^2PR4 become wider with respect to a fixed channel bit period. Note that E^2PR4 provides the closest fit to the channel frequency response for $PW_{50}/T = 2$. This means that little equalization is required to make the recording channel an E^2PR4 system.

While the PR4 channel is most widely used now, there are certain advantages to using higher-order PR systems such as EPR4 and E^2PR4. As shown in Fig. 12.13, the channel frequency response for $PW_{50}/T = 2$ is much closer to that of EPR4 or E^2PR4 than that of PR4. As a result, equalizing the channel frequency response to PR4 will require more amplification of the high-frequency components above $0.25/T$. The "boost" of high-frequency components will inevitably amplify the noise in the system and degrade the channel error rate. If the noise is additive Gaussian noise, EPR4 equalization gains 2–3 dB in SNR over PR4. Therefore, EPR4 and E^2PR4 systems allow higher channel densities, consistent with the fact that these systems have more nonzero samples per isolated pulse.

Another advantage of the EPR4 and E²PR4 systems is that they are more practical to use with the 2/3(1,7) code than PR4 if we do not want to sacrifice user bit density. While a PR4 pulse extends over two channel periods, an E²PR4 pulse extends over four channel periods. Calculations show that the E²PR4 system with the (1,7) code allows almost the same user density as the PR4 system with the 8/9(0,4/4) code (see the next section). Using the (1,7) code doubles the minimum distance between adjacent transitions, and thus decreases nonlinear distortions such as NLTS and partial erasure.

The main disadvantage of EPR4 and E²PR4 channels is the increased complexity of the ML detector required. As we will discuss in Section 12.4, the PR4 channel allows an extremely simple realization of the ML decoder compared with the extended PRML. EPR4 and E²PR4 systems also require more sophisticated schemes for the clock and gain recovery (Section 12.3).

12.2.3 Channel and user densities of PR4, EPR4, and E²PR4 channels

The standard definition of channel density is the ratio of the PW_{50} of *unequalized* isolated pulse to the channel bit period T. However, PRML equalization modifies the shape of the isolated pulse. For example, if a significant high-frequency boost is present in the PRML channel, the equalized isolated voltage pulse will have a smaller PW_{50} than the un-equalized one.

In an ideal case, the pulse width before equalization should match that after equalization, so the equalizer does not introduce excessive noise boost and the channel density is optimal for data recovery. In this case, the channel density before equalization should be close to that after equalization. Therefore, we can take the ideal equalized isolated pulses of PR channels to calculate the channel densities as given by PW_{50}/T, assuming that PW_{50} is measured at 50% of the unipolar pulse amplitude. Using the isolated pulses shown in Figs. 12.2, 12.9, and 12.10, we obtain that

$$D_{ch} \equiv \frac{PW_{50}}{T} = \begin{cases} 1.65 & \text{for PR4,} \\ 2 & \text{for EPR4,} \\ 2.31 & \text{for E}^2\text{PR4.} \end{cases}$$

For example, if PW_{50} of the isolated pulse from the magnetic head is 16.5 ns, the optimal channel bit period for a PR4 system will be closed to

$16.5/1.65 = 10$ ns. In practice, the channel density for PR4 tends to be ~2, which is ~20% higher than the optimal value of 1.65. This is possible by making a proper tradeoff between the pulse slimming with a PRML equalizer and the noise boost. The achievable bit error rate in practical PR4 channels with $D_{ch} = 2$ is usually better than 10^{-9}.

PRML channels are typically characterized in terms of user density: $D_u = D_{ch}R$, where R is the code rate. The user density represents how many user bits of information can be stored in a unit of the medium (as measured by PW_{50}). Let us compare PRML systems with the peak detection channel. Most peak detection systems use 2/3(1,7) encoding and the channel bit period is one-half of the flux change period B. According to Equation (3.17), the resolution of the channel is ~70% when $PW_{50}/B = 1$ or $PW_{50}/T_{ch} = 2$, which is acceptable. Therefore, the peak detection channel density is 2. The corresponding user density is $2 \times 2/3 = 4/3 = 1.33$. For a PR4 system with 8/9(0,4/4) encoding and a channel density of 1.65, the user density is 1.47. An E^2PR4 system with 2/3(1,7) encoding has a natural channel density of 2.31, which gives a user density of 1.54. Note that this is almost the same as that of the PR4 system with the 8/9 code. However, since the $d = 1$ constraint increases the actual distance between transitions on the medium, nonlinear distortions are reduced. The following table provides a comparison of user densities for different detection methods:

Method	User Density 2/3(1,7) code	User Density 8/9(0,4/4) code
Peak detection (70% resolution)	1.33	0.89
Peak detection (80% resolution)	1.00	0.67
PR4	1.10	1.47
EPR4	1.33	1.77
E^2PR4	1.54	2.05

The comparison of user densities has limited values in terms of judging the relative merit of these channels. More reliable comparison should take into consideration their sensitivity to noise, nonlinear distortions, equalization, etc. Typically about 30% gain in user density can be achieved using PRML instead of peak detection.

12.2.4 Principles of equalization

The goal of equalization is to modify the frequency response of a magnetic recording channel so as to match it with the frequency response of a desired PRML scheme. Consider channel spectra shown in Fig. 12.14. The

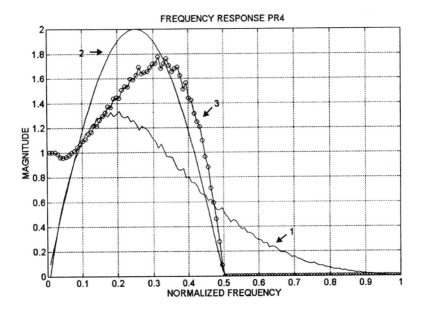

FIGURE 12.14.   Channel spectrum (curve 1), and the frequency responses of PR4 channel (curve 2) and equalizer (curve 3).

unequalized channel spectrum (curve 1) must be transformed with an equalizer so as to match the desired PR4 frequency response (curve 2). This transformation is performed with a linear filter with the frequency response given by curve 3.

The function of an equalizer can also be interpreted as that it transforms (reshapes) the sources pulse $s(t)$ into the target pulse $p(t)$.[5,7] Let $S(\omega)$ be the spectrum (Fourier transform) of $s(t)$, $P(\omega)$ the spectrum of $p(t)$, where $\omega = 2\pi f$ is the angular frequency, then the transfer function of the reshaping filter (equalizer) is

$$H(\omega) = \frac{P(\omega)}{S(\omega)}. \tag{12.7}$$

The inverse Fourier transform of $H(\omega)$ is the impulse response $h(t)$ of the filter. In time domain, the target pulse obtained by the equalizer is the convolution of the source pulse with the impulse response $h(t)$:

$$p(t) = \int_{-\infty}^{\infty} h(\tau)s(t - \tau)d\tau. \tag{12.8}$$

There are many different approaches of designing a PRML equalizer. A simple way to realize the desired frequency response is to use a *continuous time filter* (CTF) with programmable cutoff frequency and boost. By changing these two parameters, a family of frequency response functions is obtained. A more flexible solution is based on digital *finite impulse response* (FIR) *filters* or programmable analog *transversal filters*.[7]

The impulse response of a digital FIR filter has a *finite* number of nonzero samples as the name indicates. It is based on the sample version of Equation (12.8):

$$p(k) = \sum_{m=0}^{N} h(m)s(k - m), \tag{12.9}$$

where $p(k)$, $h(k)$, and $s(k)$ are the samples of $p(t)$, $h(t)$, and $s(t)$, respectively. The digital FIR filter is implemented using delay registers (denoted by D or Z^{-1}), multipliers (denoted by triangles), and adders (denoted by Σ), as shown in Figure 12.15. Each nonzero multiplier represents a *tap* (or a branch) of the filter. The more taps in the filter, the more complex it is. The sampling theorem states that if the samples are taken with a sampling frequency that is at least twice higher than the bandwidth of the channel, then the equivalent continuous signals $p(t)$, $h(t)$, and $s(t)$ can be obtained. Therefore, the PR equalization can be achieved with a digital FIR filter with sufficient sampling frequency and number of taps. Some mixed-design implementations of FIR filters use similar architecture, but analog signal levels obtained from sample-and-hold circuits instead of digital samples.

A programmable analog transversal filter realizes Equation (12.8) based on the continuous time signals $p(t)$ and $s(t)$:

$$p(t) = \sum_{m=0}^{N} h(m\tau)s(t - m\tau), \tag{12.10}$$

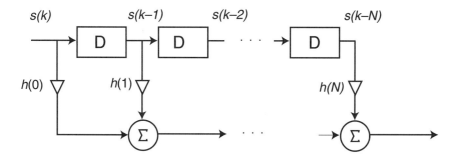

FIGURE 12.15. Structure of digital FIR filter with $N + 1$ taps.

where τ is the delay time in each delay lines, which replace the delay registers in Fig. 12.15. The architecture of the transversal filter is very similar to that of FIR filter. The source pulse $s(t)$ is taken as the input, then the output of each delay line is multiplied by some coefficient $h(m\tau)$ using an analog multiplier, and finally the outputs of all the multipliers are added together to give the target pulse $p(t)$.

To design either a digital FIR or an analog transversal equalizer properly, it is important to select the delay time and the total number of taps. In a digital FIR filter, the samples are processed at the channel clock rate. In contrast, in an analog transversal filter, the delay lines are fixed and the delay time must be $\leq 1/2f_{max}$, where f_{max} is the channel bandwidth. For example, for a PR4 signal at 100 Mflux/s, the signal bandwidth is 50 MHz, so any delay time less than 10 ns would be appropriate.

Determining the total number of taps is somewhat more complicated than choosing the delay time. The number of taps determines the total length of the impulse response of the equalizer and, therefore, the detail in the frequency response of the filter. The acceptable rule of thumb is that the total length of the delay lines should not be less than the duration of the pulse to be equalized. In most PRML channels the function of equalization is distributed between a continuous time filter at the channel front-end and a digital FIR filter. The FIR filters typically have 6–10 programmable taps.

The frequency response of the equalizer is obtained from Equation (12.3):

$$H(\omega) = \sum_{k=0}^{N} h(k)\exp(-i\omega k\tau). \tag{12.11}$$

For example, let the delay time $\tau = 5$ ns, $N = 11$ (12 taps), and the set of coefficients $h(k) = \{1\ 1\ -1\ -1\ 1\ 1\ -1\ -1\ 1\ 1\ -1\ -1\}$, then the amplitude (absolute value) of the frequency response of the equalizer is shown in Fig. 12.16. By choosing this set of coefficients we have realized a kind of "band-pass" filter at frequencies between 40 and 60 MHz.

To provide reshaping, the FIR filter coefficients must be determined from the desired frequency response. Analytically, the "reshaping" problem can be solved if the digitized waveforms of the source isolated voltage pulse and the target isolated pulse are available. Let the input and output pulses be defined by their values in the sequence of points $t_1, t_2, \ldots, t_M$, and $s_{i,j} = s(t_i - \tau_j), p_i = p(t_i)$. The quality of equalization is measured with the *mean squared error* (MSE) or distance between the equalized and target signals:

$$E(\vec{h}) = \sum_i \left(\sum_j s_{i,j} h_j - p_i \right)^2, \tag{12.12}$$

where h_j are the equalizer coefficients.

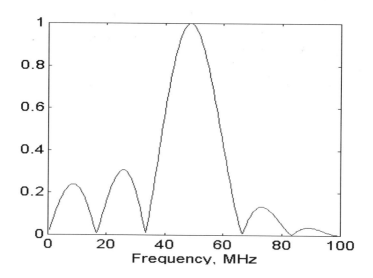

FIGURE 12.16. Example of the frequency response of an equalizer.

The best equalization is achieved when the distance is at its minimum. This occurs when

$$\frac{\partial E}{\partial h_l} = \sum_i \left[2s_{i,l} \left(\sum_j s_{i,j} h_j - p_i \right) \right] = 0, \, l = 1, \ldots, M, \qquad (12.13)$$

This results in a system of linear equations:

$$\sum_j \left[\left(\sum_i s_{i,j} s_{i,l} \right) h_j \right] = \sum_i p_i s_{i,l}, \qquad (12.14)$$

or in matrix forms:

$$\mathbf{S}^T \mathbf{S} \mathbf{H} = \mathbf{S}^T \mathbf{P}, \qquad (12.15)$$

where $\mathbf{S} = (s_{i,j})$, $\mathbf{H} = (h_i)$, $\mathbf{P} = (p_i)$. The solution to Equation (12.15) is given by the formula:

$$\mathbf{H} = (\mathbf{S}^T \mathbf{S})^{-1} \mathbf{S}^T \mathbf{P}. \qquad (12.16)$$

Therefore, if the input and output signal shapes are available, Equation (12.16) gives the set of equalizer coefficients that minimizes the MSE between the equalized and target pulses.

For example, consider the following set of coefficients:

$$h = \{0.22, -0.84, 0.12, 0.46, -0.04, 0.67, -0.36, 0.26, 0.62, 0.22, -1.0, 0.22\}$$

These coefficients were calculated so as to reshape the response of an inductive head with $PW_{50} \approx 40$ ns to the isolated pulse of the PR4 channel at 60 Mb/s. The frequency response of the resulting equalizer is shown in Figure 12.17. If the equalizer is followed by a 30-MHz-bandwidth low-pass filter, the isolated pulse shown in Fig. 12.18 (left) will be transformed into the nearly ideal PR4 isolated pulse shown in Fig. 12.18 (right).

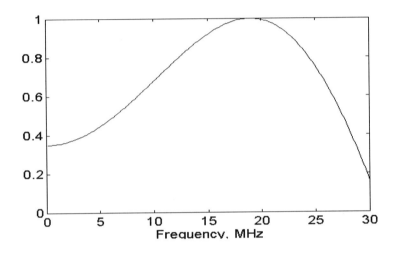

FIGURE 12.17.   Frequency response of a PR4 equalizer.

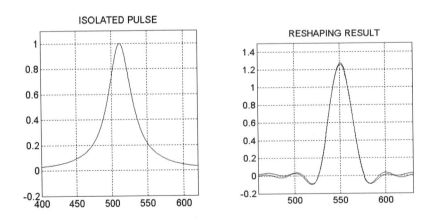

FIGURE 12.18.   Isolated pulse before (left) and after reshaping (right). The reshaped pulse is shown together with the ideal PR4 pulse.

In practice, the analytical calculation of the equalizer coefficients is time-consuming and suboptimal. Analytically calculated tap coefficients $\{h_i\}$ do not necessarily provide the best error rate of the PRML channel. The disagreement between the theory and the practice is due to the following:

1. The noise in the magnetic recording channel is correlated (e.g., medium noise), and an equalizer modifies the correlation properties of the noise. The "ideal" equalizer may create undesirable noise patterns and degrade the error rate of the ML detector. In this situation, certain "misequalization" improves the error rate performance of the PRML channel. The "optimal" equalizer allows a certain trade-off between the desirable target pulse shape and the noise boost at the equalizer output.

2. Magnetic recording channel is nonlinear. A certain residual level of nonlinear distortions is typically present in the readback signal, even after write precompensation. Therefore, the equalizer optimal for low-density readback signals (e.g., isolated pulses) becomes suboptimal for high-density signals, which often contains more nonlinear distortions. A certain trade-off between ideal equalization of isolated transitions and high-frequency patterns (dibit, tribit, and series of transitions) appears to improve the error rate of the PRML channels.

In practice, the analytically calculated equalizer coefficients are used only as the initial approximation, they are adjusted by writing a typical data pattern on the disk (e.g., random pattern) and using an adaptive procedure to minimize some "quality" measure.[8] The ultimate measure of channel quality is the bit error rate at the output of the ML detector, but a faster optimization algorithm can be based on minimizing the MSE between the received samples and their ideal values.

For example, an optimization strategy is based on the gradient optimization algorithm. If the set of coefficients at the kth step of the optimization procedure is

$$\mathbf{h}^k = \{h_1, h_2, \ldots, h_N\}, \tag{12.17}$$

the algorithm tweaks each tap coefficient h_i to minimize the MSE value $E(k)$. It changes h_i, checks the new value of MSE, and determines an update vector:

$$\nabla \mathbf{h}^k = \left[\frac{\partial E(k)}{\partial h_1}, \frac{\partial E(k)}{\partial h_2}, \ldots, \frac{\partial E(k)}{\partial h_N} \right], \tag{12.18}$$

for which MSE is decreasing. The updated set of tap coefficients is calculated as

$$\mathbf{h}^{k+1} = \mathbf{h}^k + \alpha\nabla\mathbf{h}^k. \tag{12.19}$$

The procedure is repeated until the MSE value is minimized, as schematically illustrated in Fig. 12.19.

12.3 CLOCK AND GAIN RECOVERY

The correct operation of any PRML system depends on taking samples of the readback signal in the exact "focus" positions of the eye diagram. As shown in Fig. 12.5, shifting the clock slightly from the correct position is enough to distort sample values substantially. Therefore, the correct clock signal must be recovered from the data signal.

The clock signal in a data detection channel is generated by a voltage-controlled oscillator. A clock recovery circuit, based on a phase-locked loop (PLL), adjusts the phase of the oscillator depending on the value of the phase error, as shown in Fig. 12.20. Input signal (i.e., readback data signal) is supplied to the phase detector, which calculates the phase error function by comparing the input signal with the output of the voltage-controlled oscillator. If the calculated phase error equals zero, the clock position is correct, and the oscillator frequency (and phase) stays exactly in the correct position. When the phase of the input signal and the phase of the oscillator diverge due to some reason, such as the instability of disk rotation, the phase error signal deviates from zero and the frequency of the oscillator is shifted to cancel out the phase error.

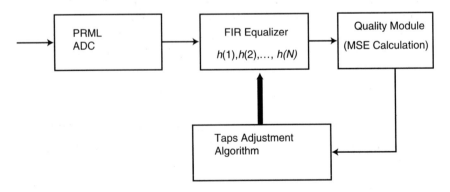

FIGURE 12.19.   Block diagram of adaptive adjustment of equalizer taps.

Input
Signal

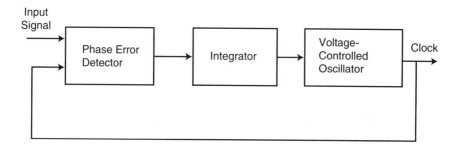

FIGURE 12.20.   Typical structure of phase-locked loop for clock recovery.

The clock recovery system solves two main problems. First, prior to reading the pattern, it is necessary to align the phase of the channel clock to the correct position of the sampling. This is referred to as the *initial phase acquisition*. The second problem is *data tracking*, i.e., following the relatively slow instabilities of the disk rotational speed. To avoid fast noisy phase shifts and to provide the stability of clock recovery, the phase error signal is passed through an integrator, which is a low-pass filter. Note that the meaning of data tracking here is different from that of tracking servo to be discussed in Chapter 15.

For the initial phase acquisition, each data sector starts with a special synchronization pattern.[5,7] For the PR4 system, the synchronization pattern in NRZ form is given by {110011001100. . . }, i.e., it consists of transitions evenly separated by two channel bit periods. The ideal PRML samples of this pattern are {+1,+1,−1,−1,+1,+1,−1,−1. . . }. The readback signal from the synchronization field resembles a sine wave, which is sampled close to its peaks, as shown in Fig. 12.21 (top). If the initial sampling phase is incorrect, as shown in Fig. 12.21 (bottom), the direction and the amount of phase shift required to reach the correct sampling point can be determined based on the first several samples. In fact, the sequence of two consecutive samples $\{s(1),s(2)\}$ determines the tabulated phase update. For example, if $\{s(1) = +1, s(2) = +1\}$, the phase update equals 0. If $\{s(1) = 0, s(2) = 1.273$ (peak of the signal)}, the phase update equals $T/2$.

The correct clock phase is obtained typically within the first several bytes of the synchronization pattern, which is followed by a special address marker. When the channel recognizes the address mark, it switches itself into data tracking mode, i.e., it recognizes the end of the synchronization pattern, as shown in Fig. 12.22.

The data tracking mode requires slow updates of the channel clock to compensate the slow instabilities of disk rotation. Assume that $s_0(k)$ is

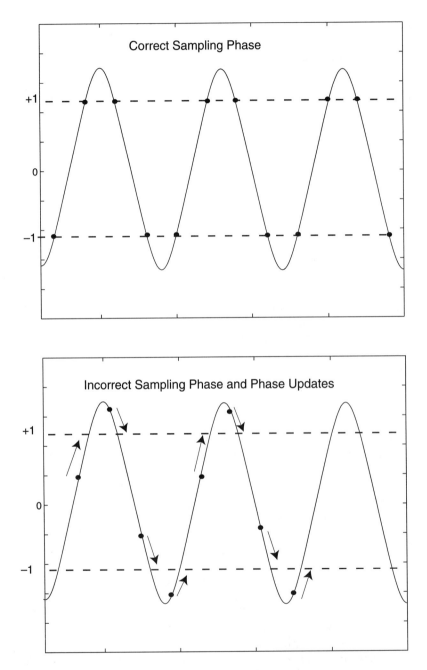

FIGURE 12.21. Synchronization pattern for PR4 channel.

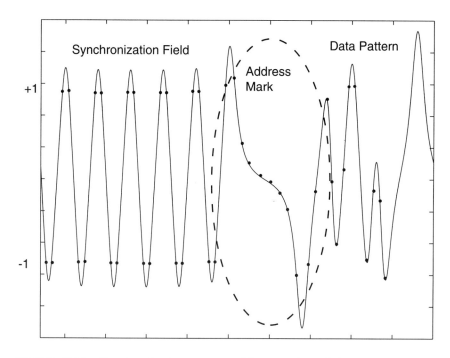

FIGURE 12.22.   Synchronization pattern, address mark, and data pattern signal.

the ideal value of the signal at the kth sampling point, and instead, some value $s(k,\tau)$ is obtained with a phase shift of τ. The resulting error is $e(k,\tau)$ $= s(k,\tau) - s_0(k)$, which depends on the value of phase shift. To minimize the error, we can use a simple gradient minimization algorithm. This can be achieved by finding the derivative of the squared error:

$$\frac{de^2(k,\ \tau)}{d\tau} = \frac{d(s(k,\ \tau) - s_0(k))^2}{d\tau} = 2e(k,\ \tau)\frac{ds(k,\ \tau)}{d\tau}. \tag{12.20}$$

If the derivative given by Equation (12.20) is known, the next phase can be calculated from the previous phase as:

$$\tau_{k+1} = \tau_k - \alpha e(k,\ \tau)\frac{ds(k,\ \tau)}{d\tau}, \tag{12.21}$$

where α is a positive convergence parameter. Equation (12.21) is reiterated and it will converge to the point of the minimum phase error.

Consider the signal and sampling points shown in Fig. 12.23, where four signal points {s1, s2, s3, s4} are sampled earlier than the correct clock positions, corresponding to ideal values {+1, 0, −1, 0}. The directions of

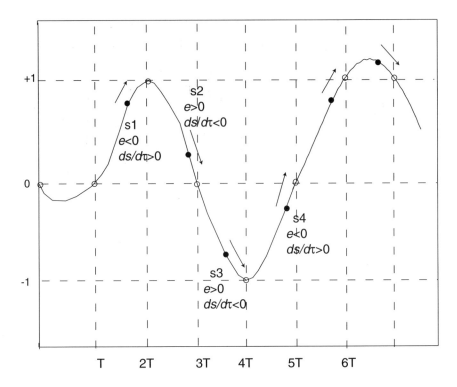

FIGURE 12.23.   Calculation of phase updates for PR4 signal.

the phase update signals will be determined from Equation (12.21). At the sample point s1, the value of the signal is less than the ideal one, so $e(k,\tau)$ < 0. However, the derivative $ds/d\tau$ at this point is positive, so the phase update signal is positive and the sample should be shifted to the right according to Equation (12.21). Similarly, for sample point s2, $e(k,\tau) > 0$, but $ds/d\tau < 0$, so the phase update signal is also positive. The same can be said about the other two sample points. Therefore, the algorithm given by Equation (12.21) correctly predicts the directions of the phase updates.

To modify the current phase based on Equation (12.21), we need to know the current error value $e(k,\tau)$ and the derivative of the signal $s(k,\tau)$. The error value can be easily calculated using the threshold-detected data as the ideal value—even if we make some errors, the overall phase error should be correct due to averaging of the individual signals by the integrator. In contrast, it is more complicated to calculate the derivative of the signal. It is usually estimated using a sequence of samples. For example, if the sequence of samples for a PR4 system is $\{+1,0,-1\}$, then the derivative in the central point is certainly negative because it is on the negative slope of a dipulse. If the sequence of samples is $\{-1,-1,+1\}$, then the derivative at the second sample point should be positive because it is on a positive slope, as shown in Fig. 12.21. A table of the expected derivatives can be constructed and used for clock recovery.

Since all the decisions about the data, samples, and clock phase updates are based on the assumption that the system amplification is correct, i.e., the "1" level is an *a priori* known value, a correct *gain control* must be provide in a PRML channel. The gain control signal is calculated from the squared error similar to that given by Equation (12.21). If a sample $s(k)$ is received instead of the ideal value $s_0(k)$, the error due to the system gain deviation is given by $e(k,\gamma) = \gamma s(k) - s_0(k)$, where γ is the system gain. The derivative of the squared error is

$$\frac{de^2(k, \gamma)}{d\gamma} = \frac{d(\gamma s(k) - s_0(k))^2}{d\gamma} = 2e(k, \gamma)s(k). \tag{12.22}$$

Therefore, the system gain should be updated according to:

$$\gamma_{k+1} = \gamma_k - \beta e(k, \gamma)s(k), \tag{12.23}$$

where β is another positive convergence parameter. The algorithm increases the system gain if the absolute value of a sample is smaller than that of the corresponding ideal value, and reduces the system gain if the opposite is true.

12.4. MAXIMUM LIKELIHOOD DETECTION

12.4.1 PRML state diagram and trellis

The maximum likelihood detection can be best understood based on the concept of *state diagram*, which describes all the possible states of the magnetic recording system and the transitions between these states.[2,3] The state diagram consists of two distinct parts: *states and transitions*. A state represents a unique physical situation at a given time, which may be specified by the current medium magnetization, and possibly its history, i.e., the medium magnetization one or several bit periods earlier. A convenient representation of the medium magnetization is provided by the NRZ data. A transition is the "event" relating the current state and the next state. There are only two types of transitions in magnetic recording: either the medium magnetization is changed between the current and the next bit periods or it is not. Note that the transition here is more broadly defined than a "magnetic transition" first introduced in Chapter 2.

State diagram is usually represented as a graph consisting of nodes (states) and oriented links between these nodes. The links represent transitions from one node to another. For example, a simple state diagram describing a memoryless magnetic recording system, such as PR1 and peak detection channel, is shown in Fig. 12.24. The meaning of the state diagram is very simple. The system may have two states, determined by current magnetization—either "0" or "1". If the current magnetization is "1" and the next magnetization is "1", the system stays in state "1" (right-hand loop). Similarly, if the current magnetization is "0" and the next magnetization is "0", the system stays in state "0" (left-hand loop). If the current magnetization is "0" and the next magnetization is "1", the system changes its state from "0" to "1" (upper arrow). If the current magnetization is "1" and the next magnetization is "0", the system goes to state "0" (lower arrow).

The NRZ state diagram for a PR4 channel is shown in Fig. 12.25a. The difference between the peak detection and the PR4 is that in the latter the state diagram has memory or is dependent on history. A state in the peak detection channel is completely described by the current magnetization, but a state in the PR4 channel is completely described by the current magnetization as well as the previous magnetization. This is because the isolated pulse in PR4 extends over two bit-periods and the presence of the previous transition affects the current channel sample value. The state diagram shown in Fig. 12.25a has four states: medium magnetization may be either "0" or "1" at the current or previous bit period. For example, if the current state is "10", it means that the current magnetization is "0" and the previous magnetization was "1". If the next magnetization is "1", the next state becomes "01".

The important feature of this state diagram is that we have *only two possible paths leading to each state*. This is because the medium magnetization can either stay in the same direction or change its sign. For example, we can arrive at state "10" only if the previous medium magnetization is "1" and current magnetization is "0". Since there are only two possible states

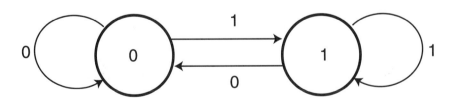

FIGURE 12.24. State diagram of a memoryless channel.

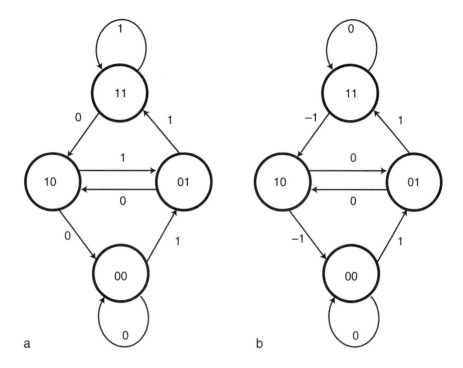

FIGURE 12.25. (a) NRZ state diagram of PR4 system. (b) PR4 state diagram with sample values.

with current magnetization being "1"—they are "11" or "01", only two paths leads to state "10".

An equivalent representation of the PR4 state diagram is shown in Fig. 12.25b, where PR4 *sample values* instead of current NRZ data are indicated along the transition paths. Each change of magnetization can generate only one sample value in the PR4 channel if we agree on the signs of magnetization. For example, change from "0" to "1" corresponds to a positive pulse, and change from "1" to "0" corresponds to a negative pulse. One has to be very careful in order not to confuse the *states* and the *samples*. First, if the magnetization between the previous state and the next state does not change for two consecutive bit periods, no magnetic transitions are generated and the samples should be equal to zero as shown by the lower and upper loops in Fig. 12.25b. If state "00" goes to state "01", the magnetization is changed from"0" to "1" and corresponds to a positive transition. This transition has samples {1,1}, so the current sample equals 1. If no magnetic transition is written next, the magnetization will change from the state "01" to state "11". This corresponds to an isolated transition and we expect to read the second sample value {1} from the isolated pulse.

To find the sample values in a state diagram of any PRML channel, we can use the corresponding PR polynomial $H(D)$. For example, the PR4 polynomial is $1 - D^2$, so the channel sample is determined by the difference of 2 NRZ bits: $s_k = a_k - a_{k-2}$. If the PR4 state changes from "11" to "10", $a_k = 0$, $a_{k-2} = 1$, thus $s_k = a_k - a_{k-2} = -1$. A *universal state diagram* can be constructed as shown in Fig. 12.26, where the number of NRZ bits to uniquely define a state is n, which is the order of polynomial $H(D)$. Therefore, PR4 and E^2PR4 channels require 2 and 4 NRZ bits, respectively. The state diagram with sample values is more useful because it visualizes how the ideal sample values in a PRML channel change with the magnetization pattern.

The equivalent representation of state diagram is the *trellis*, which is obtained by tracing the time sequence of state changes. Take the PR4 channel as an example, we first write down the four states at the current and at the next time instant. Then we trace the possible links between these states, as shown in Fig. 12.27. Typically we indicate the corresponding ideal sample value at each link. The concept of trellis is very important.

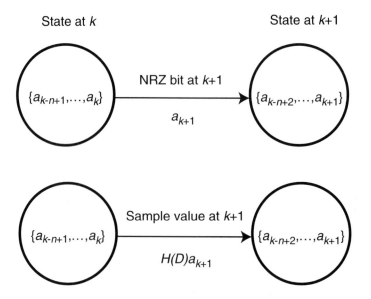

FIGURE 12.26. Universal state diagram with paths indicated by the current NRZ bits (top) or the current sample values (bottom).

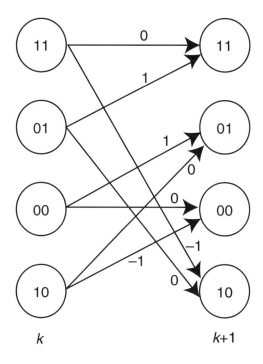

FIGURE 12.27. Trellis for PR4 system.

When the ML detector makes its decisions, it actually "extends" the trellis frame shown in Fig. 12.27 for several consecutive time instants $k, k + 1$, $k + 2$, etc., and estimates the likelihood of possible trajectories in the trellis structure.

Now consider the trellises for EPR4 and E^2PR4 channels, which are more complicated than that of PR4. In the EPR4 system we have to trace the history of three consecutive changes of magnetization which gives us $2^3 = 8$ possible states: "111", "110", "101", "100", "011", "010", "001" and "000", as shown in Figure 12.28. To find the samples corresponding to the transitions between EPR4 states, we use EPR4 polynomial (Equation 12.6). For example, when state "110" changes to state "101", the sample value is $s_k = a_k + a_{k-1} - a_{k-2} - a_{k-3} = 1 + 0 - 1 - 1 = -1$. Since we typically normalize all the samples with the maximum value, which is 2 for EPR4, so the normalized sample value is $-1/2$.

The trellis for the E^2PR4 system has $2^4 = 16$ states. However, if $d = 1$ encoding is used, the number of states in E^2PR4 will be reduced. In fact, the $d = 1$ constraint will eliminate all the states which have adjacent

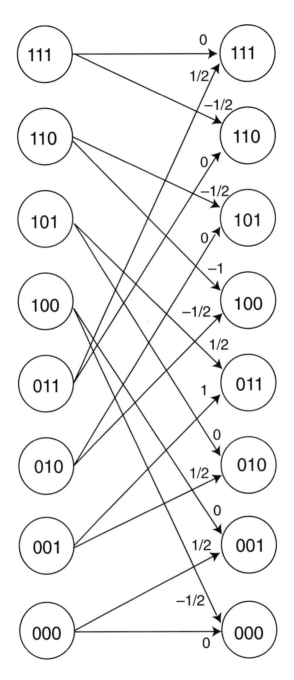

FIGURE 12.28.   Trellis for EPR4 system.

magnetic transitions. In other words, the states with NRZ data "1101", "1011", "1010", "0101", "0100", and "0010" will be eliminated. The resulting trellis is shown in Fig. 12.29, which is greatly simplified from the unconstrained case.

12.4.2 Maximum likelihood or Viterbi detection algorithm

When several transitions in a PR4 state machine are traced from state to state, the PR4 trellis is extended by adding the same diagram at each subsequent time instant. Consider the trellis extension shown in Fig. 12.30. Assume that at time instant k the exact state of the system is known to be "01". When the next channel sample comes from the ADC output, the system goes to another state. If the ideal channel sample equals 1, the system will go from state "01" at time k to state "11" at time $k + 1$. If the channel sample equals 0, the system will go from state "01" at time k to state "10" at time $k + 1$. However, the incoming channel sample is distorted by noise, so both trajectories are possible between time instants k and $k + 1$. Therefore, no decisions are taken at this point, and two states ("11" and "10") are considered as possible candidates for the system at time $k + 1$.[2,3]

Each of the possible states is assigned a certain number, or a "path metric," which corresponds to the squared difference between the channel sample $s(k + 1)$ and the ideal channel sample $s_0(k + 1)$. For example, the path metrics for states "11" and "10" at time $k + 1$ are

$$M_{k+1}("11") = [s(k + 1) - 1]^2,$$
$$M_{k+1}("10") = [s(k + 1) - 0]^2.$$

The next channel sample $s(k + 2)$ causes the system to go to a different state at time $k + 2$. As seen from Fig. 12.30, there are four alternatives at this time: state "11" may change into states "11" and "10", while state "10" may change into states "00" and "01". Therefore, all four states are to be considered at time $k + 2$. Each trajectory leading to a particular state will have an accumulated path metric. For example, the metric for state "10" at time $k + 2$ is obtained by adding the squared difference between the channel sample $s(k + 2)$ and the ideal sample (-1) to the metric $M_{k+1}("11")$:

$$M_{k+2}("10") = M_{k+1}("11") + (s(k + 2) + 1)^2.$$

When the next sample $s(k + 3)$ is received, the system is going from time instant $k + 2$ to $k + 3$. As seen from Fig. 12.30, there are eight possible

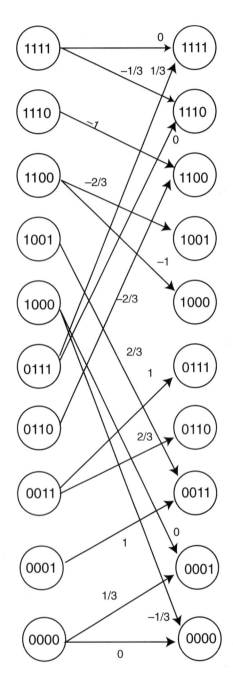

FIGURE 12.29. E^2PR4 trellis with $d = 1$ constraint.

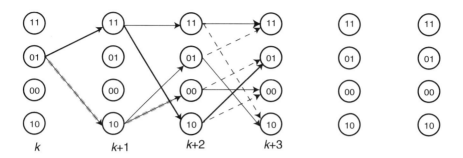

FIGURE 12.30. Candidate trajectories in PR4 trellis.

trajectories at this step, since there are four possible state at time $k + 2$. However, at time $k + 3$ two competing trajectories are converging to a state. Distance metrics (or path metrics) have been assigned the two competing trajectories, one of which should be discarded by comparing the path metrics. For example, two trajectories converging to state "11" at time $k + 3$ will have the following metrics:

$$M(1)_{k+3}("11") = M_{k+2}("11") + (s(k + 3) - 0)^2,$$
$$M(2)_{k+3}("11") = M_{k+2}("01") + (s(k + 3) - 1)^2.$$

If $M(1)_{k+3}("11") < M(2)_{k+3}("11")$, then the trajectory "01"–"11", shown in Fig. 12.30 with a dashed line, will be discarded. In this case, the accumulated path metric at time $k + 3$ for state "11" will be equal to $M_{k+3}("11") = M_{k+2}("11") + (s(k + 3) - 0)^2$. Note that after four of the eight competing trajectories are discarded, only four possible paths are continued through the trellis.

The Viterbi algorithm is based on the fact that only one surviving trajectory will remain after several consecutive steps through the trellis. The process of discarding erroneous paths from the trellis is illustrated in Fig. 12.31. As the next sample comes to the PRML channel, the trellis is extended to time instant $k + 4$. Similarly to Fig. 12.30, there are eight possible trajectories at this time and four of them will be discarded by comparing their path metrics. However, note that all the dashed trajectories coming from states "11" and "10" at time $k + 3$ are discarded at time $k + 4$. Therefore, all the trajectories leading to these states up to time $k + 3$ should also be discarded. This leaves us with only two surviving trajectories up to step $k + 3$: {"01", "11", "10", "01"} and {"01", "10", "00", "00"}.

The continuation of this process to time instant $k + 5$ is illustrated in Fig. 12.32. Now all the trajectories coming from state "00" at time $k + 4$

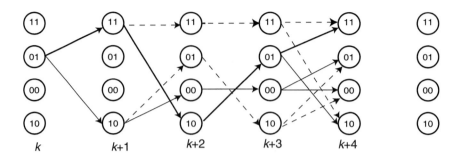

FIGURE 12.31. Extension of trellis shown in Fig. 12.30 to time instant $k + 4$.

are to be discarded. Tracing all the paths in the trellis coming to state "00", then the trajectory {"01", "10", "00", "00"} starting at time k is eliminated. This leaves only one surviving trajectory {"01", "11", "10", "01"} up to time $k + 3$, which constitutes the solutions of the Viterbi algorithm. Now we have to start from time instant $k + 3$, and repeat the process to find the next only surviving trajectory.

The Viterbi algorithm can be summarized as follows:

1. Starting from the known state, calculate the path matric for each possible trajectory in the trellis leading to the current state.
2. If two trajectories converge to the same state, select the one with a smaller path metric and discard the other. Trace the trellis and eliminate all the discarded trajectories.
3. Continue this process until only one surviving trajectory is left at some number of steps N behind the current step k. The trajectory constitutes the output of the ML detector up to step $k - N$.

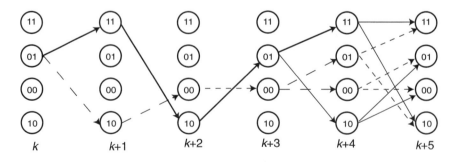

FIGURE 12.32. Extension of trellis (Fig. 12.31) to time instant $k + 5$.

It is important to understand that the ML detector can provide decisions only after some delay, i.e., after one surviving path is found. Therefore, a special "path memory" is required to trace the history of the trajectories in the trellis. In general, it is impossible to predict how fast the competing paths will converge, but the probability of disagreement decreases exponentially with the number of steps.

Note that for any PRML scheme, once a trajectory is split into competing paths, zero samples will not update the ML decisions. A sequence of zero samples will effectively continue the "parallel" paths through the trellis. Therefore, a long sequence of zero samples, or equivalently a long string of zeroes in the NRZI data pattern, will increase the delay of the Viterbi algorithm. To eliminate long strings of zeroes, RLL encoding with (0,4/4) or (1,7) codes are used.

Another problem in the ML detector is the occurrence of a catastrophic or quasi-catastrophic sequence, which creates parallel paths through the trellis, i.e., the trajectories that will never converge to the same state. These sequences exist for a periodic pattern and in general have a very low probability of occurrence. To get rid of these sequences, special scramblers (randomizers) are used.

The number of steps that should be kept in the ML memory depends on the number of states n in the trellis. Theoretically, a memory of about one hundred steps may be necessary. In practice, however, a trajectory with the minimum metrics after a reasonable number of steps is considered to be the winning trajectory. Therefore, keeping about $5(n + 1)$ steps in the ML memory is generally enough for most practical applications.

12.4.3 Interleave and sliding threshold in PR4 channel

The PR4 polynomial is $H(D) = 1 - D^2$. The current sample $s(k)$ is obtained from NRZ data $\{a_k\}$ as $s(k) = (1 - D^2)a_k = a_k - a_{k-2}$. It means that the current sample depends only on the current NRZ bit and the NRZ bit two channel periods earlier, but not on the previous NRZ bit. Since the rule holds for any time instant, we can split the sequence of samples in PR4 channel into even and odd sequences and process them independently. For each of the even or odd sequences, we obtain a greatly simplified trellis as shown in Fig. 12.33, which is often called the *interleaved* PR4 trellis.

The interleaving property of the PR4 greatly simplifies the design of PR4 systems: even and odd samples are processed simultaneously in two simple ML detectors, each running at half of the original channel data rate, and are later combined into a single NRZ data stream, as shown in Fig. 12.34.

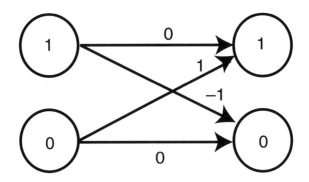

FIGURE 12.33.   Interleaved PR4 trellis, equivalent to $1 - D$ channel.

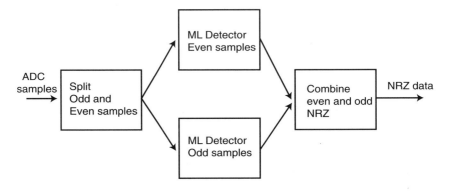

FIGURE 12.34.   Interleaving ML detection for PR4 channel.

An interesting feature of PR4 interleaving is that the ML detector for the interleaved PR4 trellis shown in Fig. 12.33 can be realized with a simple threshold-detector-type scheme. To see how the ML detection can be obtained, we need to calculate the error metrics for the interleaved PR4 trellis. The metric is calculated by squaring the difference between the channel sample $s(k)$ and the ideal sample which may cause the corresponding transition from one state to another. The table of possible transitions and the corresponding metrics of the ML detection follows:

From State	To State	Ideal Sample	Squared Error
1	1	0	$[s(k) - 0]^2 = s^2(k)$
1	0	−1	$[s(k) - (-1)]^2 = s^2(k) + 2s(k) + 1$
0	0	0	$[s(k) - 0]^2 = s^2(k)$
0	1	1	$[s(k) - 1]^2 = s^2(k) - 2s(k) + 1$

We can now perform a simple transformation. Subtract $s^2(k)$ from the error metric of every possible path and divide the resulting error by 2, we obtain the following table:

From State To State	Ideal Sample	[Squared Error $-$ $s^2(k)$]/2
1 1	0	0
1 0	-1	$s(k) + 1/2$
0 0	0	0
0 1	1	$-s(k) + 1/2$

The squared error is now reduced to the sum or difference of the sample $s(k)$ and the threshold (1/2). This is very important for the hardware implementation of the Viterbi algorithm. The actual calculation of the metrics and their updates is demonstrated in Fig. 12.35. Assume that at instant k, state "1" has an accumulated metric $M1(k)$ and state "0" $M0(k)$. When the next sample $s(k+1)$ comes, the metric is updated as follows:

$$M0(k + 1) = \min\{M0(k), M1(k) + s(k+1) + 1/2\},$$
$$M1(k + 1) = \min\{M1(k), M0(k) - s(k+1) + 1/2\}. \quad (12.24)$$

There are only three possible extensions from the current step k to the next step $k + 1$ in the trellis, as shown in Fig. 12.36. The first extension takes place when the current state is "0". In this case, both the trajectories leading from "0" to "0" and from "0" to "1" should win over the competing

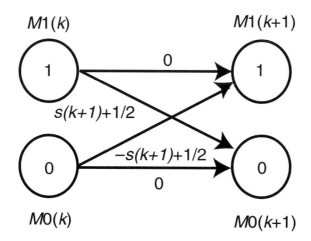

FIGURE 12.35. Metric updates for interleaved PR4 trellis.

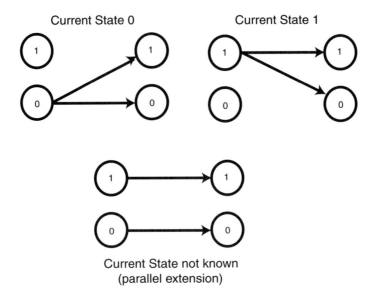

Current State 0   Current State 1

Current State not known
(parallel extension)

FIGURE 12.36.   Three possible trellis extensions in interleaved PR4 trellis.

trajectories from state "1". Writing Equation (12.24) for this case, we obtain that

$$M0(k) - s(k+1) + 1/2 < M1(k), \qquad (12.25)$$
$$M0(k) < M1(k) + s(k+1) + 1/2.$$

The inequalities (12.25) are equivalent to the following:

$$M0(k) - M1(k) < s(k+1) - 1/2. \qquad (12.26)$$

This means that it is enough to only look at the "difference metric":

$$D(k) = M0(k) - M1(k),$$

to determine whether state "0" was detected. If we now calculate the "updates" of the metrics $M0(k + 1)$ and $M1(k + 1)$ using Equation (12.24), the resulting "difference metric" at step $k + 1$ equals: $D(k + 1) = s(k+1) - 1/2$.

Similarly, we may consider the cases of the extension from state "1" and of parallel extension. In summary, we obtain the following set of simple rules for calculating the metrics and updates:

Current State = 0: $s(k+1) > D(k) + 1/2; D(k + 1) = s(k+1) - 1/2.$

$$(12.27)$$

Current State = 1: $s(k+1) < D(k) - 1/2; D(k + 1) = s(k+1) + 1/2.$

$$(12.28)$$

Current State Unknown: $D(k) - 1/2 < s(k+1) < D(k)+ 1/2;$ $\quad(12.29)$

$$D(k + 1) = D(k).$$

For example, Equation (12.27) means that if the channel sample $s(k+1)$ exceeds the current difference metric $D(k)$ by 1/2, state "0" is detected and the new difference metric is given by $s(k+1) - 1/2$.

Equations (12.27)–(12.29) allow us to realize the PR4ML channel with an extremely simple scheme called the *sliding threshold*, as shown in Fig. 12.37. Assume that $D(1) = 0$ at the first step. If the incoming sample $s(2)$

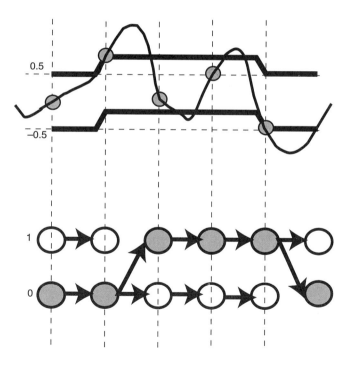

FIGURE 12.37. Realization of PR4ML detector with sliding threshold (top), and the corresponding trellis (bottom). The shaded states are the ML detector output.

> 1/2, e.g., $s(2) = 0.75$, state "0" is detected and the difference metric $D(2)$ is set to 0.25, i.e., the metric plays the role of an adjustable threshold. If the next sample is within the window from -0.25 to 0.75, the state cannot be determined, the difference metric is kept unchanged and the decision of the detector is postponed. However, once the incoming sample goes below the window, state "1" is detected and the parallel trajectory associated with state "0" in the trellis is discarded. In essence, the difference metric D is the center of a *sliding* detection window from $D - 0.5$ to $D + 0.5$. The window follows the samples, and the decisions of the ML detector are made after comparing the sample with the detection window.

Unfortunately, this simple realization of the ML detector can be used only for the interleaved PR4 system. Other types of PRML detectors cannot be simplified to such an extent, and usually require real-time metric calculation for each state (8 states for EPR4 and up to 16 states in E^2PR4).

12.4.4 Error events in maximum likelihood detection

An error occurs in the ML detector when the accumulated path metric for the wrong path through the trellis is smaller than the path metric of the correct path. While there are many possible trajectories that may diverge and converge in the trellis, only some of them are the dominant sources of errors.

When a sequence of channel samples $s_k = \{s_1, s_2, \ldots, s_N\}$ is coming to the ML detector, each of these samples equals the sum of the "ideal" noise-free sample s_k^0 and noise n_k: $s_k = s_k^0 + n_k$. Assume that two trajectories in the trellis diverge at $k = 1$ and converge back at $k = N$. The correct trajectory has noise-free samples s_k^0, while the erroneous trajectory has different sample values s_k^e. When the correct and the erroneous trajectories converge at time $k = N$, the accumulated path metrics for these two trajectories are:

$$M \text{ (Correct)} = \sum_{k=1}^{N} (s_k - s_k^0)^2 = \sum_{k=1}^{N} n_k^2$$

$$M \text{ (Wrong)} = \sum_{k=1}^{N} (s_k - s_k^e)^2 = \sum_{k=1}^{N} (s_k^0 + n_k - s_k^e)^2$$

$$= \sum_{k=1}^{N} (s_k^0 - s_k^e)^2 + 2 \sum_{k=1}^{N} ((s_k^0 - s_k^e)n_k + \sum_{k=1}^{N} n_k^2$$

The ML detector will make an error if $M(\text{Correct}) - M(\text{Wrong}) > 0$, i.e.,

$$M\,(\text{Correct}) - M\,(\text{Wrong}) = -\sum_{k=1}^{N}(s_k^0 - s_k^e)^2 - 2\sum_{k=1}^{N}(s_k^0 - s_k^e)n_k > 0. \qquad (12.30)$$

Only the second term in Equation (12.30) may become positive. Therefore, the smaller the amplitude of the first sum, the greater the probability of error of the ML detector. The first sum is called the *squared distance* between the two trajectories in the trellis.

For any type of PRML channels, certain trajectories in the trellis have a *minimum squared distance* between one another:

$$d_{\min}^2 = \sum_{k=1}^{N}(s_k^0 - s_k^e)^2. \qquad (12.31)$$

For example, the PR4 system trellis shown in Fig. 12.30 has a minimum squared distance of 2. Other sequences have squared distances equal to 4, 6, etc. A wrong trajectory or its corresponding samples sequence is usually called an "error event." The squared distance between the error event and the correct sequence of samples is called the *squared distance of the error event*.

Finding *all* the possible error events is a complicated task, which is difficult to achieve in general. However, knowing the *typical* and *dominant* error events is necessary to analyze the error performance of PRML channels. The following table describes the sequences of samples corresponding to *all* the possible transitions from a state at step k to a state at step $k + 3$ in the PR4 trellis:

To: From:	"11"	"01"	"00"	"10"
"11"	0 0 0	0 −1 0	0 −1 −1	0 0 −1
	−1 0 1	−1 −1 1	−1 −1 0	−1 0 0
"01"	1 0 0	1 −1 −1	1 −1 −1	1 0 −1
	0 0 1	0 −1 0	0 −1 0	0 0 0
"00"	1 1 0	1 0 0	1 0 −1	1 1 −1
	0 1 1	0 0 1	0 0 0	0 1 0
"10"	0 1 0	0 0 0	0 0 −1	0 1 −1
	−1 1 1	−1 0 1	−1 0 0	−1 1 0

This table is very interesting. It demonstrates that there are only two possible three-element sequences between each two states. Note that all these sequences differ by the same sequence {−1,0,1}. These are exactly the samples of a dipulse in the PR4 system. This means that a *typical* short error event in the PR4 system looks like an extra or missing dipulse and no other short error events are possible. Therefore, the ML detector in the PR4 channel is sensitive to a specific noise burst resembling a dipulse response as shown in Fig. 12.38. In comparison, a single extra pulse or two unipolar noise bursts will not confuse the ML algorithm to the same degree. In other words, a dipulselike noise burst is more likely to cause an error event in the Viterbi detector.

The theory of Viterbi detection proves that longer error events have smaller probability to occur in the trellis. The probability of the error event typically falls off almost exponentially with the length of the error event. This means that the shortest error events create the main contribution to the total error rate of the system. The dominant error events in PRML channels are thus found to be as follows:

1. For the PR4 system with 8/9 rate code, the most probable error events are {1, 0, −1}, {1, 0, 0, 0, −1}, {1, 0, 0, 0, 0, 0, −1}, etc. Of course, both polarities of dipulses are possible, e.g., {−1, 0, 1}. These error events have a squared distance of 2, which is the

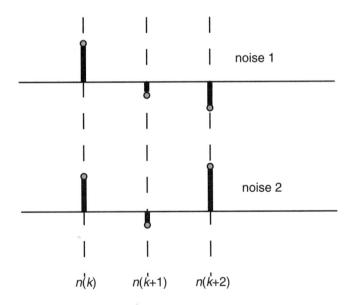

FIGURE 12.38.   Illustration of dangerous (noise 1) and "harmless" (noise 2) noise samples in PR4ML detection.

minimum squared distance. Other error events have a squared distance of 4 or higher, and their contribution to the error rate of the system is negligible.

2. For the EPR4 system with 8/9 rate code, the most probable error events are {1/2, 1/2, −1/2, −1/2}, {1/2, 1/2, 0, 0, −1/2, −1/2}, {1/2, 0, −1/2, 0, −1/2, 0, 1/2}, etc. The minimum squared distance is 1. Other error events may have a squared distance of 1.5, such as {1/2, 0, −1, 0, 1/2} and {1/2, 1/2, −1/2, 0, 1/2, −1/2, −1/2}.

3. For the E²PR4 system with 8/9 rate code, the most probable error event is {1/3, 1/3, −1/3, 0, 1/3, −1/3, −1/3} with a minimum squared distance of 6/9 = 0.67.

4. For the E²PR4 system with (1,7) code, the only minimum distance error sequence is {1/3, 2/3, 0, −2/3, −1/3}, so the minimum squared distance is 10/9 = 1.11. The next probable error event has a squared distance of 12/9 = 1.33 and is given by {1/3, 2/3, 0, −1/3, 1/3, 0, −2/3, −1/3}

The knowledge of minimum distance error events can be used to estimate the error rate of the ML detector. In the case that there is only *additive white Gaussian noise* (AWGN) in the channel, the error rate can be readily determined analytically. The error events may be caused by the noises in the samples so that the received samples are $s(k) = s_0(k) + n(k)$ where $s_0(k)$ are the ideal samples. Using the *vector representation*, we may plot the correct and wrong sequences of samples in an N-dimensional space, where N is the length of the error event, as shown in Fig. 12.39. In any real channels, we receive samples plus noise instead of ideal samples. The Gaussian noise in the N-dimensional space has an isotropic distribution and is described by a noise "sphere" in which all directions are equivalent. Therefore, any received sequence of samples is lying somewhere inside the sphere. A correct sequence will be recognized by the ML detector if the received vector is closer to the correct sequence than to the wrong sequence. According to Equation (12.30), if the distance between the correct and wrong sequences is d, the probability of an error event is simply the probability of the following event:

$$-\sum_{k=1}^{N}(s_k^0 - s_k^e)n_k > d^2/2.$$

In the case that the sample noises are AWGN, the *probability of an error event* in the ML detector is $Q(d^2/2\sigma)$, where the Q-function is defined in Chapter 11 [see Equations (11.2)-(11.5)], and σ is the standard deviation

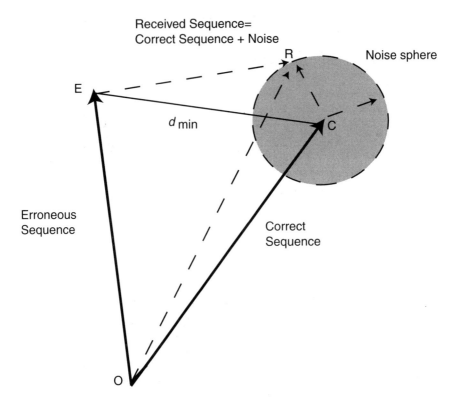

FIGURE 12.39.   Vector representation of correct/wrong sequences and noise. The ML detector makes no error if $|CR| < |ER|$.

of the noise term $-\Sigma_{k=1}^{N}(s_k^0 - s_k^e)n_k$. Note that the Q-function is a very steep function when the argument is >10.

At reasonably low error rates, the main contribution to the error rate is from the minimum distance error events because other error events have much larger distances from the correct vector and the contributions from such error events are typically several orders of magnitude smaller. Therefore, the error rate of the ML detector is determined by $Q(d_{min}^2/2\sigma)$, where d_{min} is the minimum distance in the channel trellis.

Now let us compare the error rate of the ML detector with that of a simple threshold detector. The distance between the PRML ideal samples is A, which is 1 for the PR4 system, 0.5 for the EPR4 system, and 1/3 for the E^{2}PR4 system, respectively. The threshold detector can distinguish between any two ideal samples if the sample noise is less than $A/2$. Therefore, the error rate of the threshold detector is given by $Q(A/2\sigma_s)$, where σ_s is the rms noise of each sample, which is *different* from σ! Note that, for the minimum distance error events,

$$\sigma = d_{min} \cdot \sigma_s,$$

so the ML detector error rate is

$$P_{ML} = Q\left(\frac{d_{min}^2/2}{\sigma}\right) = Q\left(\frac{d_{min}/2}{\sigma_s}\right).$$

The last expression means that if one of the sample values in the received vector exceeds half of the minimum distance, an error event will occur, which is consistent with Fig. 12.39.

For the PR4 system, $d_{min} = \sqrt{2}$, $A = 1$, so the ML detector error rate is $Q(0.7/\sigma_s)$, while that of the threshold detector is $Q(0.5/\sigma_s)$. Simply put, the PR4ML detector has an SNR gain of 3 dB over the simple threshold detector. For the EPR4 system, $d_{min} = 1$, $A = 1/2$, so the ML detector error rate is $Q(0.5/\sigma_s)$, while that of the threshold detector is $Q(0.25/\sigma_s)$. An SNR gain of 6 dB over the simple threshold detector is realized. For the E^{2}PR4 system with (1,7) encoding, $d_{min} = 10/9$, $A = 1/3$, so the ML detector error rate is $Q(0.556/\sigma_s)$, while that of the simple threshold detector is $Q(0.167/\sigma_s)$. The SNR gain in the last case is 10.5 dB.

The above results do not tell us the relative merit of the three PRML detectors, which is ultimately established by error margin analysis (see Section 12.5). Note that the value of σ_s in each case is normalized by the maximum sample amplitude just like the sample values. Consequently, for a given readback signal and noise, the σ_s value will be dependent on what PRML channel is implemented. The above results are obtained without taking noise correlation, nonlinearities, media defects, and interference signals into account. If the noise statistics is not known, the probability of errors must be estimated based on comparing the path metrics for the correct and the erroneous trajectories.

The ML detector makes an error when a wrong path through the trellis is chosen instead of a correct one. The received sequence of samples at the input of the ML detector is noisy and can be represented by $s(k) = s_0(k) + n(k)$, where $s_0(k)$ are ideal PRML sample values, and $n(k)$ are the noise samples. If an error is made during the detection process, a "wrong" sequence of samples $b(k)$ is detected. Note that an arbitrary "wrong" sequence $b(k)$ can always be represented as $b(k) = s_0(k) + m(k)A$, where $m(k)$ is the number of levels between $s_0(k)$ and $b(k)$ and A is the minimum step between ideal PRML sample values. For example, $A = 1$ for the PR4ML system. Assume that an error event has a finite length of N samples, then an error occurs in the ML detector if

$$\sum_{k=1}^{N} [s(k) - s_0(k)]^2 > \sum_{k=1}^{N} [s(k) - b(k)]^2. \tag{12.32}$$

Substituting $s(k) = s_0(k) + n(k)$ and $b(k) = s_0(k) + m(k)A$ into Equation (12.32), we obtain that

$$\sum_{k=1}^{N} [n(k)]^2 > \sum_{k=1}^{N} [n(k) - m(k)A]^2, \tag{12.33}$$

Opening the brackets in Equation (12.33), we obtain the following condition for the occurrence of an error event in the ML detector:

$$\frac{1}{\displaystyle\sum_{k=1}^{N} [m(k)]^2} \sum_{k=1}^{N} m(k)n(k) > \frac{A}{2}. \tag{12.34}$$

This means that an error occurs if a linear combination of noise samples, given by the left side of Equation (12.34), exceeds a threshold equal to $A/2$. The coefficients $m(k)$ describe how the "correct sequence" samples are different from the "wrong" sequence samples, i.e., they are the sample values of the error events. Therefore, the ML detector works like a linear noise *filter*. Remember that a typical error event for the PR4 channel with (0,4/4) code is the error sequence $m(k) = \{-1,0,1\}$, corresponding to detecting a false dipulse instead of zero signal or missing a real dipulse. It follows from Equation (12.34) that such an error occurs when $|n(k) - n(k - 2)| > 1$.

The noise correlation greatly affects the error rate of the PRML channel, and the correlated noise in multidimensional space is no longer spherical. An intuitive model of the correlated noise is shown in Fig. 12.40. The "negatively" correlated noise, corresponding to the type 1 noise in Fig. 12.38 distorts the spherical distribution into the direction of the erroneous sequence, while the "positively" correlated noise bends the noise distribution along the correct sequence of data. Obviously, the negative noise

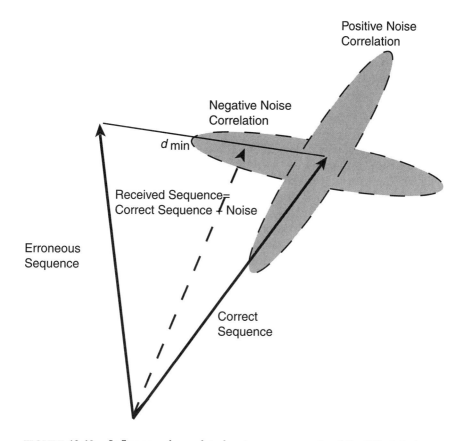

FIGURE 12.40. Influence of correlated noise on error rate of the ML detector.

correlation is not good for achieving a lower error rate and thus should be avoided.

12.5 PRML ERROR MARGIN ANALYSIS

Like any magnetic recording systems, PRML channels should provide extremely low BER (10^{-9} or better). When the system is optimized, the channel BER is low and direct error rate measurements from the output of the PRML channel are time consuming. For example, if the error rate is on the order of 10^{-8} and at least 100 errors are to be processed to obtain a reliable statistical estimate of the channel error rate, this means that at least 10^{10} bits are to be read from the channel and analyzed. Assume that the channel data rate is 100 Mbit/s, then the error rate measurement will take at least 1.5 minutes just to acquire the data.

A standard way of analyzing the error rate performance of the channel is stressing the system and checking how many errors appear at the system output. There are several ways to stress the PRML channel.[10–12] One approach is to add noise from a noise generator to the system input and measure the errors at the system output for different noise levels. The ML detector will make a large number of errors at high noise levels and the number of errors will gradually decrease when the amplitude of the external noise becomes smaller. Another way is to move the head away from the on-track position in order to increase the noise level in the readback signal. In this "off-track" test the dependence of the number of errors on the off-track position can be obtained.

While both of the described approaches are widely used, they do not fully reflect the system in normal conditions. Externally injected noise has different correlation and spectral energy distribution from head and medium noises. Off-track stressing is often used to characterize off-track system performance (Chapter 14), but it does not reflect the normal system in the on-track position. The degradation of PRML system performance is often due to NLTS, PE, and transition noises, which are different from the off-track noise and are not fully reflected in the off-track test.

A flexible way for stressing the PRML system is provided by the *sequenced amplitude margin* (SAM) method,[11] which is somewhat similar to the time margin analysis for the peak detection channel. The SAM algorithm is based on determining how close the ML detector is to making an error when it selects surviving trajectories in the trellis.

Consider a portion of the trellis for the PR4 system as shown in Fig. 12.41, where we assume that the correct data pattern is known *a priori.* There are two cases in which we can make this assumption. One is when there is an actual Viterbi detector in the system and the error rate is low so that the outputs of the Viterbi detector can be trusted. Another case is more related to a test equipment environment. In this case, the known pattern of data is written on a disk, and proper synchronization between written and read data is provided.

The correct path through the trellis is shown in Fig. 12.41 as the thick line, i.e., a transition from state "01" to state "11" has occurred, with an ideal sample value of $s_0(k+1) = 1$. In a real system, some noisy sample value $s(k+1) = s_0(k+1) + n(k+1)$ is received. At step $k+1$ we already have

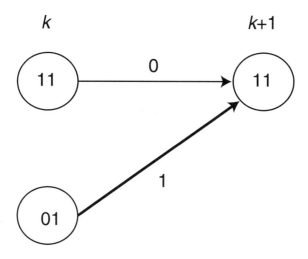

FIGURE 12.41. Part of the PR4 system trellis. The correct trajectory is shown as the thick line.

two accumulated path metrics: $M("11")$ and $M("01")$. The Viterbi detector makes a correct decision at step $k + 1$ if

$$M("01") + [s(k+1) - 1]^2 < M("11") + [s(k+1) - 0]^2,$$

or

$$B(k+1) = M("01") - M("11") + [n(k+1)]^2 - [s(k+1) - 0]^2 < 0. \quad (12.35)$$

More generally, the left side of Equation (12.35) can be rewritten as

$$B(k+1) = M("S1") - M("S2") + [s(k+1) - m1]^2 - [s(k+1) - m2]^2, \quad (12.36)$$

where $S1$ and $S2$ are the two previous states leading to the correct state, $m1$ and $m2$ are the corresponding ideal sample values, and $s(k+1)$ is the sample value.

The sample $s(k+1)$ equals $m1 + n(k+1)$. The difference between $m1$ and $m2$ is constant, at least for the standard realization of Viterbi detectors. For examples, $|m1 - m2| = 1$ for PR4, $|m1 - m2| = 1/2$ for EPR4, and $|m1 - m2| = 1/3$ for E^{2}PR4 system. Therefore,

$$B(k+1) = M("S1") - M("S2") - (m1 - m2)^2 - 2(m1 - m2)n(k+1), \quad (12.37)$$

An error will occur if $B(k+1) > 0$.

If we know path metrics for the two previous states leading to the correct state, we may easily calculate the left side of Equation (12.37).

Even if $B(k+1)$ is always less than zero, it has a distribution of values as shown in Fig. 12.42. Since the distribution (histogram) of $B(k+1)$ is known, we may integrate it from a variable threshold C to $+\infty$:

$$Q(C) = \int_C^\infty P(B)dB, \qquad (12.38)$$

where P is the probability density function (PDF) of B. If it is Gaussian with a mean value of zero, then the function is the same as the Q-function defined previously. The value of $Q(0)$ corresponds to the error rate at the nominal threshold since an error occurs if $B(k+1) > 0$. Similarly, $Q(C)$ represents the expected error rate if the decision threshold in the Viterbi algorithm is reduced to C. In this case, an error occurs if $B(k+1) > C$.

In essence, the SAM algorithm is based on obtaining a histogram of the differences of the path metrics and integrating it with a variable threshold. During the normal operation of the ML detector, the threshold is set to "0", i.e., the correct trajectory is selected when the accumulated path metric $M(\text{correct}) - M(\text{wrong}) < 0$. This nominal threshold value

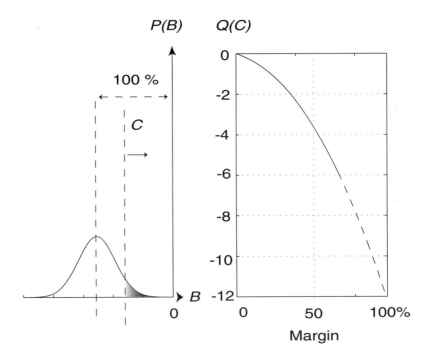

FIGURE 12.42. Illustration of the SAM plot generation. Left: distribution of values $B(k+1)$. Right: SAM plot obtained by integrating $P(B)$ with variable threshold.

corresponds to a *sequenced amplitude margin* of 100% in Fig. 12.42. Decreasing the decision threshold to C such that the correct paths are selected based on $M(\text{correct}) - M(\text{wrong}) < C$, where C is a negative number, causes the errors at the output of the ML detector to increase accordingly. A plot of the error rate vs. the decision threshold is called the *sequenced amplitude margin* (SAM) *plot*, as shown in Fig. 12.42 (right). The very small error rates are often *extrapolated* from the experimental points at the relatively large error rates to save time. The mean value of B tends to be the minimum squared distance in the trellis, so the *sequenced amplitude margin* (%) can be defined as follows:

$$\text{Margin (\%)} = \frac{C + d_{\min}^2}{d_{\min}^2}$$

Note that the definition here is different from the *operation margin*, which is understood as the "room" left to the nominal decision threshold to maintain the error rates below the specified level. If the decision threshold is $C < 0$ at the specified error rate level, then

$$\text{Operation Margin (\%)} = 1 - \text{Margin (\%)} = \frac{-C}{d_{\min}^2}.$$

For example, if the specified error rate is 10^{-10}, which is reached when the decision threshold is reduced to minus one-tenth of the minimum squared distance, then the operation margin at 10^{-10} is 10%. Obviously, a steeply descending SAM plot is desired to have a large operation margin.

A typical histogram of path metric difference is illustrated in Fig. 12.43. The minimum distance errors correspond to the main peak of the distribution. The next distance error events may create one or more additional peaks. However, the dominant source of errors in the PRML channel is generated by the minimum distance error events, so the reasonable normalization of the sequenced amplitude margin is to assign the 0% margin to the peak of this distribution, or the mean value of B, as was done previously.

Note that an equivalent SAM plot can be obtained based on Equation (12.34). If the probability density function of the left side is $p(f)$, where

$$f = \frac{\sum_{k=1}^{N} m(k)n(k)}{\sum_{k=1}^{N} [m(k)]^2},$$

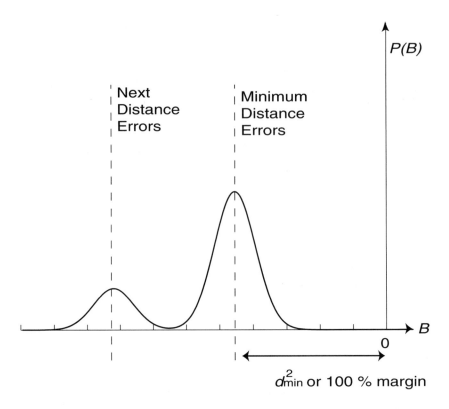

FIGURE 12.43.   Histogram of the path metric differences in the ML detector.

then the error rate of the PRML channel is

$$Q(A/2) = \int_{A/2}^{\infty} p(f)df.$$

In the case of the PR4ML channel, the 100% margin is assigned to $A = 1$, and the 0% margin is assigned to $A = 0$. Note that the standard deviation of f is not the same as that of B.

Some experimental SAM plots are shown in Fig. 12.44, where the differences between equalization result in horizontal shifts of the SAM plot. Misequalization boosts the noise in the system, and off-track position introduces additional noise into the system. Both reduce the SAM plot slope and the operation margin.

The SAM plot is a powerful tool for evaluating the performance of PRML channels. Changes in the slope and shifts of the SAM plot similar to those shown in Fig. 12.44 are important clues to determining

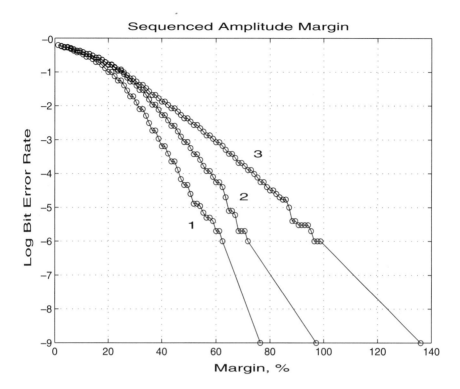

FIGURE 12.44. Examples of SAM plot: optimal equalization and on-track measurement (curve 1); misequalized signal (curve 2); off-track measurement (curve 3).

what are affecting the error rate. The SAM plot also predicts the actual BERs in the real operation conditions. For example, the SAM plots in Fig. 12.44 are calculated based on 10^6 samples and extrapolated to 10^{-9} level so that the operation margin at an error rate of 10^{-9} can be predicted.

12.6 PERFORMANCE OF PRML CHANNELS

PRML channels achieve their best performance when the readback signal is accurately equalized to the desired target shape without excessive noise amplification, and simultaneously the magnetic recording channel is maintained linear. Any deviation of the isolated pulse shape and any nonlinear distortion in the channel will greatly degrade the error rate of the PRML channel. However, these do not translate into the same impact on the ML detection process. As discussed in the previous sections, the ML detector is affected by the noise correlation properties. The noises and nonlinear distortions that do not match the probable error patterns are much less critical. For example, relatively low-frequency modulation of the system gain will not affect the performance of the ML detector.[13,14]

A simplified analysis of the error rate of the PRML system is based on considering two different noise sources. The first noise source is a random, irregular noise n, introduced by head, media, electronics, and clock jitter. The second noise source is "shape distortion" noise d. This distortion term describes regular deviations of the pulse shape from ideal due to misequalization, pulse asymmetry, and nonlinear distortions. These distortions are regular in the sense that they may be predicted. For examples, distortions caused by NLTS may be described as a deviation from an ideal dipulse shape; distortions caused by misequalization can be described as a deviation from a target isolated pulse.

Using the definition of the random and shape distortion noise terms, the total noise of each sample equals $n(k) + d(k)$. Equation (12.34) can now be rewritten as

$$\frac{\sum\limits_{k=1}^{N} m(k)n(k)}{\sum\limits_{k=1}^{N} [m(k)]^2} + \frac{\sum\limits_{k=1}^{N} m(k)d(k)}{\sum\limits_{k=1}^{N} [m(k)]^2} > \frac{A}{2}. \tag{12.39}$$

which is the condition of an error event. Note that the last term in Equation (12.39) represents the "correlation" of the shape distortion pattern with the error events in the ML detector. In other words, if the shape distortion "matches" the shape of the error event, this term will become large and the probability of error will increase.

Consider the PR4 system. The most probable error pattern for PR4 detection is given by $m(k) = \{1,0,-1\}$ or $\{-1,0,1\}$. Therefore, NLTS and PE are extremely dangerous because they match the error pattern. In fact, the reduction of dipulse amplitude by any means is "matched" with the most probable error pattern. Similarly, any misequalization that results in different signs of distortion two bit periods apart is also critical for a PR4 system.

The left side of Equation (12.39) consists of two terms, each being the linear combination of the corresponding noise component with the coefficients of the error sequence $m(k)$. Let the *probability density function* (PDF) of the first term be $p(f)$ and that of the second term be $g(f)$, and assume that the two noise terms are statistically independent, then the PDF of the left side of Equation (12.39) is given by the convolution of both distributions:

$$P(f) = \int_{-\infty}^{\infty} p(f - x)g(x)dx. \tag{12.40}$$

Therefore, the error rate of the PRML channel is

$$G(A/2) = \int_{A/2}^{\infty} P(f)df = \int_{-\infty}^{\infty} Q(A/2 - x)g(x)dx, \tag{12.41}$$

where

$$Q(A/2) = \int_{A/2}^{\infty} p(f)df,$$

Note that A is the distance between the ideal PRML sample values. There-fore, *the total error rate $G(A/2)$ of a PRML system is the convolution of the error rate function $Q(A/2)$ of a linear and ideally equalized system with the probability density function of the distortion term.* The latter includes the shape distortions due to nonlinearities and misequalization.[13]

Function $Q(A/2)$ represents the error rate due to the random noise term. It has a well-known bell-like (or "waterfall") shape, with its slope determined mainly by the standard deviation of the random noise and by the particular set of coefficients $m(k)$. Note that the standard deviation of the noise in the ML detector depends on the system equalization, which may alter the noise bandwidth and may boost noise at some frequencies.

In the presence of *misequalization,* which excludes the boost effect of equalization on random noise, the PRML signal samples at clock locations contain a discrete set of noise values $d(k)$, resulting from the superposition of the misequalization errors at current and neighboring pulses. For exam-ple, if an ideal isolated PR4 pulse has values {0,1,1,0} and the misequalized pulse samples are {0,0.9,1,0.1}, then the set of $d(k)$ is {±0.1,±0.2}. Any linear combination $d(k)$ in Equation (12.39) will result in additional errors. In the case of $m(k)$ = {1,0,−1}, the distribution $g(f)$ for a random data pattern can take these discrete values: {0,±0.05,±0.1,±0.15}.

The main impact of NLTS and PE is to decrease PRML sample amplitudes for adjacent transitions, which generates a set of $d(k)$ and additional discrete values in the distribution $g(f)$. Figure 12.45 demonstrates the histogram of $g(f)$ in a low-noise channel. The distribution was obtained using simulated signals with artificially introduced NLTS equal to 25% of the bit period. Note that $g(f)$ consists of a main peak at the zero location and a number of additional peaks. The most distant noticeable peak at approximately $f = \pm.33$ has an amplitude of about 10% of the main peak amplitude.

According to Equation (12.41), the convolution of the distribution of the random noise term with $g(f)$ gives the total error rate, which is approximately the superposition of many shifted distributions $Q(A/2)$ weighted with the amplitudes of the corresponding peaks. The result of the convolution is shown in Figure 12.46. Note that the resulting error rate at about 33% margin is close to 10% of the original $Q(A/2)$ value at 0% margin.

Nonlinear distortions are especially critical in PRML channels because they are matched to the probable error patterns. This is true for all PRML systems such as PR4, EPR4 and E^2PR4 because the most probable error

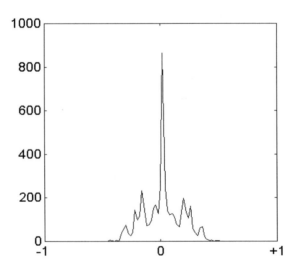

FIGURE 12.45.   Distribution $g(f)$ (not normalized) in a low-noise channel with 25% NLTS.

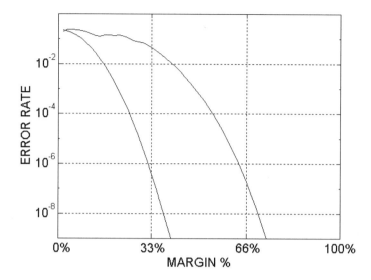

FIGURE 12.46. Convolution of $Q(f)$ (left curve) with $H(f)$ in Fig. 12.46. The result (right curve) is close to a shifted copy of $Q(f)$.

events for all of them have a dipulse shape. The minimum distance error sequences are listed in the following:

PR4:	$\{1, 0, -1\}$
EPR4:	$\{1/2, 1/2, -1/2, -1/2\}$
E²PR4:	$\{1/3, 2/3, 0, -2/3, -1/3\}$

The shape of these error events indeed matches the dipulse shape. Therefore, any nonlinear distortions, which reduce the dipulse amplitude, will increase the probability of errors in the ML detector.

A moderate amount of NLTS can be *precompensated* in the write process by intentionally delaying a transition in the presence of previous transitions. The effects of write precompensation are demonstrated in Figs. 12.47 and 12.48. The sample histograms without and with precompensation are shown in Fig. 12.47, which indicates that the separation between the sample values is greatly improved by precompensation.

The SAM plots corresponding to the histograms in Fig. 12.47 are shown in Fig. 12.48. Precompensation of NLTS greatly increases the slope of the SAM plot, just like reducing the random noise in the channel. Consequently, the operation margin of the channel increases accordingly.

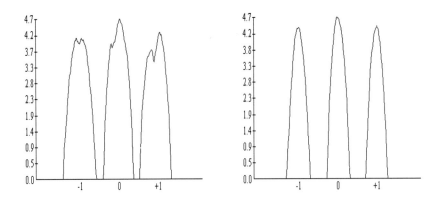

FIGURE 12.47. Sample values distributions without precompensation (left) and with precompensation (right). NLTS equals 10% of the bit period. The histograms are shown on a logarithmic scale.

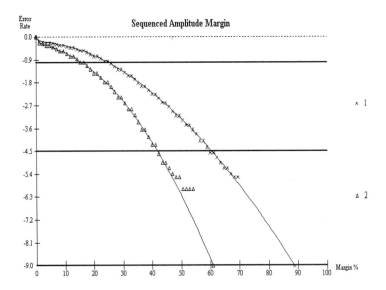

FIGURE 12.48. SAM plots for PRML channel without precompensation (right) and with precompensation of NLTS (left).

At high recording densities when NLTS approaches 20% of the bit period, the situation becomes more complicated. If all the transitions in a random pattern are precompensated the same amount that is optimal for a dipulse, the error rate performance is *not* optimized, as illustrated in Fig. 12.49. If a random data pattern is written on the disk and all the transitions in the pattern are precompensated using the same precompensation parameter that is optimal for a dibit pattern, then curve 1 is generated. Apparently, this strategy does not work well because it may *overcompensate* many transitions. If the first, second, and third transitions in a group of adjacent transitions are precompensated with individually optimized parameters and the fourth and further transitions are assigned a precompensation parameter equal to that of the third transition, the SAM plot is improved significantly (curve 3). In contrast, for a pattern consisting of isolated transitions and dibits, i.e., without three or more transitions adjacent to one another, precompensating the second transition in each dibit produces the best SAM plot (curve 2) among the three cases.

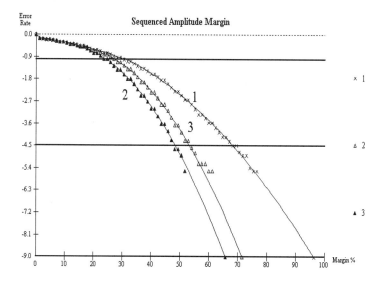

FIGURE 12.49. SAM plots with different precompensation strategies: (1) random pattern, same precompensation for all the transitions; (2) isolated dibit pattern, same precompensation; (3) random pattern, precompensation for the first, second, and third transitions individually.

The adjacent transition interactions are very important when considering NLTS and write precompensation.

As discussed in Chapter 10, the degree of interactions between transitions in a random pattern is determined by the head/medium combination. Another important nonlinear distortion is the interaction between old information and new information, which is called the hard transition shift (HTS) and will also affect the sample values like NLTS does. Figure 12.50 demonstrates an extreme case in which the sample histogram is strongly dependent on the HTS. The HTS value for this particular head and medium combination is about 12% of the bit period, while the NLTS value is close to 25% of the bit period. When the medium is effectively AC-erased and NLTS is precompensated, the channel samples are concentrated at their ideal values. However, when the medium is DC-erased and rewritten, the resulting interactions between NLTS and HTS (Chapter 10) cause strong distortions of the distributions, as shown in Fig. 12.50 (right).

The related SAM plots are shown in Fig. 12.51. In the case where NLTS is not precompensated, the operation margin is the smallest (curve 1). When the pattern is written with optimal precompensation on an AC-erased medium, the operation margin improves greatly (curve 2). If the same is done on a DC-erased medium, however, the SAM plot shifts to the right (curve 3). To compensate the influence of the HTS on the data pattern, the polarity of each transition is taken into account and the value of the HTS is subtracted from the precompensation parameter of each hard transition. The procedure shifts the SAM plot to the left and improves the operation margin (curve 4). Unfortunately, the HTS shift cannot be

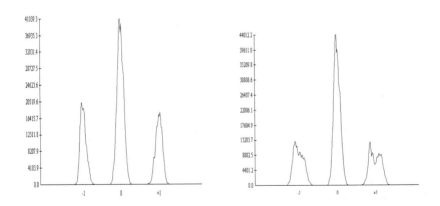

FIGURE 12.50. Sample histograms with NLTS precompensation for AC-erased medium (left) and for positive DC-erased medium (right).

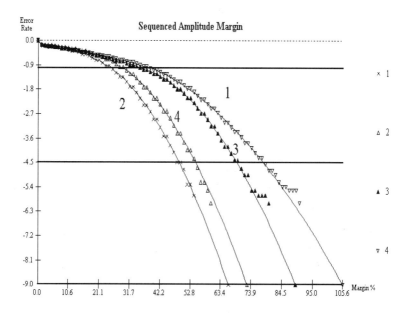

FIGURE 12.51. SAM plots for a PR4 system: (1) without NLTS precompensation; (2) with AC-erased medium and NLTS precompensation; (3) with DC-erased medium and NLTS precompensation; (4) with DC-erased medium and precompensation of both NLTS and HTS.

precompensated because the old information pattern is not known in a real detection channel.

As mentioned previously, equalization affects the error rate of the PRML channel by modifying the system SNR and introducing specific shape distortions if equalization is not ideal. System SNR considerations are very important. To match the frequency response of the channel to the PRML frequency response, a boost in certain frequencies is required of equalization. Theoretical calculations for Gaussian noise and the optimal density range predict that the SNR loss can be 2–3 dB for PR4, and 1–1.5 dB for EPR4 and E^2PR4 systems. When actual measurements are done in real recording systems, spectral distributions of medium and electronic noises are mixed and the results are usually less pessimistic for PR4 and more pessimistic for extended PRML schemes. The results may be further altered by the pulse shape, which can be asymmetrical and very different from the Lorenzian pulse. Therefore, the actual degradation of SNR caused by equalization is almost impossible to predict, and should be measured experimentally.

Shape distortions introduced by an equalizer can be destructive or beneficial. Some of the shape distortions are more destructive than the others. For example, if the isolated PR4 pulses samples are $\{\ldots, 0, 1.1, 0.9, 0, \ldots\}$, the dipulse samples are $\{\ldots, 0, 1.1, -0.2, -0.9, 0, \ldots\}$; if the isolated pulse sample are $\{\ldots, 0, 0.9, 1, 0.1, 0, \ldots\}$, then the dipulse samples are $\{\ldots, 0, 0.9, 0.1, -0.9, -0.1, 0, \ldots\}$. The first dipulse will cause less PRML errors because its first sample is pushed up to 1.1. In contrast, the second dipulse will be more subject to errors because its first sample is pushed down to 0.9 and the third sample is -0.9. Like random noises and nonlinear distortions, pulse shape distortions due to misequalization will be more prone to errors if they reduce the dipulse amplitude.

A real recording channel is somewhat nonlinear, so the optimal equalization is actually pattern dependent. Consider the SAM plots shown in Fig. 12.52. At first, an isolated pulse is captured and the equalizer coefficients are calculated so as to reshape it to the targeted isolated pulse. Next, the equalizer thus obtained is applied to a data pattern consisting of isolated transitions. Finally a SAM plot (curve 1) is generated. Obvi-

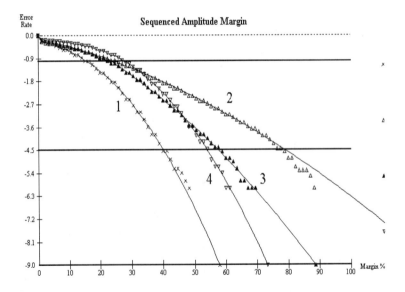

FIGURE 12.52. SAM plots for PR4ML channel: (1) optimal equalization for isolated transitions, pattern consisting of isolated transitions; (2) optimal equalization for isolated transitions, random pattern with NLTS precompensation; (3) equalization adjusted for random pattern, random pattern with NLTS precompensation; (4) equalization adjusted for random pattern, pattern consisting of isolated transitions.

ously, the equalizer works well because the operation margin is large. However, if the same equalization is applied to a random pattern with the optimal write precompensation, a much worse SAM plot (curve 2) is generated. If the equalizer is adjusted for the random pattern with NLTS precompensation so as to optimize the system error rate, an improved SAM plot (curve 3) is obtained, but this equalizer is not optimal for the pattern consisting of isolated transitions (curve 4). In short, different data patterns have different optimal equalizations, which must be carefully considered in the design and characterization of PRML channels.

References

1. H. Kobayashi and D. T. Tang, "Application of partial response channel coding to magnetic recording system," *IBM J. Res. Develop.*, July 1970, p. 368; R. W. Wood and D. A. Petersen, "Viterbi detection of class IV partial response on a magnetic recording channel," *IEEE Trans. Commun.*, **34**, 454, 1986.

2. E. R. Kretzmer, "Generalization of a technique for binary data communication," *IEEE Trans. Commun. Technol.*, **14**, 67, 1966; G. D. Forney, Jr., "The Viterbi algorithm," *Proc. IEEE*, **61**(3), 268, 1973.

3. B. Sklar, *Digital Communications: Fundamentals and Applications.* (Englewood Cliffs, NJ: Prentice Hall, 1988).

4. P. H. Siegel and J. K. Wolf, "Modulation coding for information storage," *IEEE Communications Magazine*, December 1991, (12), 68.

5. R. D. Cidecyan, F. Dolvio, R. Hermann, W. Hirt, and W. Schott, "A PRML system for digital magnetic recording," *IEEE Journal on Selected Areas in Communications*, **10**(1), 38, 1992.

6. T. Howell, *et al.*, "Error rate performance of experimental gigabit per square inch recording components," *IEEE Trans. Magn.*, **26**(5), 2298, 1990.

7. J. D. Cocker, R. L. Galbraith, G. J. Kerwin, J. W. Rae, and P. Ziperovich, "Implementation of PRML in a rigid disk drive," *IEEE Trans. Magn.*, **27**(6), 4538, 1991.

8. H. Thapar and A. Patel, "A class of partial response systems for increasing storage density in magnetic recording," *IEEE Trans. Magn.*, **23**(5), 3666, 1987.

9. J. Hong, R. Wood, and D. Chan, "An experimental 180 Mb/sec PRML channel for magnetic recording," *IEEE Trans. Magn.*, **27**(6), 4532, 1991.

10. A. Taratorin, "Method and apparatus for measuring error rate of magnetic recording devices having a partial response maximum likelihood data detection channel," U.S. Patent 5,355,261, 1996.

11. T. Perkins "A window margin like procedure for evaluating PRML channel performance," *IEEE Trans. Magn.*, **31**(2), 1109, 1995.

12. A. Kogan *et al.*, "Histograms of processed noise samples for measuring error rate of a PRML data detection channel," U.S. Patent 5,490,091, 1997.

13. A. Taratorin, "Margin evaluation of PRML channels: non-linear distortions, misequalization and off-track noise performance," *IEEE Trans, Magn.*, **31**(6), 3064, 1995.

14. P. Ziperovich, "Performance degradation of PRML channel due to nonlinear distortions," *IEEE Trans. Magn.*, **27**, 4825, 1991.

CHAPTER 13

Decision Feedback Channels

As the channel density (the ratio of PW_{50} to the channel bit period) exceeds 1.5, the inter-symbol interference (ISI) becomes a critical factor: the readback voltage pulses from adjacent transitions overlap, and the transitions can no longer be detected reliably with peak detection channels. The PRML channels cancel ISI by equalizing the input readback to a target waveform so that higher channel densities become possible. The PR4ML channel density is approximately 2.0, and the extended PRML channels (EPR4 and E^2PR4) are achieving channel densities of around 2.5. However, the extended PRML channels are somewhat complicated, and require sophisticated ML detector, equalizers, and accurate timing and gain recovery circuits.

To achieve even higher channel densities (>2.5) in PRML channels, it may require either excessive boosting of noise in the equalizer, or unacceptable hardware complexity and power consumption. Therefore, there are considerable interests in alternative data detection channels. Decision feedback equalization (DFE) channels are designed for channel densities above 2.5 and allow simpler channel architecture than PRML channels. The DFE channels are relatively new, so very few hardware implementations are available and most estimates of DFE performance are obtained by modeling.

13.1 PRINCIPLE OF DECISION FEEDBACK

The main idea of decision feedback is to cancel ISI using a feedback circuit. This approach is different from partial response channels, where the input readback pulse is equalized to a target waveform.

The signal from the isolated transition appears before the head actually travels over the transition center, as shown in Fig. 13.1. This part of

433

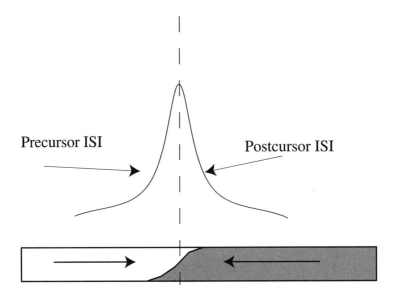

FIGURE 13.1.   Precursor and postcursor ISI.

signal is called a *precursor* ISI if we think of a transition as a "cursor."[1,2] The tail of the readback signal extending after the transition location is referred to as a *postcursor* ISI. It is important to realize that precursor ISI is *noncausal* because the signal appears *before* the actual transition. Postcursor ISI is *causal* because the signal appears *after* the transition.

To cancel ISI with a feedback circuit, the ISI should be causal. A precursor ISI occurs before we have the knowledge of the cursor, so we cannot cancel it with a feedback circuit. Therefore, the first step in decision-feedback (DFE) channels is to minimize the precursor ISI and to "translate" the pulse energy into the postcursor time domain. This operation can be done with a simple linear filter. For example, subtract the delayed voltage pulse $p(t - \tau)$ by the attenuated original pulse $\alpha p(t)$, resulting in $-\alpha p(t) + p(t - \tau)$, where $\alpha < 1$, as shown in Fig. 13.2.

Once the precursor ISI is removed, the postcursor ISI can in principle be removed with a feedback loop, if the presence of the transition is somehow detected. The main concept of the decision feedback equalization channels is to remove the precursor ISI and the postcursor ISI separately while detecting the transitions,[1,3] as illustrated in Fig. 13.3. The channel has two different equalizers: the forward equalizer that removes precursor ISI, and the feedback equalizer that removes the postcursor ISI. Once a signal comes to the feedback loop, the presence or absence of a

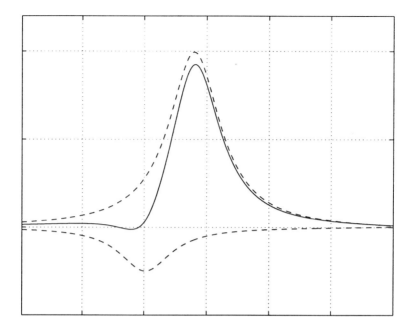

FIGURE 13.2. Example of equalization in decision feedback channel, $PW_{50}/T = 2.5$, $\alpha = 0.25$, $\tau = 2T$. where T is channel bit period.

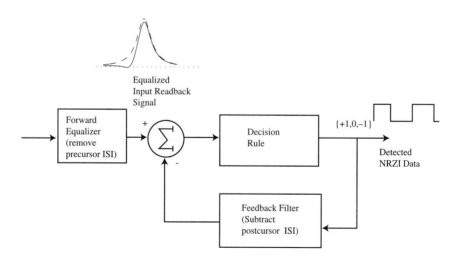

FIGURE 13.3. Block diagram of decision feedback channel.

transition has to be determined with a certain decision rule. If a transition is detected correctly, a binary "1" is fed into the feedback filter, which has a step response similar to the postcursor tail of the isolated voltage pulse shown in Fig. 13.2, and the postcursor ISI is subtracted from the incoming input signal. The subtraction may affect several channel bits depending on the length of the postcursor tail. If no transition is detected, a binary "0" is fed into the feedback filter, and the incoming signal is unaltered. It is worthy to note that the complexity of the feedback filter increases *linearly* with the length of the postcursor tail. In contrast, the complexity of PRML channels increases *exponentially* with the length of the ISI tail.

   One of the main advantages of the decision feedback detection is that the forward equalizer in DFE channels has less amplification that the PRML equalizer. For example, Fig. 13.4 illustrates the typical frequency responses of the DFE and PR4 equalizers at a channel density of 2.5. The DFE equalizer has a nearly flat frequency response compared to PR4 equalizer. Therefore, the DFE channel has a much less noise boost and

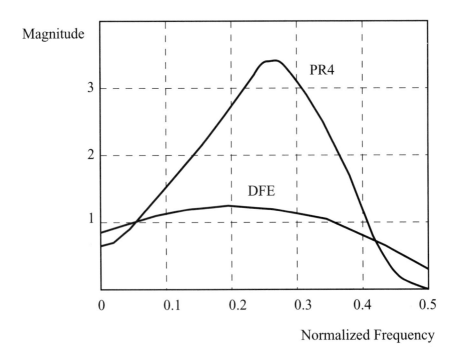

FIGURE 13.4. Frequency responses of PR4 and DFE forward equalizers at a channel density of 2.5.

provides an advantage in achievable SNR, at least for the case of additive Gaussian white noise. The typical SNR gain is 2–3 dB at a channel density of 2.5.

While the basic idea of DFE channel is straightforward, its efficient implementation of the channel is not that simple. The detection algorithm (decision rule) must operate in real time: the presence of absence of a transition must be decided immediately in order to compensate the post-cursor ISI. The detector of transitions is typically reduced to a simple slicer, i.e., the presence or absence of a transition is decided by comparing the signal value with a threshold, similar to the threshold detector in peak detection channels. Unfortunately, the slicer outputs are less reliable than the decisions of the ML detector. More seriously, a single bit error in the detected data will cause several output bits after the error event to be in error. If the slicer makes an error, the feedback loop will subtract a nonexistent postcursor ISI tail from the incoming signal, thus making the next decision more likely another error. Likewise, the second error may propagate to the third bit. In other words, a single error bit may result in a burst of error bits. This is often called the *error propagation* problem of DFE channels.

Based on theoretical modelling results, the overall performance of standard DFE channels is either equivalent to or marginally better than PRML channels. This is because the anticipated SNR gains are offset by the losses caused by error propagation in the feedback loop.[4-6] However, if DFE channels are used with $d = 1$ constraint modulation codes, for example, the (1,7) code, then an efficient channel implantation can be achieved. The DFE channel for the (1,7) code is also referred to as the multi-level DFE (MDFE), or DFE17 channels.[3,4]

13.2 FIXED-DEPTH TREE SEARCH WITH DECISION FEEDBACK AND (1,7) CODE

The basic idea of MDFE channel derives from a detection channel called the *fixed-depth tree search with decision feedback* (FDTS/DF) channel, first proposed by Moon and Carley.[3-5] The FDTS detector is quasi-optimal detector that replaces the slicer detector in the DFE channel as shown in Fig. 13.3 and makes the decisions more reliable. It is essentially a truncated version of the maximum likelihood (ML) (Viterbi) detector. The truncated search in the trellis achieves quasi-optimal performance of the ML detector, but with less complexity. To understand the principle of FDTS/DF, consider a block diagram of the discrete FDTS/DF channel shown in Fig. 13.5. The

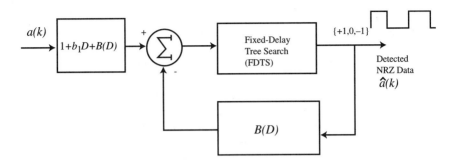

FIGURE 13.5.   Discrete-time FDTS/DF channel.

channel has a forward equalizer with a special transfer function, which simplifies the final derivation of the FDTS detection. This particular transfer function, however, does not present a limitation to the final algorithm.

The digital samples of the input analog readback waveform are obtained and then sent to the discrete time FDTS/DF channel. The samples of the input signal are equalized to a polynomial $1 + b_1D + B(D)$, where D is the delay operator (Section 12.2), and $B(D)$ is some polynomial in the form of $B(D) = b_2D^2 + b_3D^3 + \ldots$. This means the output of forward equalizer for the incoming NRZ pattern (a_k) can be calculated as:

$$s_k = [1 + b_1D + B(D)]a_k.$$

Similar to PRML channels, the impulse response of the forward filter equals to the samples of a dibit: $\{1,b_1,b_2, \ldots \}$. The polynomial $B(D)$ corresponds to the causal postcursor ISI, and is removed with the feedback filter being $B(D)$. In other words, the ISI from the samples starting from b_2 are removed, but the ISI from the first two samples is not removed in this implementation.

As mentioned previously, FDTS is a truncation of the Viterbi algorithm. For example, Fig. 13.6 demonstrates the tree search with a depth of 2. Assume that at time $k - 1$ the detected NRZ bit equals to $+1$, then the detected NRZ bit at time instant k can be either $+1$ or -1, corresponding to positive magnetization or negative magnetization. Similarly, at the next time instant, each of the two trajectories split into two alternative trajectories, resulting in a total of four sequences of states $\{a_{k-1},a_k,a_{k+1}\}$:

$$\{+1,+1,+1\},\{+1,+1,-1\},\{+1,-1,+1\}, \text{ and } \{+1,-1,-1\}.$$

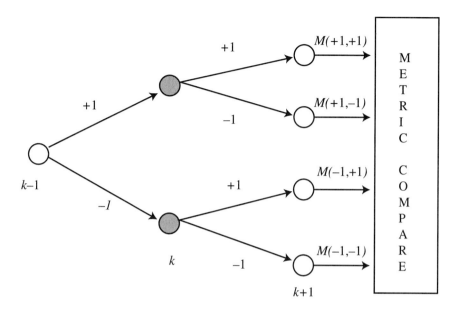

FIGURE 13.6.   Example of FDTS with a depth of 2.

For each of these states, the total path *metric* is calculated similarly to the Viterbi algorithm (Section 12.4). If s_k is the detected channel sample value, and s_k^0 is the ideal channel sample, the path metric is calculated as

$$M(a_k, a_{k+1}) = (s_k - s_k^0)^2 + (s_{k+1} - s_{k+1}^0)^2. \qquad (13.1)$$

However, the ideal channel samples s_k^0 are equal to the uncompensated residual of the ISI as given by the terms $(1 + b_1 D)a_k = a_k + b_1 a_{k-1}$. Therefore,

$$M(a_k, a_{k+1}) = [s_k - (a_k + b_1 a_{k-1})]^2 + [s_{k+1} - (a_{k+1} \qquad (13.2)$$
$$+ b_1 a_k)]^2.$$

Once all four path metrics $M(a_k, a_{k+1})$ are calculated at time $k + 1$, the smallest one is selected, and the corresponding decision bit a_k appears at the FDTS output. Note that the decision about bit a_k is done one channel clock later. This is why only the residual ISI starting from the b_2 sample is compensated in the feedback loop.

The above implementation of FDTS/DF allows considerable simplification if the $d = 1$ coding constraint is used. In this case the third path in Fig. 13.6 is eliminated, since it violates the $d = 1$ constraint: the sequence of NRZ states $\{+1,-1,+1\}$ has two adjacent transitions. Therefore, only the path metrics of the remaining three states are to be calculated. The ideal samples for these paths are listed below:

Path	Samples
$\{+1,+1,+1\}$	$\{1 + b_1, 1 + b_1\},$
$\{+1,+1,-1\}$	$\{1 + b_1, -1 + b_1\},$
$\{+1,-1,-1\}$	$\{-1 + b_1, -1 - b_1\},$

The ideal samples can be represented as the points in a two-dimensional *decision space* (s_k, s_{k+1}), as shown in Fig. 13.7, where the value of b_1 has been set to 1. The accumulated square error criterion given by Equation (13.2) and the Viterbi algorithm is equivalent to determining which ideal points, shown as bullets in Fig. 13.7, is the closest to a detected point (s_k, s_{k+1}). Line L1 is equidistant from the points $(-1+b_1, -1-b_1)$ and

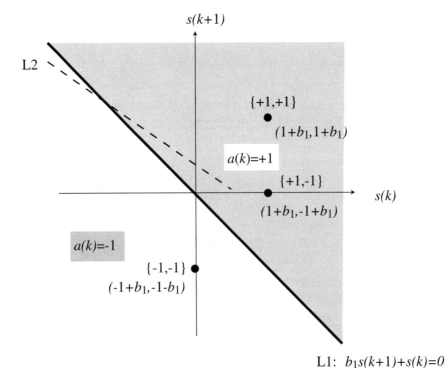

FIGURE 13.7. Decision space for three search with the $d = 1$ constraint. Note that $a(k) \equiv a_k, s(k) \equiv s_k$.

$(1+b_1, -1+b_1)$, and is described by the equation $b_1s_{k+1} + s_k = 0$. Therefore, if the linear combination of channel samples $b_1s_{k+1} + s_k > 0$, the
decision is made in favor for $a_k = +1$. Similarly, line L2 is equidistant
from the points $(-1+b_1, -1-b_1)$ and $(1+b_1, 1+b_1)$, and defines the decision boundary between the two ideal samples.

It is apparent from Fig. 13.7 that the first decision boundary given
by $b_1s_{k+1} + s_k > 0$ will cover the major fraction of all error events for a_k
$= +1$. Theoretical modeling indicates that the probability of error due to
using the L1 decision boundary instead of the combined L1 and L2 boundaries is at least 10 orders of magnitude below the bit error rate of the
channel. Therefore, the structure of FDTS detector for $d = 1$ constrained
codes can be implemented as a linear filter $(b_1 + D)$ followed by a slicer,[4]
as shown in Fig. 13.8. The equivalent FDTS detector simply calculates
$b_1s_{k+1} + s_k$, and then compares it with zero.

The equivalent FDTS/DF channel in Fig. 13.8 can be transformed
by moving the $(b_1 + D)$ filter to the forward equalizer path and by
simultaneously adding the same filter to the feedback path, as shown in
Fig. 13.9. In both cases, the equivalent polynomial before the slicer is

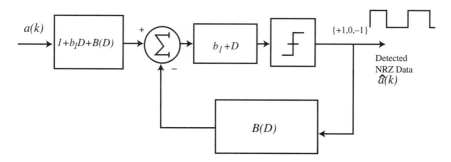

FIGURE 13.8.   Block diagram of equivalent FDTS/DF channel.

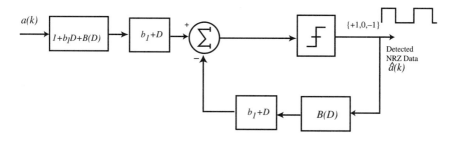

FIGURE 13.9.   Multilevel DFE (MDFE) channel.

$(1 + b_1 D)(b_1 + D)$. Note that the latter is very similar to the standard DFE channel shown in Fig. 13.3, but the slicer detector in Fig. 13.9 now performs binary decisions on a *multilevel* signal with ideal samples given by $(1 + b_1 D)(b_1 + D)a_k$. The channel in Fig. 13.9 is referred to as the *multilevel decision feedback* (MDFE) channel, or DFE-17 channel. The MDFE channel achieves the performance of the $d = 1$ constrained FDTS/DF channel with a two-tap transversal filter!

13.3 EQUALIZATION FOR MDFE CHANNEL

The transfer functions $F(D)$ and $H(D)$ of the forward and the backward (feedback) equalizers in the MDFE channel must be selected properly[3,4]:

$$\text{Forward equalizer: } F(D) = [1 + b_1 D + B(D)]\cdot(b_1 + D), \quad (13.3)$$
$$\text{Backward equalizer: } H(D) = B(D)\cdot(b_1 + D).$$

Expanding the polynomials, we obtain that

$$F(D) = b_1 + (1 + b_1^2)D + b_1 D^2 + (b_1 + D)\cdot(b_2 D^2 + b_3 D^3 + \ldots), \quad (13.4)$$
$$H(D) = (b_1 + D)\cdot(b_2 D^2 + b_3 D^3 + \ldots).$$

Recall that the PR4 channel is described by the polynomial $1 - D^2$, which results in ideal samples of a dibit as $\{1,0,-1\}$. Similar to the PRML channels, the coefficients of the transfer function of the forward equalizer can be interpreted as the ideal samples of an equalized isolated dibit. Therefore, the ideal samples of a dibit at the output of the forward equalizer in MDFE channel are:

$$\{s_0, s_1, s_2, s_3, \ldots\} = \{b_1, 1 + b_1^2, b_1 + b_1 b_2, b_1 b_3 + b_2, \ldots\} \quad (13.5)$$

Similarly, Equation (13.4) gives the tap values of the feedback filter:

$$\{h_2, h_3, \ldots\} = \{b_1 b_2, b_1 b_3 + b_2, \ldots\}, \quad (13.6)$$

which corresponding to $D^2, D^3, \ldots$ delay lines, respectively. There are no lower-order delay lines D^0 and D in the feedback filter. Starting from the second (nonzero) tap of the feedback filter, its values are the same as the samples of an equalized dibit at the output of the forward equalizer: $h_3 = s_3$, $h_4 = s_4, \ldots$, etc. It means that starting from the second tap of the feedback equalizer, they must be set to the corresponding samples of the dibit to ensure that the postcursor ISI tail is subtracted from the incoming signal. The value of the first tap of the feedback equalizer equals to $b_1 b_2$, which is $s_2 - s_0$ from Equation (13.5).

Once the coefficients of the forward equalizer are determined and the sequence of the dibit samples is given by $\{s_0, s_1, s_2, s_3, \ldots\}$, then the tap values of the feedback equalizer are set as $\{s_2 - s_0, s_3, s_4, \ldots\}$. Since the feedback equalizer coefficients are determined by the samples of the equalized dibit at the output of the forward equalizer, the design of the forward equalizer is critical to the overall MDFE performance.

Consider a typical equalized dibit shown in Fig. 13.10. It has certain nonzero precursor ISI, a typical pulse covering three first samples $\{s_0, s_1, s_2\}$, and a decaying postcursor tail to be canceled by the feedback filter. An optimal forward equalizer must satisfy the following constraints:

1. Minimize the precursor ISI by forcing all precursor dibit samples to zeros.
2. Maximize the second dibit sample as given by $1 + b_1^2$. As we will see in the next section, the value of the second sample determines the error rate of the MDFE channel.

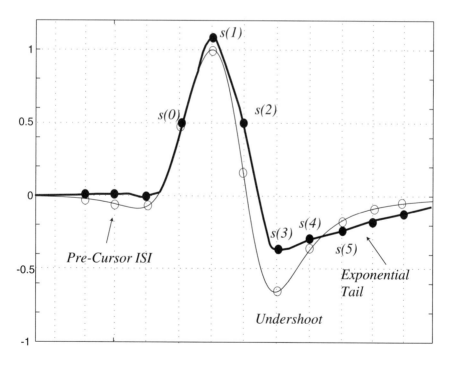

FIGURE 13.10. Optimization of impulse response of forward equalizer. Thin line: a dibit signal; thick line: equalized signal.

3. Try to match samples s_0 and s_2. Although this is not absolutely necessary, it forces the first tap of the feedback equalizer to zero. This makes the feedback loop less critical, i.e., the ISI for the second sample does not have to be canceled). The less the number of taps of the feedback equalizer, the easier the circuit design.

4. Minimize the postcursor undershoot and make the postcursor tail to decay exponentially. The minimization of undershoot is advantageous since smaller values will be subtracted from the incoming signal in the feedback loop. The exponential decay of the tail is desirable for circuit design: the feedback equalizer can have a small number of taps, followed by an integrator circuit with an approximately exponential response. This structure is equivalent to a feedback equalizer with a larger number of taps.

The forward equalizer is usually implemented in analog domain. The coefficients of the forward equalizer are calculated with numerical methods. Surprisingly, a simple way of canceling the precursor ISI, described by a filter $-\alpha p(t) + p(t - \tau)$ shown in Fig. 13.2 gives a good approximation to the optimal equalizer design.

13.4 ERROR RATE OF MDFE CHANNEL AND ERROR PROPAGATION

To determine the error rate of the MDFE channel, consider the block diagram of the channel shown in Fig. 13.9. If the input sequence of NRZ bits is given by $\{a_{k-1}, a_k, a_{k+1}\}$, the channel sample before the $b_1 + D$ filter at time $k + 1$ is

$$s_{k+1} = a_{k+1} + b_1 a_k + R(a_{k-1}, a_{k-2}, \ldots) + n_{k+1},$$

where R is the postcursor ISI, and n_{k+1} is the noise term. The sample is then filtered by the $b_1 + D$ filter, and the ISI is subtracted by the feedback filter. Therefore, the signal level coming to the slicer is[3,4]

$$s_{k+1} = b_1 a_{k+1} + (1 + b_1^2) a_k + b_1 a_{k-1} + b_1 n_{k+1} + n_k. \tag{13.7}$$

If we ignore error propagation, i.e., assume that all the previous decisions were correct, then the errors are caused by noises.

There are eight possible combinations of $\{a_{k-1}, a_k, a_{k+1}\}$ bits. However, two of these combinations given by $\{+1, -1, +1\}$ and $\{-1, +1, -1\}$ violate the $d = 1$ constraint. The remaining six combinations generate various ideal sample values at the slicer according to Equation (13.7):

$$
\begin{aligned}
\{a_{k-1}, a_k, a_{k+1}\} &= \{+1, +1, +1\} & s_{k+1} &= 2b_1 + (1 + b_1^2), \\
\{a_{k-1}, a_k, a_{k+1}\} &= \{+1, +1, -1\} & s_{k+1} &= (1 + b_1^2), \\
\{a_{k-1}, a_k, a_{k+1}\} &= \{+1, -1, -1\} & s_{k+1} &= -(1 + b_1^2), \\
\{a_{k-1}, a_k, a_{k+1}\} &= \{-1, -1, +1\} & s_{k+1} &= -(1 + b_1^2), \\
\{a_{k-1}, a_k, a_{k+1}\} &= \{-1, -1, -1\} & s_{k+1} &= -2b_1 -(1 + b_1^2), \\
\{a_{k-1}, a_k, a_{k+1}\} &= \{-1, +1, +1\} & s_{k+1} &= (1 + b_1^2),
\end{aligned}
\tag{13.8}
$$

where the noise term has been set to zero. As seen from Equation (13.8), there are only four distinct sample levels at the slicer input: $\pm[2b_1 + (1 + b_1^2)]$ and $\pm(1 + b_1^2)$.

When the noise term $b_1 n_{k+1} + n_k$ is present at the slicer input, the "+1" bit will be confused with "−1" if the total noise exceeds one-half of the minimum distance $d_{\min}$ between two input sequences. In this case, it is the distance between the two closest sample levels: $\pm(1 + b_1^2)$, so $d_{\min} = 2(1 + b_1^2)$. Assume that the noise term n_{k+1} is Gaussian with a standard deviation of σ and the two adjacent noise samples are uncorrelated, then the variance of the total noise at the slicer σ_Σ^2 is given by:

$$
\sigma_\Sigma^2 = \langle (b_1 n_{k+1} + n_k)^2 \rangle = \langle b_1^2 n_{k+1}^2 + 2b_1 n_k n_{k+1} + n_k^2 \rangle = (1 + b_1^2)\sigma^2,
$$

where $\langle \rangle$ stands for average or expectation value. Therefore, the raw error rate in the MDFE channel without error propagation is determined as:

$$
P_e \approx Q\left(\frac{d_{\min}/2}{\sigma_\Sigma}\right) = Q\left(\frac{\sqrt{1 + b_1^2}}{\sigma}\right).
\tag{13.9}
$$

The term $1 + b_1^2$ equals to the second sample of the equalized dibit. Therefore, the larger the second sample, the better error rate will be achieved by MDFE.[5]

Following the same argument, the advantage of MDFE over standard DFE can be easily demonstrated. In the latter the $d = 1$ constraint is removed, so the two closest levels at the slicer input are given by the two sequences prohibited in MDFE:

$$
\begin{aligned}
\{a_{k-1}, a_k, a_{k+1}\} &= \{+1, -1, +1\}, & s_{k+1} &= 2b_1 - (1 + b_1^2) = -(1 - b_1)^2, \\
\{a_{k-1}, a_k, a_{k+1}\} &= \{-1, +1, -1\}, & s_{k+1} &= -2b_1 + (1 + b_1^2) = (1 - b_1)^2,
\end{aligned}
\tag{13.10}
$$

which gives a minimum distance of $d_{\min} = 2(1 - b_1)^2$. Evidently, the standard DFE channel has a smaller minimum distance than the MDFE channel, so the MDFE channel has a lower error rate.

Note that the dependence of the minimum distances on b_1 are opposite for the standard DFE and MDFE, so the equalization targets for the two channels are different. The MDFE equalizer maximizes the second dibit sample given by $1 + b_1^2$, while the standard DFE equalizer minimizes the first sample b_1.

The estimate of the MDFE channel error rate as given by Equation (13.9) ignored the error propagation in the feedback loop.[6,7] The estimation of the error propagation is complicated because the current decision $\hat{a}_k$ of the MDFE channel depends on the current channel sample s_k and a set of the possible feedback errors, given by the difference between correct and incorrect channel bits:

$$\{e_{k-2}, e_{k-3}, \ldots\} = \{\hat{a}_{k-2} - a_{k-2}, \hat{a}_{k-3} - a_{k-3}, \ldots\}.$$

The error propagation is usually modeled as a Markov chain. The Markov chain[6,7] is a special probabilistic model in which the probability of the current state of the system, such as the probability of error in the MDFE channel, depends only on the previous state. For the MDFE channel with additive Gaussian white noise, several simulation results are available in the literature. The conclusions of these simulations can be summarized as follows[7]:

1. The main contribution to the MDFE error rate is generated by short error bursts. For example, the probability of having two errors is typically an order or magnitude smaller than the probability of a single error.
2. The log probability $L(n)$ of a burst error of length n due to the feedback decays asymptotically with n. Figure 13.11 presents a simulated probability of burst errors in the MDFE channel with Gaussian white noise.
3. The total degradation of MDFE error rate caused by error propagation is moderate, typically not more than an order of magnitude.

In spite of the great promise of the MDFE channel, it has not yet been used in commercial disk drives. To evaluate the practicality of the MDFE channel, we must compare it with PRML channels, which are the dominant designs at present and in the near future. Fig. 13.12 presents the simulated PR4 and MDFE error rates at a user density of 2.5. Apparently the MDFE channel outperforms the PR4 channel slightly at the given user density. This result is expected because of the smaller noise boost of the MDFE forward equalizer. The simulation of MDFE in Fig. 13.12 included the effect of error propagation, and the noise was modeled as the mix of the electronics and media noises. However, accurate comparison of the MDFE

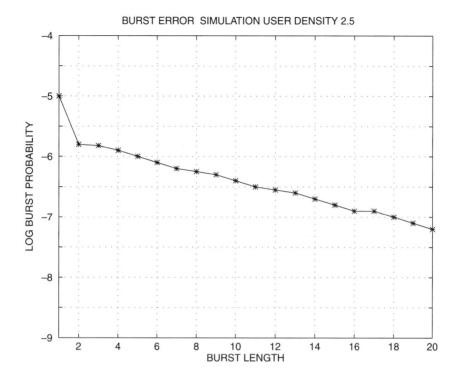

FIGURE 13.11.   Probability of burst errors in MDFE channel.

and PRML channels using hardware channels and magnetic spin-stand
are yet to be performed.

13.5 TIMING AND GAIN RECOVERY IN MDFE CHANNEL

Several schemes have been proposed for MDFE timing recovery, most of
which are decision-directed and are based on the symmetry of the sequence of samples at the MDFE slicer.[4,7,8,9]

Consider the input sequence of bits $\{a_{k-1}, a_k, a_{k+1}\} = \{+1, -1, -1\}$, corresponding to the presence of a transition. Note that we always have at
least two consecutive bits of the same polarity for $d = 1$ constraint code,
i.e., sequences like $\{a_{k-1}, a_k, a_{k+1}\} = \{+1, -1, +1\}$ are not allowed. From
Equation (13.7), the ideal MDFE slicer levels for the second and third bit
are equal to $s_k = +(1 + b_1^2)$ and $s_{k+1} = -(1 + b_1^2)$, or simply $+A$ and $-A$

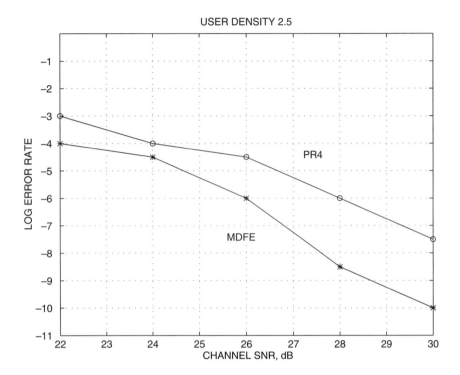

FIGURE 13.12.   Simulated performance of MDFE and PR4 channels.

with $A = 1 + b_1^2$. For the transition of the opposite polarity $\{a_{k-1}, a_k, a_{k+1}\} = \{-1, +1, +1\}$, the consecutive slicer levels are equal to $-A$ and $+A$. Therefore, if the sampling phase is correct, the sum of two consecutive samples at the slicer should vanish: $s_k + s_{k+1} = 0$.

If the sampling phase is "early" for $\{a_{k-1}, a_k, a_{k+1}\} = \{+1, -1, -1\}$, the first sample will be larger than A, and the second sample larger than $-A$, and $s_k + s_{k+1} > 0$. Conversely, if the sampling phase is "late," $s_k + s_{k+1} < 0$. For a transition of the opposite sign, $\{a_{k-1}, a_k, a_{k+1}\} = \{-1, +1, +1\}$, the situation is symmetrical: if the sampling phase is early, $s_k + s_{k+1} < 0$; if it is late, $s_k + s_{k+1} > 0$. Therefore, $\Delta_t = (s_k + s_{k+1})a_k$ is indicative of phase error and can be used for timing phase recovery when a transition is detected $(a_k \neq a_{k-1})$.

Similarly, the gain update signal can be easily derived from the difference of two adjacent samples, which should be $(s_{k+1} - s_k)a_k = 2A$ nominally. Therefore, the gain update signal is determined by the difference between the real value and the nominal value:

$$\Delta_g = (s_{k+1} - s_k)a_k - 2A.$$

Like PRML channels, timing and gain recovery is essential for the MDFE channel to function properly.

References

1. B. Sklar, *Digital Communications, Fundamentals and Applications,* (Englewood Cliffs, NJ: Prentice Hall, 1988); J. M. Cioffi, W. L. Abbott, H. K. Thapar, C. M. Melas, and K. D. Fisher, "Adaptive equalization in magnetic-disk storage channels," *IEEE Communications Magazine,* February 1990, p. 14.

2. J. W. Bergmans, *Digital Baseband Transmission and Recording,* (Boston, MA: Kluwer Academic Publishers, 1996).

3. J. J. Moon and L. R. Carley, "Performance comparison of detection methods in magnetic recording," *IEEE Trans. Magn.,* **26**, 3155, 1990; J. G. Kenney, L. R. Carley, and R. Wood, "Multi-level decision feedback equalization for saturation recording," *IEEE Trans. Magn.,* **29**, 2160, 1993.

4. J. G. Kenney and R. Wood, "Multi-level decision feedback realization of FDTS/DF," *IEEE Trans. Magn.,* **31**, 1115, 1995.

5. L. R. Carley and J. G. Kenney, "Comparison of computationally efficient of FDTS/DF against PR4ML," *IEEE Trans, Magn.,* **27**, 4567, 1991.

6. V. Krachkovsky, Y. Lee, and L. Bin, "Error propagation evaluation for MDFE detection," *IEEE Trans. Magn.,* **33**, 2770, 1997.

7. J. Ashley, M. Blaum, B. Marous, and C. M. Melas, "Performance and error propagation of two DFE channels," *IEEE Trans. Magn.,* **33**, 2773, 1997.

8. J. Hong, Y. Lee, H. Mutoh, Q. Sun, H. Ueno, J. Wang, and R. Wood, "An experimental MDFE detector," *IEEE Trans. Magn.,* **33**, 2776, 1997.

9. Y. Lee, L. Ong, J. Wang, and R. wood, "Timing acquisition for DFE detection," *IEEE Trans. Magn.,* **33**, 2761, 1997.

CHAPTER **14**

Off-Track Performance

In previous chapters we have usually assumed that the read/write heads are more or less on-track, i.e., the head and data track are aligned properly. The on-track performance of a magnetic recording system determines how close we can pack data bits (linear data density) and how fast we can transfer data bits (data rate). In reality, however, the read/write head is often off-track. Therefore, we need to consider how the off-track position of the read/write head affect the performance of magnetic disk drives and how the data track configuration can be best designed, which are the subjects of this chapter. We will consider how to control the off-track position of the read/write head in the next chapter. The off-track performance of a magnetic recording system is extremely important because it determines how close data tracks may be written on disk surfaces (data track density).

14.1 STRUCTURE OF MAGNETIC TRACK

When a track of information is written on a magnetic medium, it has a certain width that is determined by the geometry of the write bubble *along the cross-track direction.* The certain width of magnetic medium enclosed by the write bubble is successfully written with magnetic transitions, resulting in a magnetic data track as the disk spins. At larger distances from the track center (outside the write bubble), the head field decreases below the medium coercivity and is not enough to write magnetic transitions. However, as shown in Fig. 14.1, the side fringing magnetic field extends from the edges of the head gap. The side field has a magnitude large enough to remagnetize the medium, but its orientation is different from the longitudinal direction and its field gradient is low. As a result, a small strip of medium close to the track edges is remagnetized, forming so-

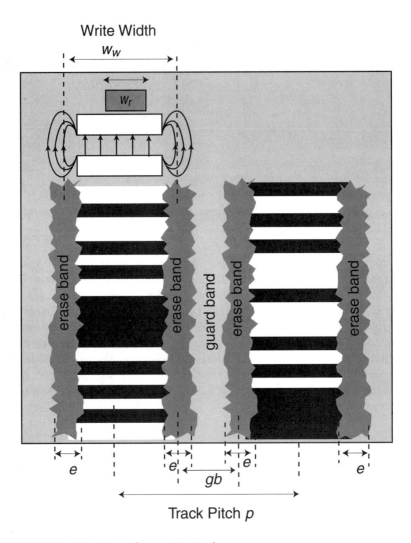

FIGURE 14.1. Structure of magnetic track.

called *erase bands*. The side fringing field will erase any old information within the erase bands, and the magnetization in the erase bands do not contribute to readback signal.

In general, the magnetic track geometry are described by the following parameters:

  e—*Erase band*: the width of poorly written or demagnetized area on each side of the track.

w_w—*Written track width*: the recorded track width plus 50% of erase bands from each side of the track. Note that **this definition differs from that in Section 5.5.3 [Equation (5.20)] by an erase band! We** choose to use this new definition here to be consistent with Lee and Bonyhard's original paper.[1]

p—*Track pitch*: the distance between the centers of two adjacent tracks.

gb—*Guard band*: the distance between the centers of the erase bands of two adjacent tracks. Note that $p = w_w + gb$. The guard band is added to avoid the interference (unwanted signal) from adjacent tracks.

w_r—*Read track width*: the effective width of recorded track which contributes to readback signal. This definition is the same as that in Section 5.5.3. [Equation (5.21)].

14.2 TRACK MISREGISTRATION AND OFF-TRACK PERFORMANCE

Recording heads generally do not stay at the center of data track, but rather "close" to the center of the data track, resulting in so-called track misregistration.[1,2] The *write-to-write track misregistration* is the misregistration between the center lines of a recorded track and an adjacent track, resulting in track-to-track "squeeze." The *write-to-read track misregistration* is the misregistration between the center line of a recorded track and that of the read head.

Let us consider the situation that no adjacent tracks are written around a data track. When a read head is offset from the track center, its readback signal partly comes from the data track, and partly from the medium outside the data track. Obviously, the larger the offset between the head gap center and the track center, the less signal is read by the head and the more medium noises are included in the readback signal. An example of the readback signal of an inductive head vs. its off-track position is shown in Fig. 14.2. As the head offset increases, the readback pulse is increasingly distorted by noises and becomes more difficult to detect.

A *track profile* test as shown in Fig. 14.3 is often used to characterize the performance of head and medium at different offsets, in which the track-average amplitude (TAA) of the readback signal is measured and plotted vs. the read head center offset from the track center. The track profile shown in Fig. 14.3 was obtained for a single track and a peak detection channel, in which TAA gradually decreases and reaches about

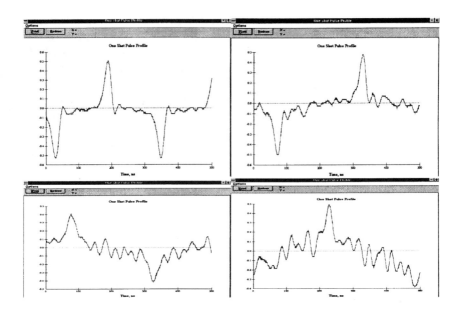

FIGURE 14.2. Readback signal at 50 (upper left), 80 (upper right), 100 (lower left), and 140 (lower right) microinches offsets from the track center, respectively.

30% of the maximum amplitude at approximately 150 μin. off the track center.

Also shown in Fig. 14.3 is the BER of peak detector vs. the read head offset, which is often called a *bathtub curve* after its shape. The bathtub curve indicates how much the read head can be allowed to go off-track at a specified error rate. The bathtub curve can also be used to predict the actual on-track BER by extrapolating the slope of the error rate dependence on the off-track position to the zero offset. As mentioned previously, the on-track BER is very small, and a long measurement time is needed to obtain reliable statistics. In Fig. 14.3 the error rate at the track center is better than 10^{-9} and is only about 10^{-3} at 150 μin.

Another way of examining the off-track performance is to measure the bit-shift plot (Chapter 11) or sequenced amplitude margin (SAM) plot (Chapter 12) at different offset values. For example, as shown in Fig. 14.4, the bit-shift plot at the track center predicts a time window margin of ~4.8 ns at an error rate of 10^{-9}. At an offset of 20 μin., the margin is reduced to 3.9 ns; at 50 μin. the margin is only ~2 ns; at 120 μin. there is no time margin left at an error rate of 10^{-4}.

Now let us consider the situation that an adjacent track is written next to the data track. In this case, the off-track performance deteriorates

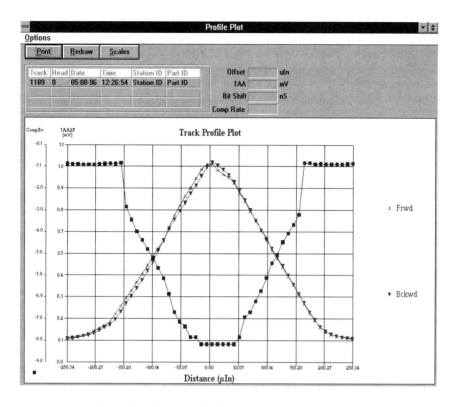

FIGURE 14.3.   Example of track profile (x and triangle) and bathtub curve (square).

dramatically because the read head picks up the interference signal from the adjacent track instead of medium noises. Figure 14.5 illustrates a track profile when an adjacent track is written at a distance of 200 μin. from the data track center. The right half of the bathtub curve shifted upward greatly due to the interference from the adjacent track. The error rate is as high as 10^{-2} when the head is moved about 40 μin. off the track center. However, the performance in the opposite direction of the data track is not affected since no interference signal exists there.

Generally speaking, track misregistration often leads to smaller read-back signals and larger interferences or noises than those obtained with on-track heads. To reduce the negative impact of the track misregistration on the SNR performance of recording heads, a guard band is inserted between adjacent tracks. Furthermore, a scheme called "write wide read narrow" is commonly employed, particularly in the disk drives utilizing inductive write/MR read dual-element magnetic recording heads. In this

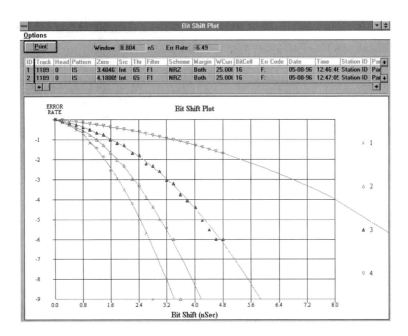

FIGURE 14.4. Bit-shift plots for the read head offset values of 0, 20, 50, and 120 μin., respectively.

case, the read track width is intentionally designed to be smaller than the written track width. The design of guard band and write/read track widths will be presented next, along the lines of Lee and Bonyhard.[1]

14.3 OFF-TRACK CAPABILITY AND 747 CURVE

The *off-track capability* (OTC) is defined as the write-to-read track misregistration ($\text{TMR}_{w,r}$) allowed at a specified error rate limit. This is closely related to so-called "k parameter," meaning that k portion of read track width can be allowed to be out of the written data track w_w. As shown in Fig. 14.6, their relation is

$$kw_r = 0.5w_r + \text{OTC} - 0.5w_w$$
$$= \text{OTC} - 0.5(w_w - w_r). \tag{14.1}$$

For the purposes of off-track capability analysis, all the other sources of noises are negligible in comparison with the old information (in the

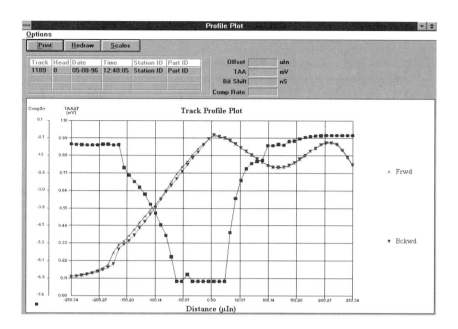

FIGURE 14.5.   Track profile and bathtub curve with an adjacent track.

guard band) and the interference signal (from the adjacent track) read by the read head. We also assume that signal or noise voltage is proportional to the fraction of read head in the data track or in the old information region, respectively.

The off-track capability depends on the distance between the data track and the adjacent track. When two tracks are separated so that their erase bands do not overlap, as shown in Fig. 14.6, the OTC is determined by the erase band, the old information in the guard band, and the possible interference signal from the adjacent track. If the adjacent track is far enough, there is no interference and the OTC should be a constant independent of track pitch. If the adjacent track is written closer and closer to the data track, at some point the erase bands of the two tracks will approach each other and the head will read less old information. As a result, the OTC in this case will increase. However, if the adjacent track is written too close to the data track, these tracks will overlap and the interference signal (or cross-talk) from the adjacent track will decrease the OTC value. Therefore, the dependence of OTC on the track pitch is nonmonotonic. The shape of the OTC vs. track pitch curve often resembles the nose of the Boeing 747 jet plane, so it is often called the "747 curve."

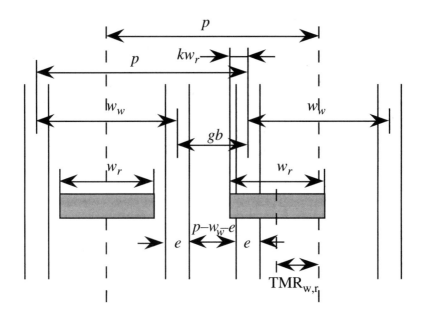

FIGURE 14.6.   Geometry of write wide read narrow scheme.

To illustrate how the 747 curve comes about, we will analytically derive the OTC with decreasing track pitch in five regimes.[1,2]

1. $p \geq w_w + kw_r + 0.5e$. In this case, the track pitch is large enough so that the read head does not read the adjacent erase band. The portion of the read width that reads the old information is $kw_r - 0.5e$. As shown in Fig. 14.6, the noise to signal voltage ratio is

$$k' = \frac{V_n}{V_s} = \frac{kw_r - 0.5e}{w_r - [kw_r + 0.5e]}$$

Since

$$OTC_1 \equiv OTC_\infty = kw_r + 0.5\,(w_w - w_r),$$

The OTC in this case is the sum of the genuine off-track capability kw_r and the extra width due to the write wide, read narrow scheme. At the specified error rate, k' is usually a constant. Therefore, in case (1), the k parameter and OTC shall be constant, independent of track pitch p.

Now what is the effect of erase band? If the k parameter is k_0 in the absence of erase band ($e = 0$), then

$$k' = \frac{k_0 w_r}{w_r - k_0 w_r} = \frac{k_0}{1 - k_0}. \tag{14.2}$$

In the presence of the erase band, the k' value for a specified error rate is unchanged:

$$\frac{k_0}{1 - k_0} = \frac{kw_r - 0.5e}{w_r - [kw_r + 0.5e]} = \frac{OTC_1 - 0.5(w_w - w_r + e)}{w_r - OTC_1 + 0.5(w_w - w_r - e)}.$$

Therefore,

$$k_0 = \frac{OTC_1 - 0.5\,(w_w - w_r + e)}{w_r - e},$$

$$OTC_1 = 0.5(w_w - w_r) + k_0 w_r + (0.5 - k_0)e, \tag{14.3}$$

$$OTC_1 = 0.5(w_w - w_r) + \frac{k'}{k' + 1}(w_r - e) + 0.5e.$$

From Equation (14.1), we obtain that

$$k = k_0 + (0.5 - k_0)e/w_r. \tag{14.4}$$

This means that the erase band increases the OTC by an amount of $(0.5 - k_0)e$, while write-wide increases the OTC by $0.5(w_w - w_r)$.

2. $w_w + kw_r - 0.5e + \Delta \le p \le w_w + kw_r + 0.5e$. This is the case when the adjacent track is written close enough so that the head reads two adjacent erase bands and the old information band, as shown in Fig. 14.7. In this case,

$$V_n = p - w_w - e,$$
$$V_s = w_r - [OTC_2 - 0.5(w_w - w_r)] - 0.5e, \tag{14.5}$$

$$\frac{k_0}{1 - k_0} = \frac{p - w_w - e}{w_r - OTC_2 + 0.5(w_w - w_r) - 0.5e},$$

$$OTC_2 = w_r + 0.5(w_w - w_r) - 0.5e - \frac{1 - k_0}{k_0}(p - w_w - e).$$

The OTC is a linear function of p with a negative slope $-(1 - k_0)/k_0$.

3. $w_w + e \le p \le w_w + kw_r - 0.5e + \Delta$. This is the case when the read head reads not only two erase bands and old information band, but also picks up the interference from the adjacent track, as shown in Fig. 14.8. The old information noise is

$$V_{oi} = p - w_w - e.$$

The interference noise from the adjacent tracks is

$$V_a = [OTC - 0.5(w_w - w_r)] - V_{oi} - 1.5e.$$

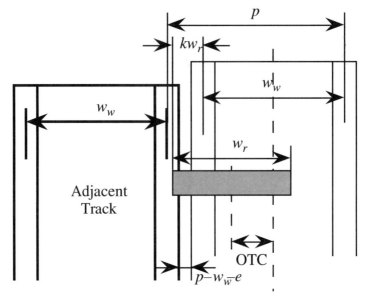

FIGURE 14.7.   Track squeeze case (2).

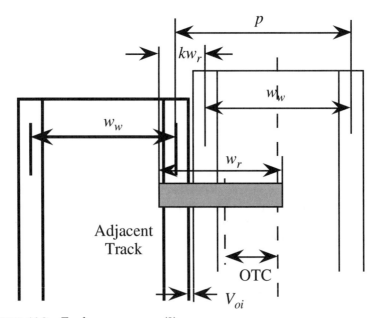

FIGURE 14.8.   Track squeeze case (3).

The real readback signal is

$$V_s = w_r - [\text{OTC} - 0.5(w_w - w_r)] - 0.5e.$$

Assuming that V_{oi} and V_a are uncorrelated, then the total noise is

$$V_n = \sqrt{V_{oi}^2 + V_a^2},$$

and the noise-to-signal power ratio is

$$k'^2 = \frac{V_n^2}{V_s^2} = \frac{V_{oi}^2 + V_a^2}{V_s^2}. \tag{14.6}$$

Let

$$\begin{aligned}
\text{OTC} &= uw_w, \\
V_{oi} &= \gamma_{oi}w_w, \\
w_r &= mw_w, \\
e &= \gamma_e w_w, \\
\eta &= 0.5(1 - m),
\end{aligned} \tag{14.7}$$

and substituting into Equation (14.6), we obtain that

$$\begin{aligned}
k'^2[m - u + \eta - 0.5\gamma_e]^2 &= \gamma_{oi}^2 + [u - \eta - \gamma_{oi} - 1.5\gamma_e]^2, \\
k'^2 u^2 - 2k'^2(m + \eta - 0.5\gamma_e)u &+ k'^2(m + \eta - 0.5\gamma_e)^2 \\
= \gamma_{oi}^2 + u^2 - 2u(\eta + \gamma_{oi} + 1.5\gamma_e) &+ (\eta + \gamma_{oi} + 1.5\gamma_e)^2.
\end{aligned}$$

Therefore,

$$Au^2 + Bu + C = 0, \tag{14.8}$$

where

$$\begin{aligned}
A &= 1 - k'^2, \\
B &= 2k'^2(m + \eta - 0.5\gamma_e) - 2(\eta + \gamma_{oi} + 1.5\gamma_e), \\
C &= \gamma_{oi}^2 + (\eta + \gamma_{oi} + 1.5\gamma_e)^2 - k'^2(m + \eta - 0.5\gamma_e)^2.
\end{aligned}$$

In this case,

$$\text{OTC}_3 \equiv uw_w, \tag{14.7}$$

where u is the solution to the quadratic equation (14.8).

4. $w_w \le p < w_w + e$. This is the case when the erase band of the adjacent track has replaced the old information, as shown in Fig. 14.9. In this case, the interference noise is

$$V_a = [\text{OTC} - 0.5(w_w - w_r)] - (p - w_w) - 0.5e.$$

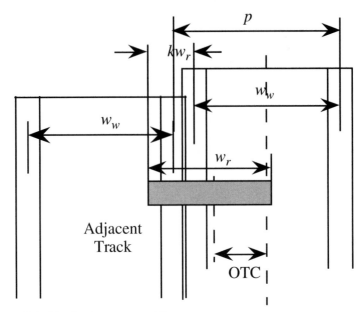

FIGURE 14.9   Track squeeze case (4).

The readback signal is

$$V_s = w_r - [\text{OTC} - 0.5(w_w - w_r)] - 0.5e.$$

Then

$$k' = \frac{V_a}{V_s} = \frac{[\text{OTC} - 0.5(w_w - w_r)] - (p - w_w) - 0.5e}{w_r - [\text{OTC} - 0.5(w_w - w_r)] - 0.5e}.$$

Therefore, the OTC in this case is

$$\text{OTC}_4 = \frac{1}{1 + k'} (p - w_w + 0.5e + k'w_r - 0.5k'e) + 0.5(w_w - w_r), \qquad (14.9)$$

which is a linear function of p with a positive slope $1/(1 + k') = 1 - k_0$.

5. $w_w - \text{OTC}_\infty \le p < w_w$. This is the case when the adjacent track moves into the data track so that part of the written data is wiped out, as shown in Fig. 14.10. In this case,

$$V_a = [\text{OTC} - 0.5(w_w - w_r)] - (p - w_w) - 0.5e,$$
$$V_s = w_r - V_a - e,$$
$$k' = \frac{V_a}{V_s} = \frac{[\text{OTC} - 0.5(w_w - w_r)] - (p - w_w) - 0.5e}{w_r - V_a - e}.$$

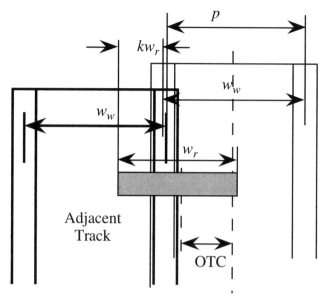

FIGURE 14.10. Track squeeze case (5).

Therefore, the off-track capability is

$$\text{OTC}_5 = p - w_w + 0.5e + 0.5(w_w - w_r) + \frac{k'}{k' + 1}(w_r - e), \quad (14.10)$$

which is a linear function of p with a slope of 1.

A minimum track pitch, $p_{\min}$, beyond which the on-track error will exceed the specified limit, can be obtained by setting $\text{OTC}_5 = 0$. This leads to

$$p_{\min} = w_w - 0.5e - 0.5(w_w - w_r) - \frac{k'}{k' + 1}(w_r - e).$$

Since

$$\text{OTC}_1 = 0.5(w_w - w_r) + \frac{k'}{k' + 1}(w_r - e) + 0.5e,$$

it can be shown that

$$p_{\min} = w_w - \text{OTC}_\infty. \quad (14.11)$$

To sum up all the cases discussed above, the plot of OTC as a function of track pitch p, is usually like the curve shown in Fig. 14.11. It indeed resembles the nose of a Boeing 747 jet plane, so it is called a 747 curve.

An experimental 747 curve is shown in Fig. 14.12. The so-called "747 test" is commonly used in disk drives to find the optimal track pitch that achieves the maximum allowable OTC at a specified error rate.

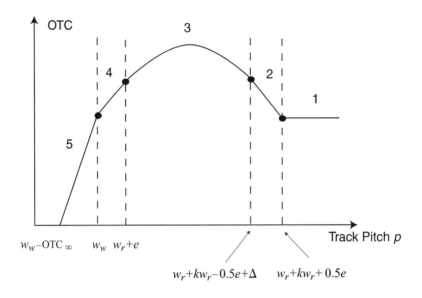

FIGURE 14.11.   Schematic 747 curve, plotted as OTC vs. adjacent track pitch.

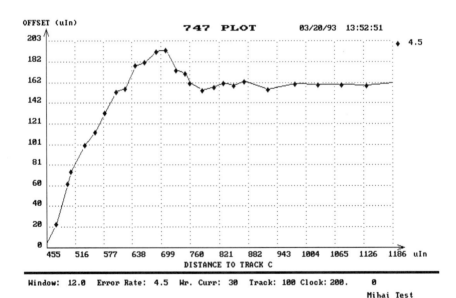

FIGURE 14.12.   An experimental 747 curve.

References

1. J. K. Lee and P. Bonyhard, "A track density model for magnetoresistive heads considering erase bands," *IEEE Trans. Magn.*, **26**, 2475, 1990.
2. M. P. Vea and T. Howell, "A soft error rate model for predicting off-track performance.," *IEEE Trans. Magn.*, **31**, 820, 1995.

CHAPTER 15

Head-Disk Assembly Servo

When a magnetic disk drive is powered up, magnetic write/read heads are moving constantly relative to magnetic disks in a three-dimensional space. The vertical spacing between a head and a disk is determined by air-bearing design and disk rotation speed, while the radial position of the head is determined by the torque or force exerted by an actuator. This chapter discusses how to measure and control the radial position of a recording head on a spinning disk.

To write/read data properly, the head has to be accurately positioned at the desired data track. Such a control task is accomplished by a head *servomechanism*. A servomechanism is a feedback system that consists of a sensing element, an amplifier, and a servomotor (actuator), used in the automatic control of a mechanical device. In this sense, magnetic disk drives are rightfully called mechanical data strorage devices.

15.1 HEAD SERVOMECHANISM

A typical head servomechanism is illustrated schematically in Fig. 15.1. When the center of the recording head is aligned to the center of the target data track, it is called to be *on track*. When the head is moved *off*

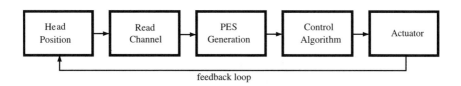

FIGURE 15.1. Schematic recording head servomechanism.

467

track, a position error signal (PES) is generated by the read head, which is the sensing element in the servomechanism. The PES are processed by the control algorithm, which determines how the actuator current responds to the PES. The actuator then modifies the head position.

The head servomechanism fulfills mainly two tasks: *seeking* and *tracking.* If the target track with desired data is far from the initial track where the head is flying, the head must be moved to the target track first. This seeking motion is schematically shown in Fig. 15.2. The head radial velocity (along the cross-track direction) is increased until nearing the midpoint between the initial track and the target track. Then the head radial velocity is decreased until reaching the target track. The total time span is called the *seek time,* which should be minimized by the control algorithm. More elaborate control trajectory can be found in the literature.[1,2] The head velocity when reaching the target track is usually not zero; it takes additional *settling time* to position the head on track. After this point, the head can read data reliably. Once the head is on track, it may move off track because of many mechanical and thermal disturbances. To keep the head on track, i.e., to make PES zero, the actuator responds to the PES through a negative feedback loop. The requirement of this tracking mechanism is to keep the write-to-write track misregistration and write-to-read track misregistration much less than the track pitch.

In magnetic disk recording, data tracks are equally spaced concentric circles. Obviously, tracking can be performed without knowing the track number. However, to perform seeking, both PES and track number must

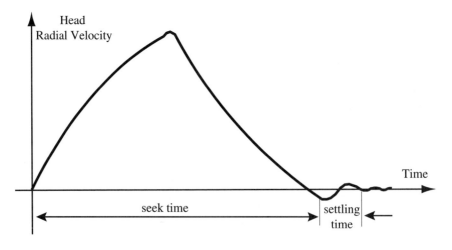

FIGURE 15.2.   Head radial velocity vs. time during typical seeking.[3]

be known. To generate PES and track number on spinning disks, special servo data are written on disks before shipping. There are mainly two schemes being used: *dedicated servo* and *sector servo*. Dedicated servo uses a dedicated disk surface and a read-only head to generate PES and track number. The servo head is mechanically coupled with the data heads in the drive, so positioning the servo head simultaneously locate all the data heads on disk surfaces. For small disk drives with a few disks, sector servo scheme is used instead, which will be discussed in the next section. In this case, sectors of servo information are shared in the data tracks on all the surfaces involved. Typically an overhead of about 7–10% of the recording area needs to be devoted to servo sectors instead of user data.

The relative merit of dedicated servo and sector servo depends on how many disks are used in a disk drive. In the case of dedicated servo, the overhead is listed in Table 15.1. For disk drives with four disks, employing dedicated servo results in an overhead of $1/(5 \times 2) = 10\%$. Therefore, disk drives with four or fewer disks usually employ sector servo schemes. The principle of servo data pattern and PES generation are similar in sector servo and dedicated servo schemes, so we will focus on sector servo hereafter.

15.2 SECTOR SERVO DISK FORMAT

The sector servo information and user data are typically organized in a manner schematically shown in Fig. 15.3, where four concentric *data bands* (*zones*) are recorded on one disk surface, denoted as A, B, C, and D, respectively.[4] There are 60 servo sectors which are synchronized in each data zone, but there are more user data sectors in outer zones. Therefore, data sectors may span one to two servo sectors and are not synchronized with the servo sectors. In this example, there are a total of 156 data sectors.

Zone recording means that each data zone uses a different recording frequency, and higher frequencies are used for outer zones than for inner zones. Therefore, more data are recorded in outer zones than in inner zones in spite that the angular velocity of the disk rotation is constant. The zone recording scheme is more efficient than the traditional single-

TABLE 15.1. Servo Overhead in Dedicated Servo

Number of disks	3	4	5	6	7
Overhead	16.7%	12.5%	10%	8.3%	7.1%

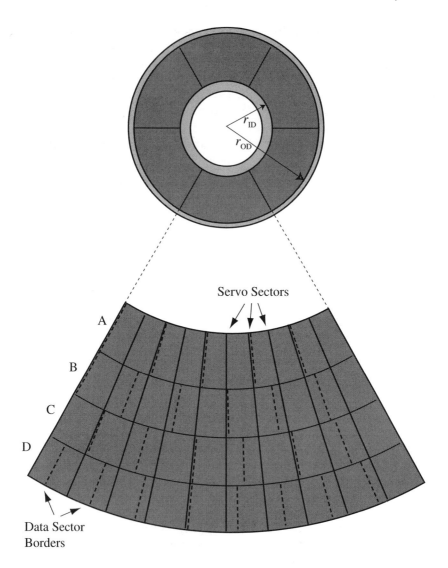

FIGURE 15.3.   Representative format of zone recorded sector servo disk, with 60 servo sectors (solid radial lines) in this example. There are four concentric data zones, and there are 30, 36, 40, and 50 data sectors (dashed lines) in zone A, B, C, and D, respectively. Note that the data zone is smaller than the physical disk size.

zone recording. On the other hand, more data zones require more sophisticated disk controllers. Therefore, a compromise of zone number is chosen
in disk drive design.

As shown in Fig. 15.3, the outer diameter ($2r_{OD}$) of the data zone is
smaller than the physical size of the disk, while the inner diameter ($2r_{ID}$)
of the data zone is greater than the hole diameter. In single-zone recording,
only the innermost track records a maximum linear density D_t of the disk,
and the number of bits per track is independent of track radius. If the
track density is D_t, then the data capacity (the total data bits recorded)
on a single disk surface is

$$C = D_l \cdot 2\pi r_{ID} \cdot D_t \cdot (r_{OD} - r_{ID}) = 2\pi D_l D_t r_{OD}^2 b (1 - b), \quad (15.1)$$

where $b = r_{ID}/r_{OD}$. The maximum capacity is achieved if $b = 1/2$. In other
words, when the OD radius is twice the ID radius, we obtain that

$$C_{max} = \frac{1}{2} \pi D_l D_t r_{OD}^2. \quad (15.2)$$

If the zone recording scheme is employed, and the OD and ID radius
of the jth zone are r_{j+1} and r_j, respectively, then, the total capacity of a
disk surface is

$$C = 2\pi D_l D_t \cdot \sum_{j=1}^{n} r_j (r_{j+1} - r_j). \quad (15.3)$$

The capacity is the maximum when the width of each zone is constant:[5]

$$r_{ID} = r_{j+1} - r_j = \frac{r_{OD} - r_{ID}}{n}, \quad r_{ID} = \frac{r_{OD}}{n + 1}, \quad (15.4)$$

where n is the number of the data zones. Therefore, the maximum capacity
is

$$\begin{aligned}
C_{max} &= 2\pi D_l D_t \cdot \frac{r_{OD} - r_{ID}}{n} \sum_{j=1}^{n} r_j \\
&= 2\pi D_l D_t \cdot \frac{r_{OD} - r_{ID}}{n} \cdot n \frac{r_{ID} + n r_{ID}}{2} \quad (15.5) \\
&= (n + 1) \pi D_l D_t r_{ID} (r_{OD} - r_{ID}) \\
&= \frac{n}{n + 1} \pi D_l D_t r_{OD}^2.
\end{aligned}$$

If there are 10 data zones, the zone recording scheme gives a maximum
capacity which is 1.82 times that of the single-zone recording, without

increasing disk diameter, linear density, or track density. The catch is that
it may not be practical to require $r_{ID} = r_{OD}/(n + 1)$, particularly when n
≥ 4 or when there is an inner landing zone. If $r_{ID} = br_{OD}$, but the width
of each zone is still equal, then the maximum capacity is modified as

$$
C_{max} = 2\pi D_l D_t \cdot \frac{r_{OD} - r_{ID}}{n} \cdot n \frac{r_{ID} + r_{ID} + (n - 1)\dfrac{r_{OD} - r_{ID}}{n}}{2}
$$

$$
= \pi D_l D_t r_{OD}^2 (1 - b)[2b + \frac{n - 1}{n}(1 - b)].
$$

(15.6)

For example, if $n = 10$, $b = 1/2$, then the maximum capacity is 1.45 times
that of single-zone recording. This maximum data capacity is about 80%
of the maximum obtained without the constraint of ID radius.

15.3 POSITION ERROR SIGNAL

The data recorded in servo sectors must be properly positioned and encoded to generate PES signal and track number (ID), and a common
approach is shown in Fig. 15.4. The servo sector is composed of preamble,
timing synchronization mark, track address information, and servo pattern which generates PES. The preamble consists of equally spaced magnetic transitions (low-density square wave recording pattern) extending

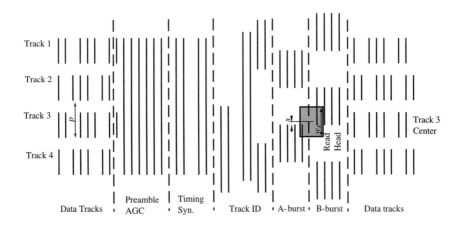

FIGURE 15.4.   Example of servo sector pattern. The solid lines represent magnetic
transitions, and the two outer dashed lines denote the extent of the servo sector.

from ID to OD. This marks the start of a servo sector, and enables the automatic gain circuit (AGC) to hold the gain constant within the servo sector. The latter function allows PES to be computed faster. Timing synchronization marks are necessary for the PES-generating demodulator circuit to function (to be discussed in Fig. 15.5). Track ID (address, not inner diameter) is the encoded servo track number.

In a simple scheme, the servo pattern consists of two kinds of transitions, called A-burst and B-burst, as shown in Fig. 15.4. These transitions are a full track pitch (p) wide while the data track width (w_w), mainly determined by write head geometry, is usually 70–80% of track pitch, and the guard band width is 20–30% of track pitch. The center of the nth data track is aligned with the edge of the neighboring nth A-burst pattern and nth B-burst pattern. The sector servo pattern can be demodulated with a circuit as shown in Fig. 15.5, which measures the integrated area of A-burst readback signal and of B-burst readback signal, denoted as $S_A(n)$ and $S_B(n)$, respectively. The readback signal, multiplied by the limiting amplifier signal gives the total integrated area $S_A(n) + S_B(n)$. If we invert the limiting amplifier signal when gate 2 goes high, then the multiplication gives $S_A(n) - S_B(n)$. Therefore, a position error signal at the nth track can be constructed as

$$PES(n) = \frac{S_A(n) - S_B(n)}{S_A(n) + S_B(n)}. \tag{15.7}$$

If the head is off the center of data track n toward data track $n - 1$ by a distance of x, $|x| \leq w_r/2$ (w_r is the read track width), then

$$S_A(n) \propto \left(\frac{w_r}{2} - x\right), \quad S_B(n) \propto \left(\frac{w_r}{2} + x\right),$$

$$PES(n) = -\frac{2}{w_r}x, \quad \text{if } |x| \leq \frac{w_r}{2}. \tag{15.8}$$

This is a linear function of x. To keep the recording head on track, the control algorithm will change the actuator current so that $x = 0$.

If the read head is near the center of track 2 instead of track 3, the slope of the PES will be positive. Therefore, the *PES* as a function of head radial position x, extending from ID to OD, is shown in Fig. 15.6. If the head track width is the same as the track pitch, then a saw-tooth waveform results. In reality, the data head track width is smaller than the track pitch, the *PES* flattens out when the data head is moving inside A-burst or B-burst:

$$x - x_n > w_r/2, \text{ but } x - x_{n+1} < -w_r/2,$$

where x_n is the position of the nth track center as shown in Fig. 15.6.

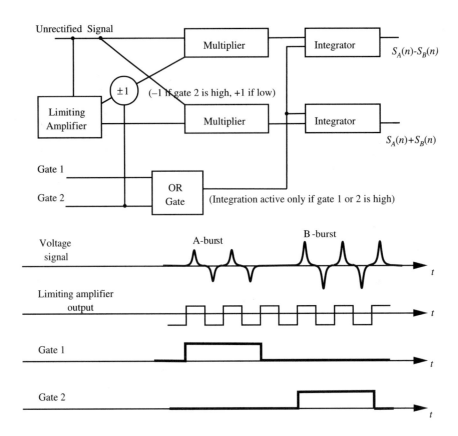

FIGURE 15.5.   Servo demodulator circuit. The limiting amplifier and the multiplier essentially rectify the voltage signal. The gate pulses can be timed by the timing synchronization marks and have the identical duration. $S_A(n)$ is the integrated area of the A-burst signal, $S_B(n)$ is that of the B-burst signal.

The PES is a periodic function with a period of $2p$. In other words, it cannot tell us the head radial position uniquely unless within the same period. To resolve this uncertainty, the sector track address is encoded in the servo data pattern. A unique track ID pattern is read by the data head when it is on that track. To resolve the uncertainty of the head position when the PES flattens out, the servo sector can be modified by adding C-burst and D-burst patterns, as shown in Fig. 15.7. The A-burst and B-burst patterns are the same as in Fig. 15.4, used to generate the normal PES. The C-burst and D-burst patterns, offset from the former by a quarter of the PES signal period ($2p$), are used to generate the so-called quadrature

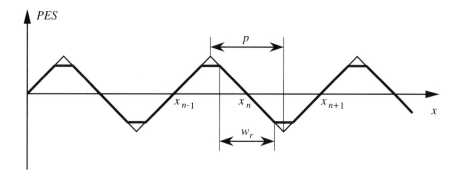

FIGURE 15.6. *PES* vs. head radial position x, where x_n is the radial coordinate of the nth track center. The thin line denotes the PES when $w_r = p$, and the thick line denotes the PES when $w_r < p$.

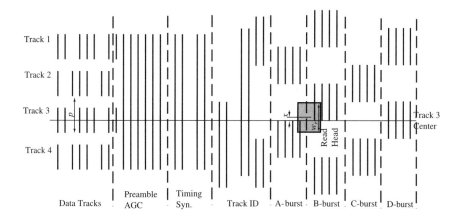

FIGURE 15.7. Normal and quadrature servo pattern. Quadrature means 90° phase shift, or $p/2$ shift. Here the period is $2p$.

PES, as shown in Fig. 15.8. Now head position can be unambiguously determined at any location, by using the normal PES signal when $|x - x_n| \leq p/4$ and the quadrature PES signal when $|x - x_n - p/2| \leq p/4$.

Servo sectors are written in the factory, and cannot be altered by the users. To write the servo data pattern, accurate radial position increments of exactly half of track pitch and accurate phase coherence of written magnetic transitions are required. This task typically requires a special servowriter consisting of a laser interferometer for radial positioning and

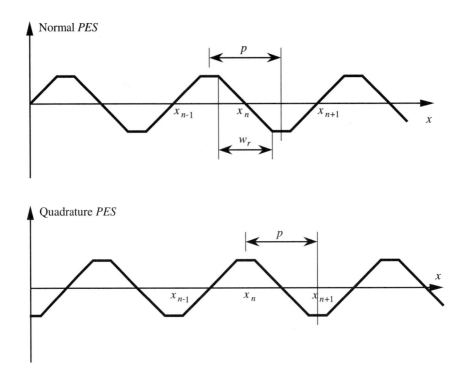

FIGURE 15.8.   Normal and quadrature PES vs. head radial position.

a clock head for phase information.[6] The read/write heads in a disk drive are used to write the servo data pattern with the assistance of the laser interferometer and the clock head.

The design of head servomechanism ultimately aims to achieve two goals: (1) to reduce the access time as much as possible without excessive power consumption and cost, (2) to reduce the write-to-write and write-to-read track misregistration. The former is mainly governed by the actuator and the control algorithm. The latter is affected by many factors such as the bandwidth of the head servo loop, the signal-to-noise ratio and the linearity of PES signal, the precision of the servowriter, the radial runout of data tracks, and cooling air.

There have been many innovations and new developments in head servomechanism including actuator designs, servowriting, and PES signal generation. These are beyond the scope of the text, and the readers are referred to the literature.

References

1. M. L. Workman, "Head-positioning servomechanism," in Magnetic Storage Handbook, edited by C. D. Mee and E. D. Daniel (New York, NY: McGraw-Hill, 1996).

2. R. K. Oswald, "Design of a disk file head-positioning servo," *IBM J. Res. Develop.,* November 1974, p. 506.

3. A. R. Hearn, "Actuator for an eight-inch disk file," *IBM Disk Storage Technology,* published by IBM, San Jose, 1980.

4. R. K. Oswald, "Position error signals," Institute of Information Storage Technology Short Course, Santa Clara University, Dec., 1993.

5. A. B. Marchant, *Optical Recording,* (Reading, MA: Addison-Wesley, 1990), p. 260.

6. C. Lee, "Servowriters: a critical tool in hard disk manufacturing," *Solid State Technology,* 207, May 1991.

CHAPTER 16

Fundamental Limitations of Magnetic Recording

The evolution of magnetic information storage technology is one of the examples that the performance of a product improves exponentially, yet its cost drops exponentially. The areal density of magnetic recording has gained more than 6 orders of magnitude since its introduction in the late 1950s, and it has been growing at an annual rate of ~60% since the early 1990s. The state-of-the-art products have areal densities of about 5 Gbit/in.^{2} and data rates up to 30 Mbyte/s. If the trend continues, the areal density of 40 Gbit/in.^{2} will be reached in 4–5 years. The projected data rate of magnetic recording is rising at ~40% annually, and will exceed 100 Mbyte/s even sooner.

The traditional and so far highly successful approach to continuously improve disk drive performance is based on a *scaling* principle. For example, in order to gain a twofold increase in areal density, the relevant head and medium dimensions such as head gap, track width, medium grain size, medium thickness, and magnetic spacing (between head and medium) are reduced by the square root of two. The resulting loss of the signal amplitude is compensated by more sensitive read head design, including MR and GMR heads. Less traditional approaches such as proximity recording, contact recording and keepered recording, have also been proposed, and practiced with various degrees of success.

Curiously enough, the "death" of magnetic recording was predicted many times for the last 15 years, whenever there were perceived technical obstacles. One of these predictions in the early 1970s claimed that the areal density of magnetic recording would be ~100 Mbit/in.^{2} because of the difficulty to achieve the extremely low flying heights needed. However, the prediction crumbled as the new disk polishing processes and head-slider designs were developed. The flying height in the modern drives is as low as 20–30 nm, and contact recording appears feasible.

While the areal densities in disk drives are limited by practical engineering considerations such as SNR and fly height, there are several

479

ultimate limits to conventional longitudinal magnetic recording on continuous magnetic medium. These limitations will be considered in this chapter. One such limitation is related to the magnetic grain size and the thermal stability of written bits. Another limitation is the achievable recording speed, which could be limited by the properties of the magnetic materials of the head and medium and by the write electronics.

16.1 SUPERPARAMAGNETISM AND TIME-DEPENDENT COERCIVITY

As we will learn in this section, the thermal effect is a fundamental limit in magnetic recording, and it is a double-edged sword! On one hand, magnetic recording media must be able to retain recorded magnetization pattern over the desired storage time, typically greater than 10 years. However, if spontaneous magnetization reversal occurs to individual medium grains, the magnetization pattern will decay and the recorded information will be lost unexpectedly. This must be avoided in the design of magnetic disk drives. On the other hand, the thermal effect makes medium coercivity time dependent. The faster the magnetization is switched, the higher the coercivity. This does not help high-speed recording at all.

16.1.1 Superparamagnetism

The spontaneous magnetization reversal is caused by thermal energy fluctuations. The probability of magnetization reversal per unit time can be approximated by the Arrhenius equation:

$$f = f_0 \exp\left(-\frac{\Delta E}{kT}\right), \tag{7.3}$$

where k is the Boltzmann's constant, T is the absolute temperature, ΔE is the energy barrier, associated with the switching process, and f_0 is the so-called "attempt frequency" to cross the barrier. The attempt frequency is determined from the precession frequency of magnetic materials and spin-lattice relaxation time. The estimated value of f_0 is approximately 10^9 Hz. The Arrhenius equation indicates that the mean time for the spontaneous magnetization reversal to occur is

$$\tau = \tau_0 \exp\left(\frac{\Delta E}{kT}\right), \tag{16.1}$$

where $\tau_0 = 1/f_0 = 10^{-9}$ s.

What is the energy barrier ΔE associated with the spontaneous magnetization reversal process? It is mainly determined by the magnetic anisotropy energy of the ferromagnetic particles or grains undergoing reversal. Consider a prolate spheroidal particle with semiaxes a and b, as shown in Fig. 16.1. In the absence of external magnetic field, the particle magnetization will be aligned along the easy axis, with its energy being at the minimum.[1] When a field H is applied, the magnetization rotates in the direction of applied field. The *energy density* (energy per unit volume) of the particle is given by:

$$E/V = K\sin^2\phi - \mu_0 H M_s \cos(\theta - \phi),\qquad(16.2)$$

where M_S is the saturation magnetization of the particle, V is the particle volume, and K is the uniaxial anisotropy constant. If the shape anisotropy is dominant, then

$$K = \frac{1}{2}(N_b - N_a)\mu_0 M_s^2,\qquad(16.3)$$

where N_a and N_b are the demagnetizing factors along the main axes of the ellipsoid. In the absence of an applied field, the energy barrier is $\Delta E = KV$, which is the difference between the maximum energy ($\phi = \pi/2$) and minimum energy ($\phi = 0$).

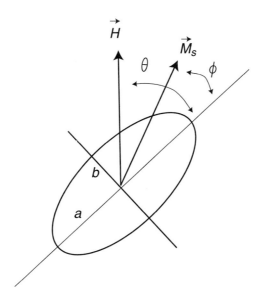

FIGURE 16.1. Single-domain prolate spheroidal particle.

If a magnetic particle is sufficiently small, its energy barrier eventually becomes comparable to thermal energy and its mean time of spontaneous reversal becomes much smaller than the observation time. Consequently, its magnetization reverses easily due to thermal energy fluctuation, just like magnetic dipoles in a *paramagnetic* material. However, the magnetic moment of the magnetic particle here is much larger because it contains thousands of atoms. Therefore, such magnetic particles are called *super-paramagnetic*, a term first coined by Bean. The magnetization of a collection of identical superparamagnetic particles can be described by the Brillouin function of a paramagnet, which has no magnetic hysteresis.

If a bit of information is to be stored by the spheroidal particles, safely for $t > 10$ years (3×10^8 s), then $f < 3.33 \times 10^{-9}$ Hz. This means that $\Delta E > 40kT$, or $V > 40kT/K$. For a typical thin-film medium, the effective anisotropy constant K is about 2×10^6 erg/cm^3. At the room temperature $T = 300$ K, $kT = 4.14 \times 10^{-14}$ erg. Therefore, the particle volume V should be greater than 828 nm^3. For a cubic grain this means that its linear dimension should be larger than ~9 nm. This limit is often called the *superparamagnetic limit* in magnetic recording, although its numerical values are dependent upon the magnetic materials and engineering assumptions used. More sophisticated methods based on micromagnetic modeling agree that magnetic bits may become thermally unstable at the room temperature if the medium grain size is smaller than 10–12 nm. Based on SNR considerations, a magnetic bit must contain at least 100 thermally stable grains (Chapter 9). The ultimate areal density of the conventional longitudinal magnetic recording due to superparamagnetic limit is approximately 40–70 Gbit/in.^{2}, depending on the assumptions of medium thickness and material parameters.

Obviously, if the anisotropy constant K can be raised, then the medium grain size can be reduced further. Therefore, the search for high anisotropy magnetic media is now a hot topic. However, we will see later that this approach is limited by the write capability of recording heads.

16.1.2   Magnetic viscosity or aftereffect

When the mean time of spontaneous reversal due to thermal fluctuation is comparable to observation time, the measured magnetization is a function of measurement time. This phenomenon is generally referred to as *magnetic viscosity* or *magnetic aftereffect.*

How does the magnetization M of a collection of independent particles vary with time because of thermal fluctuation? The probability of the magnetization *not* having hopped the energy barrier is

$$P(t) = e^{-t/\tau},$$

while the probability of the magnetization having hopped is

$$1 - P(t) = 1 - e^{-t/\tau}.$$

Therefore, the magnetization is expected to vary exponentially:

$$M(t) = M_0 e^{-t/\tau} + M_\infty (1 - e^{-t/\tau}),$$

where M_0 and M_∞ are the initial and final saturated magnetization, respectively. In the presence of a constant opposing magnetic field, $M_\infty = -M_0$, then

$$M(t) = M_0[2\exp(-t/\tau) - 1] = M_0\left[2\exp\left(-f_0 t \exp\left(-\frac{\Delta E}{kT}\right)\right) - 1\right]. \quad (16.4)$$

Note that the "exponential of exponential" term in Equation (16.4) implies that the magnetization decay is extremely sensitive to the thermal effect. As the ratio of $\Delta E/kT$ decreases, the transition from the stable regime for information storage to the rapid decay regime is very abrupt, like jumping of a cliff!

An illustration of the "cliff" effect is given in Fig. 16.2, which is plotted based on Equation (16.4) with an assumption of $\Delta E = KV = Kd^3$, where d is the linear dimension of the magnetic grain, $T = 300$ K, $kT = 4.14 \times 10^{-14}$ erg, and $K = 2 \times 10^6$ erg/cm^3. The linear grain size is varied from 10 to 9 nm, with steps of 0.1 nm. A sharp decay of magnetization is observed below $d = 9.2$ nm within 10 days from the initial recording. The extreme sensitivity of the magnetization decay to the grain volume presents a serious problem in ultra-high-density magnetic recording.

The ultra-high-density medium consists of very small grains, but they have a distribution of anisotropy and size. As a result, the grains with smaller volumes or anisotropy constants will decay faster than the rest of the medium. Therefore, the observed dynamics of magnetization decay is more like the superposition of exponential decays with different rates. Experimentally observed magnetization decay is approximately linearly dependent on the logarithm of time:

$$M(t) = C - S\log t, \quad (16.5)$$

where C is a constant, and S is known as the *magnetic viscosity coefficient*,[2–5] which is often given as the percentage of the magnetization decay per decade of time. A "high-viscosity" material will exhibit a large decay.

It can be argued that Equation (16.5) is only valid in a limited range of time, for experimental observation time is always finite. The magnetization decays for CoCrPt/CrTi thin-film media with selected medium thicknesses are shown in Fig. 16.3, where the 10-nm-thick medium exhibits a large decay with time, while the 20-nm-thick medium shows no sign of

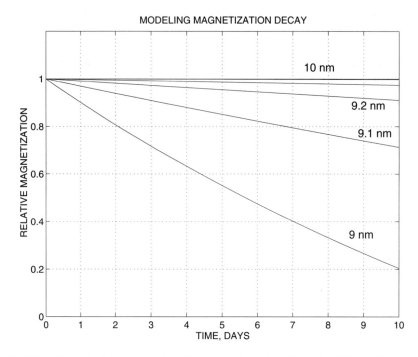

FIGURE 16.2. Model of magnetization decay based on Equation (16.4). The grain size is varied from 10 to 9 nm.

magnetization decay. It is very difficult to measure the exact grain sizes of thin magnetic media. By a rough estimate, however, the grain size of the 10-nm-thick medium is on average 20–30% smaller than that of the 20-nm-thick medium.

The magnetization decay is very sensitive to the value of the energy barrier ΔE, which can be lowered by the presence of a magnetic field (to be shown in the next section). Therefore, the demagnetizing fields from adjacent transitions will exacerbate the decay, and high-frequency magnetization patterns such as dibits and transition bursts will decay faster than isolated transitions.

To understand the energy barrier reduction due to a magnetic field, consider the switching process of a single domain particle in Fig. 16.1. Assuming that an external magnetic field H is applied along the easy axis (EA) and opposite to the initial magnetization, i.e., $\theta = \pi$, Equation (16.2) can be simplified as

$$E/V = K\sin^2\phi + \mu_0 M_S H\cos\phi. \tag{16.6}$$

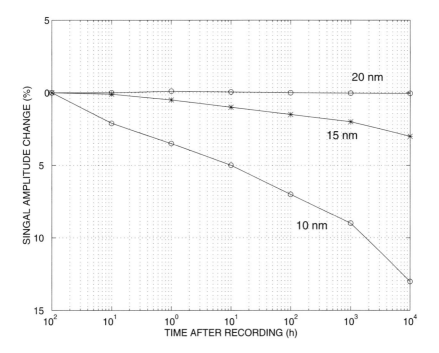

FIGURE 16.3.   Experimental measurements of magnetization decay for thin film magnetic media at selected thicknesses.

There are two local minimums of energy at $\phi = 0$: $E_{min} = \mu_0 M_S HV$ and $\phi = \pi$: $E_{min} = -\mu_0 M_S HV$, respectively. In order to switch the magnetization from $\phi = 0$ to $\phi = \pi$, the magnetization must jump the peak energy barrier, which can be found by letting the derivative of Equation (16.6) be zero:

$$\frac{1}{V}\frac{dE}{d\phi} = 2K\sin\phi\cos\phi - \mu_0 M_S H\sin\phi = 0, \tag{16.7}$$

$$\cos\phi = \frac{\mu_0 M_S H}{2K}.$$

Substituting into Equation (16.6), we obtain the energy peak:

$$E_{max}/V = K\left[1 + \left(\frac{\mu_0 M_S H}{2K}\right)^2\right] \tag{16.8}$$

and the energy barrier is calculated as:

$$\Delta E = E_{max} - E_{min} = KV\left[1 + \left(\frac{\mu_0 M_s H}{2K}\right)^2\right] - \mu_0 M_s HV$$

$$= KV\left(1 - \frac{\mu_0 M_s H}{2K}\right)^2 = KV\left(1 - \frac{H}{H_0}\right)^2,$$

(16.9)

where the ratio $H_0 = 2K/\mu_0 M_s$ is often referred to as the "intrinsic" coercivity (or anisotropy field as in Chapter 1). When $H = H_0$, the energy barrier $\Delta E = 0$, meaning that the magnetization reversal takes place when external field $H \geq H_0$, even in the absence of thermal energy.

The above derivation assumed a perfect alignment of applied field and EA, which is not realistic for magnetic media with a distribution of grain orientations. The dispersion of EAs and the interactions among the particles lead to a modified version of Equation (16.9):

$$\Delta E = KV\left(1 - \frac{H}{H_0}\right)^m,$$

(16.10)

where power m takes a value between 3/2 and 2.

Equation (16.10) demonstrates that external magnetic fields may decrease the energy barrier and accelerate thermal decay. This is responsible for the difference observed between the magnetization decay of isolated transitions and that of high-frequency patterns, as shown in Fig. 16.4. The transitions recorded at high linear densities are subject to demagnetizing fields, and thus decay faster than isolated transitions.

16.1.3 Time-dependent coercivity

The thermal effect not only causes magnetization decay, but also makes coercivity time dependent.[2-5] The time-dependent *effective coercivity* $H_c(t)$ is defined as the field that causes half of the magnetization in a sample to reverse at time t after the field reversal. In other words, the magnetization at time t vanishes. Combining Equations (16.4) and (16.10), we obtain that

$$H_c(t) = H_0\left\{1 - \left[\frac{kT}{KV}\ln\left(\frac{f_0 t}{0.693}\right)\right]^{1/m}\right\},$$

(16.11)

which is sometimes called Sharrock's law as it is first popularized by M. P. Sharrock based on his studies of particulate recording media. The

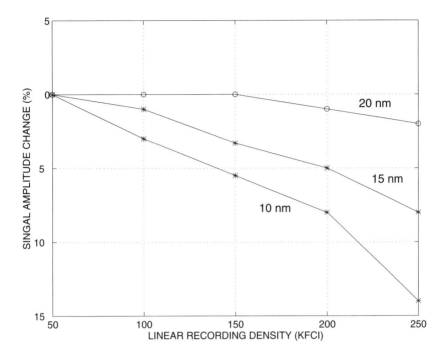

FIGURE 16.4.   Readback signal decay vs. linear densities for medium thicknesses of 20, 15, and 10 nm, respectively.

equation has also been verified for thin-film disk media in a number of experimental measurements.[6]

Figure 16.5 shows the dependence of the effective coercivity on the writing time. The coercivity becomes larger if writing is to be accomplished in a short time (on the order of 1 ns). This can be understood based on the thermal effect discussed previously. The attempt frequency for the magnetization to hop over the energy barrier is $f_0 \approx 10^9$ Hz.

When the channel bit period is as small as several nanoseconds, very few attempts of magnetization reversal due to thermal fluctuation can occur in the time frame. Therefore, magnetization reversal is driven by the external field only, and the effective coercivity is equal to the intrinsic coercivity H_0 . At a longer time scale, e.g., ~1 ms, ~10^6 magnetization reversal attempts due to thermal fluctuations can occur, so a lower magnetic field will be sufficient to cause the magnetization reversal. In other words, the effective coercivity at a long time is smaller than the intrinsic coercivity.

The time dependence of coercivity has serious implications for high-data-rate recording. The coercivity of magnetic media is traditionally

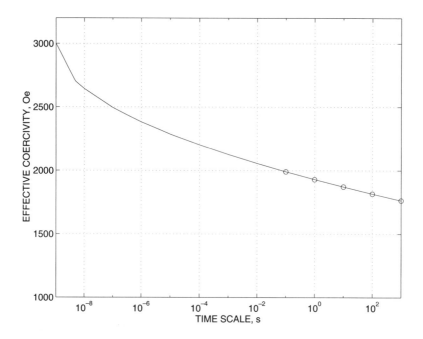

FIGURE 16.5.   Typical time dependence of effective coercivity.

measured using vibrating sample magnetometer (VSM) at a time scale of several seconds, which is neither the effective coercivity during high-frequency writing (~1 ns) nor the effective coercivity at the expected life of archival storage (~10 years). If the thermal effect is small ($KV \gg 40kT$), the effective coercivity is close to the intrinsic coercivity in all three cases. However, if the thermal effect is not small, on the one hand, higher write fields than the coercivity, measured by VSM are required to reverse magnetization in high-frequency recording. On the other hand, effective coercivity values lower than that measured by VSM are expected during archival storage. No wonder the thermal effect was called a double-edged sword.

16.2 DYNAMIC EFFECTS IN MEDIUM, HEAD, AND ELECTRONICS

The areal recording densities and data rates of magnetic disk drives are both increasing continuously at a fast pace. The areal density of conventional magnetic recording is limited by the superparamagnetic effect, what limits the date rate?

The data rates of disk drives have been rising at ~40% annually in the 1990s. The state-of-the-art disk drives may operate at a user data rate of ~30 MB/s, which translates into a channel data rate of 270 Mbit/s using 8/9 modulation code. The corresponding bit period is 3.7 ns. The bit period for a user date rate of 60 MB/s is 1.85 ns, and the corresponding channel data rate and recording frequency are 540 Mbit/s and 270 MHz respectively. As the recording frequency approaches GHz range and the disk velocity goes up, a number of frequency-dependent *dynamic* effects in magnetic media, write heads and channel electronics are becoming limiting factors to the continual increase in data rates.

16.2.1 Medium switching time

The most fundamental constraint to data rate could be the magnetic medium switching time, which is determined by at least two limiting mechanisms. One is the time-dependent coercivity, which was analyzed in the previous section. Another limiting mechanism is the gyromagnetic switching speed,[6-8] which will be discussed here.

When a magnetic field is applied to an isolated magnetic grain, its magnetization reversal is not instantaneous, but rather a damped gyromagnetic precession process. The magnetization reversal dynamics is described by the Landau-Lifshitz-Gilbert equation:

$$\frac{d\vec{M}}{dt} = -\gamma(\vec{M} \times \vec{H}) + \frac{\alpha}{M_s}\left(\vec{M} \times \frac{d\vec{M}}{dt}\right), \tag{16.12}$$

where $\vec{M}$ is the magnetization of the grain, $\vec{H}$ is the applied magnetic field, γ is the gyromagnetic ratio, M_s is the saturation magnetization, and α is the *damping constant* (with no dimension). Equation (16.12) is mathematically equivalent to the Landau-Lifshitz equation:

$$\frac{d\vec{M}}{dt} = -\gamma'(\vec{M} \times \vec{H}) + \frac{\lambda}{M_s}[\vec{M} \times (\vec{M} \times \vec{H})],$$

where

$$\gamma' = \frac{\gamma}{1 + \alpha^2}, \lambda = \frac{\gamma\alpha}{1 + \alpha^2} = \gamma'\alpha,$$

and λ is often called the *relaxation frequency*, which should have a unit of Hz/Oe if the Landau-Lifshitz equation is used in the Gaussian unit system.

The damping constant describes the loss of energy of the grain and is analogous to "friction." When the damping constant $\alpha = 0$, the magnetization of the grain rotates continuously around the applied field with the precession frequency, and no switching will occur. This case never happens in reality because there are always some damping force, no matter how small it is. When damping is nonzero, the grain loses energy and the magnetization will align to the applied field after a number of precession cycles. The switching processes are schematically illustrated in Fig. 16.6. If $\alpha \ll 1$, the switching process will take many precession cycles, so the switching time is long. If $\alpha \gg 1$, the magnetization will rotate directly toward the applied field without making many precessions, but the switching time is long too because the "friction" to magnetization motion is too large. The shortest switching time is obtained at a moderate damping constant $\alpha = 1$. At the so-called *critical damping* condition, the switching time is equivalent to $1/\pi$ of precession cycle time[9]:

$$\tau_{min} = \frac{2}{\gamma H},$$
(16.13)

where H is the applied field, $\gamma \approx 17.5$ MHz/Oe. The observed damping constant for magnetic materials is generally in the range between 0.01 and 0.1. At present, there is no precise measurement of the damping constant in magnetic media. Some data obtained from ferromagnetic reso-

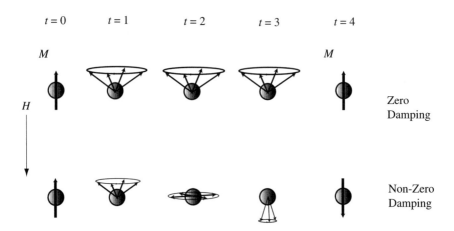

FIGURE 16.6. Simplified illustration of the gyromagnetic switching processes. Magnetic field H is switched downward at $t = 0$. Magnetization cannot be switched to the field direction without damping.

nance studies indicate that the damping constant for Co alloys could be close to 0.01.[6]

The Landau-Lifshitz-Gilbert equation describes the magnetization reversal of a single domain particle. However, there are usually some interactions between grains in magnetic media, such as the demagnetizing field and intergranular exchange coupling. Therefore, complicated micromagnetic models have been developed to simulate the magnetization reversal in longitudinal thin-film media.[8] For $\alpha \approx 0.01 - 0.05$ and the typical parameters of modern magnetic media, the medium switching time is estimated to be approximately 0.5 ns. This corresponds to a maximum recording frequency of ~1 GHz. However, given the uncertainty in the values of the damping constant, the actual medium switching time could be several times smaller.

The impact of medium switching time on the recording process is now masked by head field rise time. As discussed in Chapter 5, the rise time of a typical writing head is on the order of several nanoseconds, significantly larger than the medium switching time. However, there are indications that the medium switching time could be deteriorated by the field rise time. The larger the head field rise time, the smaller the magnetic field applied to the grain over a short time scale, meaning that a gyromagnetic precession cycle takes longer. Generally speaking, a larger and faster field pulse will switch the medium magnetization more quickly.

The problem of medium switching is an area of active research. The available measurements and models agree that the medium switching time for a typical magnetic thin-film medium could be on the order of hundreds of picoseconds (0.1–0.5 ns). This limits the achievable recording frequency to several GHz. In other words, the maximum channel data rate due to the medium switching limit is several hundreds of megabytes per second.

16.2.2 Write head and electronics

The problem of the head field rise time was already addressed in Chapters 5 and 10. The total write field rise time is approximately determined as[10,11]:

$$\tau_f = \sqrt{\tau_h^2 + \beta^2 \tau_c^2}, \tag{5.10}$$

where $\beta \leq 1$ is the coefficient depending on the write current, τ_h is the intrinsic head rise time and τ_c is the write current rise time. As was discussed in Section 10.2.5, the field rise time introduces complicated pattern-dependent transition shifts into the recorded data and limits the achievable data rate of magnetic recording.

Even if an ideal step of write current ($\tau_c = 0$) is applied to the write head, its intrinsic flux response time will not be zero. Currently the intrinsic head rise time is ~3 ns at best. To make τ_h small, the yoke length should be as short as possible. This requires multilevel coils with closely packed coil turns under the head yoke. The saturation of the head apex slows down the rise of head field, and should be avoided. This means that high-saturation soft magnetic materials should be employed in high-data-rate write heads.

The eddy currents in the yoke oppose the change of yoke magnetization and slow down the magnetization reversal. The eddy currents effectively reduce the magnetic yoke thickness to its skin depth, which is equal to

$$\delta = \sqrt{\frac{2\rho}{\omega\mu\mu_0}},$$

where ρ is the yoke resistivity, ω is the angular frequency of write current, and μ is the (relative) DC magnetic permeability of the yoke material. The minimization of the eddy currents can be achieved either by using high-resistivity yoke materials, or by lamination of the yoke with insulating materials. Lamination is expensive in manufacturing, but may be the only feasible solution to create effective high frequency write head. The choice of the DC permeability is determined by two conflicting requirements: On one hand, it must be large enough to give a sufficiently large head efficiency; on the other hand, it must be small enough to give a sufficiently large skin depth at high frequencies.

Similar to medium switching time, the gyromagnetic motion of yoke magnetization as governed by the Landau-Lifshitz-Gilbert equation also contributes to the head field rise time. This is the ultimate limit to the achievable write head frequency, and depends mainly on the magnetic properties of head yoke material.

On another front, write drivers are being continuously improved to shorten the current rise time τ_c. The peak-to-peak write driver output voltage is

$$V_{p-p} = 2(V_s - 2\Delta V),$$

where V_s is the power supply voltage, and ΔV is the voltage drop across the active devices in the write driver. The base-to-peak write current I_w can be determined from:

$$V_{p-p} = 2\left(L\frac{dI}{dt} + IR\right) = 4L\frac{I_w}{\tau_c} + 2I_wR, \tag{16.14}$$

where L is the sum of head and interconnect inductance and R is the total series resistance at the output of the write driver. Therefore, a smaller value of τ_c requires a higher peak-to-peak voltage, and subsequently a higher power supply voltage. However, a higher power supply voltage is incompatible with high-speed silicon VLSI and drives up power consumption. The state-of-the-art write drivers have a current rise time close to 1.5–2 ns.

Finally, it should be mentioned that read heads such as MR heads or GMR heads are generally faster than write heads, so they are not limiting factors to achieving high data rates. However, this scenario may change if new materials and devices such as spin-dependent tunneling junctions are used for read heads.

References
1. E. C. Stoner and E. P. Wolfarth, "A mechanism of magnetic hysteresis in heterogeneous alloys" *Phil. Trans. R. Soc. Lond.,* Ser. A, **240**, 599, 1948.
2. M. P. Sharrock. "Time-dependence of switching fields in magnetic recording media," *J. Appl. Phys.,* **76**(10), 6413, 1994.
3. M. P. Sharrock, "Time-dependent magnetic phenomena and particle-size effects in recording media," *IEEE Trans. Magn.,* **26**, 193, 1990.
4. S. H. Charap, P. Lu, and Y. Ho, "Thermal stability of recorded information at high densities," *IEEE Trans. Magn.,* **33**, 978, 1997.
5. P. Lu and S. H. Charap, "Magnetic viscosity in high-density recording," *J. Appl. Phys.,* **75**(10), 5768, 1994.
6. S. M. Stinett, W. D. Doyle, P. J. Flanders, C. Dawson, "High speed switching measurements in CoCrPtTa thin film media," *IEEE Trans. Magn.,* **34**, 1828, 1998.
7. J. Zhu and H. N. Bertram, " Micromagnetic studies of thin metallic films," *J. Appl. Phys.,* **63**(8), 3248, 1988.
8. Q. Peng and H. N. Bertram, "Micromagnetic studies of recording dynamics and switching speed in longitudinal thin film media," *IEEE Trans. Magn.,* **33**, 2677, 1997.
9. S. Chikazumi and S. H Charap, *Physics of Magnetism,* Krieger, 1964, p. 336–360.
10. K. B. Klassen, R. G. Hirko, and J. Van Peppen, "High speed magnetic recording," *IEEE Trans. Magn.,* **34**, 1822, 1998.
11. K. B. Klaassen and R. G. Hirko, "Non-linear eddy-current damped thin film write head model," *IEEE Trans. Magn.,* **32**, 3254, 1996.

CHAPTER 17

Alternative Information Storage Technologies

This book is primarily devoted to magnetic recording, particularly magnetic disk drives. There are many variations of magnetic information storage technology, such as magnetic floppy disk drives, magnetic tape drives, and removable hard disk drives. However, there are even more varieties of alternative information storage technologies commercialized or under development. These include optical disk recording,[1] near-field optical recording,[2] holographic recording,[3] semiconductor flash memory,[4] magnetic RAM,[5] scanning probe recording,[6] electron trapping optical memory (ETOM),[7] etc. In this chapter we will briefly discuss some of these technologies as representative alternatives to magnetic recording.

There were many historical predictions that magnetic recording would be replaced by other technologies. All have turned out to be wrong so far. It is now widely believed that magnetic recording will remain dominant in the foreseeable future. However, we are indeed dealing with a broad field called "information storage," and we should always keep an open mind toward alternative or new technologies. A good analogy of this is the transportation technologies. Automobiles are the dominant means today, but in long run, we should keep an eye on small planes, bullet trains, jumbo jet liners, and even flying carpets in science fictions. In this sense, the following sections serve the purpose of keeping us in a good perspective.

17.1 OPTICAL DISK RECORDING

Optical recording generally refers to the recording systems that use light beams to record or recover information. For example, compact-disk-read only memory (CD-ROM) using light to read information is regarded as an optical recording device. Most optical recording systems employ disks

495

or tapes as storage media. Holographic recording belongs to optical re-cording, but the storage media are very different from optical disk re-cording in nature, so it will be treated separately in Section 17.2. Optical tape recording is very similar to optical disk recording except that the latter employs flexible tape media, so we will focus on optical disk recording hereafter.

The principle of optical disk recording can be generally illustrated in Fig. 17.1. The light is focused by an objective lens to form the *optical stylus,* i.e., a tightly focused spot of light. Diffraction effects related to the wave-like nature of light limit the minimum size of the focused spot. The tightest possible focus is achieved using a coherent light source such as a laser. The smallest possible diameter is

$$D \equiv 0.56 \frac{\lambda}{NA}, \tag{17.1}$$

where NA is the numerical aperture of the objective lens, and λ is the wave length of the laser.

17.1.1 CD-ROM

The read-only optical disks, including audio compact disk (CD) and CD-ROM, were first introduced by Philips and Sony in 1980 and 1985, respec-tively. Digital information is replicated into these disks in factories and cannot be altered by users. CD-ROM is a direct extension of audio CD. The main difference between audio CD and CD-ROM is the formatting of their digital data: the former is formatted for audio applications while the latter is for computer applications. CD-ROM must have superior data reliability than audio CD. Nowadays a CD-ROM drive can play audio CDs with appropriate software, but not vice versa.

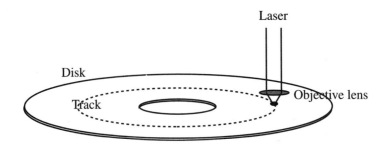

FIGURE 17.1.   Schematic of typical optical disk recording device.

The CD-ROM technology is very important and provides the techno-logical base for more advanced optical recording devices. The substrate manufacturing methods for read-only disks are used to make substrates for write-once and erasable optical disks. A CD is a plastic (polycarbonate) disk with an outer diameter of 120 mm (4.7 in), an inner diameter of 15 mm, and a thickness of 1.2 mm. As illustrated in Fig. 17.2, a CD is basically composed of an injection-molded substrate, an aluminum reflector layer (~100 nm thick), and a plastic protection layer (~10–20 μm). The protec-tion layer is usually a UV-cured acrylate, on which the CD title information is printed.

The principle of CD readback[8] is shown in Fig 17.3. The reflected light beam from a bump will interfere destructively with the reflected light beam from grooves because of a 180° phase difference, leading to a weak signal. On the contrary, the absence of a bump will lead to a strong reflected signal. The effective change in optical reflectivity is typically

FIGURE 17.2. Schematic cross section of CD structure (not to scale).

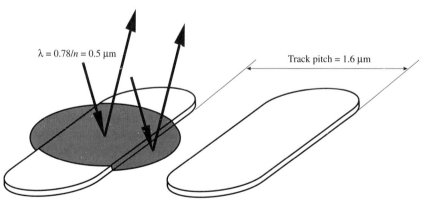

FIGURE 17.3. CD readback principle. Note that the wavelength in the substrate is the laser source wavelength divided by the substrate refractive index.

~70%. Therefore, we can detect information recorded on the disk. For example, a bump can be encoded as a "1", while no bump can be encoded as a "0".

By design the light beam enters and leaves through a 0.7-mm-diameter circle for each bit from the substrate side instead of the protection layer side, as shown in Fig. 17.4. As a result, surface debris has to be three orders of magnitude larger than the readout spot before the beam is obscured. This feature is very important in that bare CDs are frequently handled by users and may catch dust and finger prints. In contrast, floppy magnetic disks (including Iomega's zip™ disks) must be carefully protected from users with elaborately designed cartridges.

The Iomega zip™ disk drive is an advanced type of floppy disk drive, which has become a popular backup storage and distribution media in recent years. The major technological breakthrough over old floppy disk drive is that the zip™ disk drive employs a "centered disk stabilization" scheme such that a flexible disk can rotate steadily at as much as 3000 RPM.[9] The old floppy disks typically rotate at 300 RPM.

The data capacity of the CD-ROM has remained ~680 MB because of the compatibility and standard considerations. However, the data transfer rate of CD-ROM has steadily risen with increasing disk rotation speeds (from 530 RPM to 12,720 RPM). For example, the data transfer rate of the "24X" CD-ROM now available can be as much as 24 times of that of the first-generation "1X" CD-ROM. It is expected that CD-ROMs will be gradually superseded by the digital video disk or digital versatile disk (DVD), which will be discussed in the next section.

One of the major applications of CD-ROM and DVD is information distribution. They are compared with other information distribution

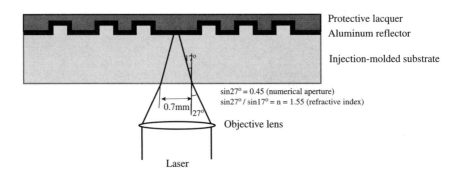

FIGURE 17.4.   The light beam enters the CD through a 0.7-mm-diameter circle. The numerical aperture of the objective lens is 0.45.

TABLE 17.1. Comparison of Selected Information Distribution Media.

Media	Information per unit	Equivalent book pages
Book	500 pages	500
Microfilm	4000 pages	4,000
Microfiche	300 pages	300
CD-ROM	680 MB	240,000
DVD-5	4.7 GB	1,700,000
3.5 in. floppy disk	1.4 MB	500
Iomega zip™ disk	100 MB	36,000

media in Table 17.1. In calculating equivalent book pages, we have assumed that 1 MB = 360 text pages. Obviously a graphics page will take much more than 3 KB of disk space. Nevertheless, it is worth noting that a CD-ROM disk could contain the information of nearly 500 average books, while a DVD-5 disk[1] could contain 3400 average books. More importantly, digital information can be accessed and searched much faster than books.

17.1.2 DVD

When CD-ROM was developed in the 1980s, a storage capacity of ~680 MB was regarded as almost unlimited. This abundance of storage capacity proved to be short-lived. By the middle 1990s, a new generation of optical disk technology is badly needed to address the large data storage requirements of video, image, multimedia, database, data backup, and other applications. Although the core technologies for DVD, the new generation of optical disks, had been attained for a while, several issues such as manufacturing standard delayed DVD commercialization until late 1997.

Both DVD-read-only (DVD-ROM) and DVD random access rewritable (DVD-RAM) are now commercially available for computer applications. The former is essentially a down-scaling of CD-ROM, while the latter employs rewritable phase-change optical recording. At the same time, DVD-audio and DVD-video[10] have been developed for the music and movie industries. The physical format of DVD-ROM and CD-ROM are

---

[1]DVD-5, DVD-9, and DVD-18 refer to different DVD-ROM formats that have data capacities of 4.7 GB (single-sided, single-layer), 8.5 GB (single-sided, two-layers), and 17 GB (double-sided, four layers), respectively.

TABLE 17.2.   DVD-ROM vs. CD-ROM

	CD-ROM disk	DVD-ROM disk
Pit length (μm)	0.83	0.4
Track pitch (μm)	1.6	0.74
Laser wavelength (μm)	0.78 (infrared)	0.635–0.65 (red)
Numerical aperture	0.45	0.6
Linear bit density (b/in.)	43,000	96,000
Track density (track/in.)	16,000	34,000
Areal density (Gb/in.^{2})	0.68	3.3
Data rate (Mb/s)	≤14.4 (12X)^a	≤21.6
Average access time	125 ms (12X)	180 ms
Disk physical diameter	120 mm (5 in.)	120 mm (5 in.)
Disk structure	one 1.2-mm substrate	two 0.6-mm substrates
		4.7 (single-sided, one layer)
Capacity (GB)	0.68 (single-side)	8.5 (single-sided, two layers)
		9.4 (double-sided, one layer)
		17 (double-sided, four layers)

^{a}The data rate and access time of CD-ROM depend on disk rotation angular velocity: 1X: 200–530 RPM, 4X: 800–2120 RPM, 12X: 2400–6360 RPM. The angular velocity varies even in the same mode because CD-ROM disks are designed to maintain a constant linear velocity relative to the optical head from outer diameter to inner diameter.

compared in Table 17.2.[11] The substrate thickness of DVD is 0.6 mm so that the back-to-back bonded structure can be used for double-sided recording. A single substrate can take up to two layer of recording layers. Other than employing multilayer recording layers, the gain in data capacity of DVD-ROM over CD-ROM comes from adopting a shorter wavelength, a larger numerical aperture, smaller ECC redundancy and sector redundancy, a better modulation code, etc. These technical improvements allow each layer of DVD-ROM to store up to 133 minutes of full-motion MPEG-2 video,[II] a feat that has eluded CD-ROM.

How is dual-layer recording accomplished in optical recording? This is made possible by a *dual-focus* objective lens and a dual-layer disk consisting of a top semi-reflective layer and bottom fully reflective layer. For example, a Panasonic hologram integrated lens combines refraction (near lens center) and diffraction (near lens edge) to form two focal points. With such a lens, both recording layers can be read simultaneously, but the signals must be separated electrically at photodetector output.

[II]The second video compression standard issued by the Moving Picture Experts Group (MPEG).

   While the application of DVD-ROM are well defined, the role of rewritable DVD in computer applications is less clear. A comparison between DVD-RAM and magnetic hard disk drives is given in Table 17.3. At present the leading-edge optical and magnetic drives claim similar areal densities. However, the greatest drawbacks of optical disk systems are their larger access times and smaller data rates than those of magnetic hard disk systems. In addition, magnetic disk drives can stack multiple disks more easily and thus are more advantageous in realizing higher overall volume density and data capacity.

   As announced by the IBM Almaden research center in December 1997, magnetic recording with an areal density of ~11.6 Gbits/in.^{2} has been demonstrated in a laboratory scale. It is believed that the conventional magnetic hard disk recording can be extended to as much as 40–60 Gbits/in.^{2} recording densities. Consequently, magnetic disk recording will continue to maintain its advantage over optical disk recording. Nevertheless, optical disk recording offers greater reliability than magnetic disk recording because there are neither head crashes nor head/disk wear in optical disk recording. Thus optical disk recording can readily employ removable media. In that sense, DVD-RAM may compete well with removable magnetic disk drives such as Iomega zip™ and jazz™ drives.

17.1.3 Phase-change optical recording

Rewritable phase-change optical disk recording is based on phase-change media that can change from crystalline phase to amorphous phase, and vice versa, with >10^6 read/write cycles. The phase-change leads to a change in optical reflectivity typically greater than 15%. Normally, a written mark is an amorphous spot in the matrix of the crystalline phase.

TABLE 17.3. DVD-RAM vs. Leading-Edge Magnetic Disk Drives in 1997[12]

	DVD-RAM (SD-W1001)	Magnetic disk (Travelstar 4GT)	Magnetic disk (Cheetah 9)
Areal density (Gb/in.^{2})	~1.8	2.6	0.94
Linear density (b/in.)	N/A	211,000	135,000
Track density (tracks/in.)	N/A	12,500	6,932
Data rate (Mb/s)	≤10.8	≤83.2	≤170
Average access time	270 ms (write) 170 ms (read)	20 ms	5 ms
Disk physical diameter	120 mm (5 in.)	65 mm (2.5 in.)	95 mm (3.5 in.)
Capacity (GB)	2.6 (single-sided, one layer)	4.1 (three disks)	9.1 (eight disks)

Typical write/erase/read processes of the phase-change media are illustrated in Fig. 17.5, where T_m is the melting point of the media, and T_x is the crystallization temperature. The write laser pulse is very short such that the melted crystalline phase is quenched into an amorphous phase. The erasure bias power should be just high enough to crystallize the amorphous spot. The read bias power should be as low as possible to keep amorphous marks from degrading, but high enough to give a good carrier-to-noise ratio (CNR). (See Equation (17.16).)

There are many alloys researched for phase-change media. The following alloys have a higher reflectivity in crystalline phase than in amorphous phase: GeTe, Sb_2Te_3, $GeTe$-SB_2Te_3 (tie-line), etc. The current media include $Ge_2Sb_2Te_5$, $GeSb_2Te_4$, and $GeSb_4Te_7$. The following alloys have a lower

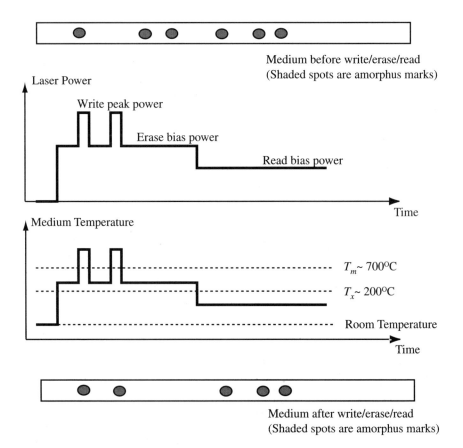

FIGURE 17.5.   Write/erase processes in erasable phase-change optical disks.

reflectivity in the crystalline phase than in the amorphous phase: In_3SbTe_2, InSeTl, InSb-GaSb (tie line), etc. An example of phase-change optical disk[13] is shown in Fig. 17.6, where the $GeTe-Sb_2Te_3-Sb$ layer is the phase-change layer, and the Al layer serves as heat sink and reflector.

17.1.4 Magneto-optic recording

So far the majority of rewritable optical disk drives are not DVD-RAM but rather magneto-optic (M-O) disk drives, which are based on the magneto-optic Kerr effect (MOKE). Magneto-optic drives in 5 1/4 in. (130-mm) form factor have been introduced since the late 1980s, but are expected to be more expensive than DVD-RAM because DVD-RAM can leverage the DVD-ROM technology and components but M-O drives require more special components.

The MOKE is illustrated in Fig. 17.7. When the magnetization is perpendicular to the magnetic film and the light beam is at normal incidence, the polar Kerr effect is observed. When the magnetization is in the film plane and parallel to the light (oblique) incidence plane, the

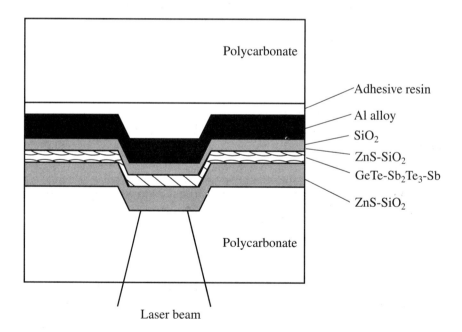

FIGURE 17.6. Example of phase-change optical disk structure (not to scale).

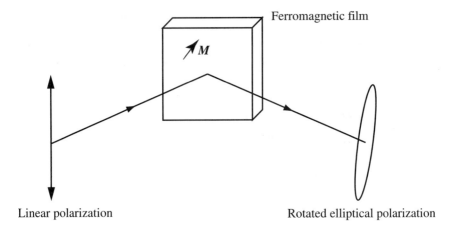

Linear polarization   Rotated elliptical polarization

FIGURE 17.7.   Magneto-optic Kerr effect.

longitudinal Kerr effect is observed. M-O recording utilizes polar Kerr effect so M-O media must possess a large perpendicular anisotropy to align magnetic domains normal to the disk plane.

A light beam can generally be represented by a *polarization vector:*

$$\begin{pmatrix} E_x \\ E_y \end{pmatrix},$$

which gives the electric field of a light wave in terms of two orthogonal (linear) polarization states. Each optical element in an optical head can be represented by a 2×2 *Jones matrix* of complex quantities,[III] which operates on the polarization vector by matrix multiplication to show how the element alters the light amplitude and polarization. At an interface between two optical media the reflected electric field is related to the incident electric field through Fresnel coefficients:

$$\begin{pmatrix} E_{p,r} \\ E_{s,r} \end{pmatrix} = \begin{pmatrix} r_{pp} & r_{ps} \\ r_{sp} & r_{ss} \end{pmatrix} \begin{pmatrix} E_{p,i} \\ E_{s,i} \end{pmatrix}, \tag{17.2}$$

where the subscript s stands for the component perpendicular to the incidence plane, and p stands for the component parallel to the incidence

[III]The convention is first introduced by Jones in 1941, thus it is called the Jones matrix.

plane. The Jones matrix for the polar M-O Kerr reflection at a magnetic surface is

$$\begin{pmatrix} r_{pp} & r_{ps} \\ r_{sp} & r_{ss} \end{pmatrix} = r \begin{pmatrix} 1 & \mp ke^{i\kappa} \\ \pm ke^{i\kappa} & 1 \end{pmatrix}, \tag{17.3}$$

where $r = r_{ss} = r_{pp}$ is the reflection coefficient, which is the square root of reflectivity R, and the alternate signs correspond to the two possible states of magnetization: up or down. Note that $r_{sp} = -r_{ps}$. The Fresnel coefficients are related to the Kerr rotation angle θ_k and the ellipticity η_k:

$$r_{ps}/r = \mp ke^{ik} \equiv \tan\theta_k + i\eta_k,$$
$$\theta_k \approx k \ll 1, \quad \eta_k \approx k\sin\kappa \ll 1. \tag{17.4}$$

The write, erase, and read processes in M-O recording are shown in Fig. 17.8. To write a downward domain in a DC magnetized (upward) M-O disk, a laser beam heats a focused area of the M-O disk above the Curie temperature (or to a certain temperature when the coercivity is less than the electromagnet bias field). Upon cooling down a magnetic domain along the bias field direction is written. This mechanism can write magnetic domains only in one direction because the electromagnet bias field cannot be switched at the data rate. Therefore, a written track needs to be erased with a constant upward bias field before new data can be written over old data. The erase power should be similar to the write power. To read back the information written on the M-O disk, a polarized laser beam (with a lower power than the write/erase power) is incident on the magnetic domain. The upward and downward magnetization give rise to opposite Kerr rotations which are distinguished by the analyzer and photodetector. The operation of the M-O read head will be further discussed in Section 17.1.5.

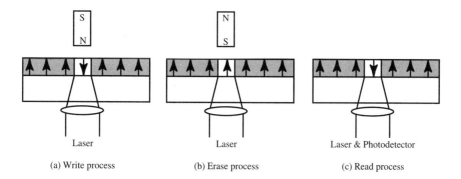

(a) Write process   (b) Erase process   (c) Read process

FIGURE 17.8. Magneto-optic write, erase and read processes.

The first generation M-O drives did not employ direct overwrite schemes, and the write-after-erasure procedure slows down the data transfer during M-O write process. There have been many schemes devised to achieve direct overwrite in M-O recording, including light modulation, field modulation, and exchange-coupled multilayer disk approaches.[14,15]

The most commonly researched and used magneto-optic media are the rare earth-transition metal (RE-TM) amorphous films. For example, many commercial M-O disks are made of GdTbFeCo alloys with a Kerr rotation angle of ~0.3° (0.005 radian) at 0.633 μm wavelength. These films are called *ferrimagnet* because they are composed of two subnetworks of magnetic dipoles: the transition metal magnetization is aligned along one direction while the rare earth magnetization along the opposite direction. Their dependence upon temperature is shown in Fig. 17.9. At the compensation temperature, the rare earth magnetization cancels the transition metal magnetization, so the net magnetization is zero. Near the compensation temperature, the coercivity approaches infinity. At the Curie temperature, both magnetizations disappear.

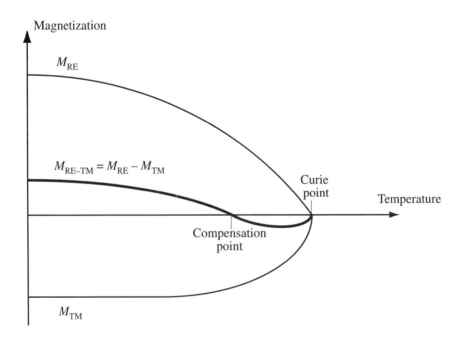

FIGURE 17.9.   Magnetization vs. temperature of RE-TM films with a compensation point.

To write, read, and erase effectively in M-O recording, the design rules for GdTbFeCo M-O media[16] are as follows:

1. Determine the Curie temperature by selecting the Co/Fe ratio. The Curie temperatures of Co and Fe are 1130°C and 770°C respectively. The Curie temperature should be much higher than room temperature, but low enough to be written or erased by laser beam.
2. Determine the compensation temperature by selecting RE/TM ratio. The compensation temperature should be close to room temperature so that the coercivity is high at room temperature.
3. Determine the magnetization-coercivity product $M_s H_c$ by selecting the Tb/Gd ratio. It is generally necessary to have either a large $M_s H_c$ product or a small domain wall energy to achieve a small domain size in M-O media.
4. Determine the anisotropy constant K_u by selecting the terbium/gadolinium (Tb/Gd) ratio and the sputtering conditions. Terbium has a high anisotropy while Gd has a low anisotropy. The M-O media must have a perpendicular anisotropy so that magnetization will be oriented out of the film plane.

17.1.5 Optical head

One of the most critical components in an optical disk drive is the optical head consisting of objective lens, diode laser, polarizer, analyzer, photodetector, etc. The Jones matrices of selected optical elements are listed in Table 17.4.

TABLE 17.4. Jones Matrices of Optical Elements.

Optical element	Jones matrix
Perfect linear polarizer aligned in the x-direction	$\begin{pmatrix} 1 & 0 \\ 0 & 0 \end{pmatrix}$
Quarter-wave plate (QWP) with its slow axis aligned in the x-direction	$\begin{pmatrix} e^{i\pi/2} & 0 \\ 0 & 1 \end{pmatrix}$
Counterclockwise rotation of the polarization or clockwise coordinate transformation by an angle of θ	$\begin{pmatrix} \cos\theta & -\sin\theta \\ \sin\theta & \cos\theta \end{pmatrix}$
Normal-incidence reflection that just reduces the overall amplitude of electric field (R is the reflectance or reflectivity)	$\begin{pmatrix} \sqrt{R} & 0 \\ 0 & \sqrt{R} \end{pmatrix}$

The operation of an optical head for CD-ROM or phase-change media is illustrated in Fig. 17.10. In a read process, the light beam transmits through a polarizing beam-splitter cube (PBSC) which is aligned with the p-polarization (E_x) of the incident beam, a coordinate rotation corresponding to the 45° alignment of the quarter-wave plate (QWP), a quarter-wave retardation, a reflection from the recording layer, a second pass through the QWP, a reverse rotation back to the coordinates of the PBSC, a deflection from the PBSC, and finally a s-polarizer. The polarization vector of the light beam detected by the photodetector is

$$\begin{pmatrix} 0 & 0 \\ 0 & 1 \end{pmatrix}\begin{pmatrix} \frac{1}{\sqrt{2}} & \frac{1}{\sqrt{2}} \\ -\frac{1}{\sqrt{2}} & \frac{1}{\sqrt{2}} \end{pmatrix}\begin{pmatrix} i & 0 \\ 0 & 1 \end{pmatrix}\sqrt{R}\begin{pmatrix} i & 0 \\ 0 & 1 \end{pmatrix}\begin{pmatrix} \frac{1}{\sqrt{2}} & -\frac{1}{\sqrt{2}} \\ \frac{1}{\sqrt{2}} & \frac{1}{\sqrt{2}} \end{pmatrix}\begin{pmatrix} 1 & 0 \\ 0 & 0 \end{pmatrix}\begin{pmatrix} E_x \\ E_y \end{pmatrix}$$

$$= \sqrt{R}\begin{pmatrix} 0 \\ E_x \end{pmatrix}.$$

This means that the detector receives all the light that initially passes through the polarizer and is then reflected from the medium surface. The optical power at the detector is equal to the reflectivity at the medium surface multiplied by the polarized laser power. Therefore, the optical power signal in optical disk recording based on reflectivity modulation is

$$P_{p-p} = P_l \Delta R, \tag{17.5}$$

where $P_l = E_x^2$ is the polarized laser power, and ΔR is the change in reflectivity between bit "1" and "0".

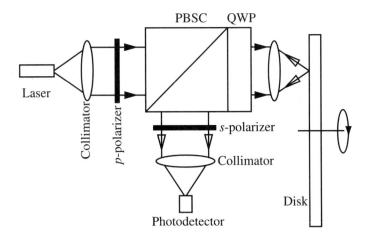

FIGURE 17.10.   Schematic of optical head operation.

The optical head is also an optical isolator. The polarization vector of the light transmitted back to the laser is

$$\begin{pmatrix} 1 & 0 \\ 0 & 0 \end{pmatrix}\begin{pmatrix} \frac{1}{\sqrt{2}} & \frac{1}{\sqrt{2}} \\ -\frac{1}{\sqrt{2}} & \frac{1}{\sqrt{2}} \end{pmatrix}\begin{pmatrix} i & 0 \\ 0 & 1 \end{pmatrix}\sqrt{R}\begin{pmatrix} i & 0 \\ 0 & 1 \end{pmatrix}\begin{pmatrix} \frac{1}{\sqrt{2}} & -\frac{1}{\sqrt{2}} \\ \frac{1}{\sqrt{2}} & \frac{1}{\sqrt{2}} \end{pmatrix}\begin{pmatrix} 1 & 0 \\ 0 & 0 \end{pmatrix}\begin{pmatrix} E_x \\ E_y \end{pmatrix} = 0.$$

This indicates that no light is reflected back to the laser, a desirable feature for reducing laser noise.

An M-O head is similar to that shown in Fig. 17.10, but a "leaky" PBSC and a different analyzer angle are used, and the QWP is not necessary. The leaky PBSC has a reduced efficiency of transmitting p-polarization compared with the standard PBSC in Fig. 17.10. In other words, the leaky PBSC transmits a fraction ($f < 1$) of p-polarized light and reflects $(1 - f)$ of the p-polarization, and reflects 100% of any s-polarized light. A perfect standard PBSC would form a well-polarized optical stylus, but it would also perfectly polarize the reflected beam in the orthogonal direction, erasing any information about the sense of Kerr rotation. Therefore, a standard PBSC cannot be used for M-O disks. In this case, the polarized beam reflects off the M-O medium surface, deflects from the leaky beam splitter, and finally passes through an analyzer oriented at θ_a (relative to the s-polarization direction). The polarization vector at the detector is

$$\begin{pmatrix} 0 & 0 \\ 0 & 1 \end{pmatrix}\begin{pmatrix} \cos\theta_a & -\sin\theta_a \\ \sin\theta_a & \cos\theta_a \end{pmatrix}\begin{pmatrix} \sqrt{1-f} & 0 \\ 0 & 1 \end{pmatrix}r\begin{pmatrix} 1 & \mp ke^{i\kappa} \\ \pm ke^{i\kappa} & 1 \end{pmatrix}\begin{pmatrix} \sqrt{f} & 0 \\ 0 & 0 \end{pmatrix}\begin{pmatrix} E_x \\ E_y \end{pmatrix}$$

$$= \sqrt{f}\,rE_x\begin{pmatrix} 0 \\ \sqrt{1-f}\sin\theta_a \pm ke^{i\kappa}\cos\theta_a \end{pmatrix}.$$

The incident laser beam is predominantly polarized in the x-direction ($E_x \gg E_y$). The case of $\theta_a = 0$ would correspond to an analyzer aligned in the y-direction (an orientation that completely extinguishes the incident polarization in the absence of any Kerr rotation). The light beam intensity at the detector is

$$P = fRP_l[(1 - f)\sin^2\theta_a + k^2\cos^2\theta_a \pm \sqrt{1-f}\,k\cos\kappa\sin(2\theta_a)], \qquad (17.6)$$

where P_l is the laser output power through the collimator and $R = r^2$ is the reflectivity of the M-O medium. The peak-to-peak M-O optical power signal is

$$P_{p-p} = 2P_lRf\sqrt{1-f}\,k\cos\,\kappa\sin(2\theta_a), \qquad (17.7)$$

which will diminish if $f = 1$ (standard PBSC) or $\theta_a \rightarrow 0$. It is natural to set $\theta_a = 45°$ so that the signal is maximized ($\kappa \sim 0$):

$$P_{p-p} \cong 2P_l Rf\sqrt{1 - f}k. \tag{17.8}$$

The ultimate optical signal which gives data information is the detector current in the photodetector which converts detected optical power P into an electrical current I. If the quantum efficiency is η (A/W), and the built-in current gain is G, then the detector current is

$$I = G\eta P. \tag{17.9}$$

If the noise equivalent power (NEP) at the detector is defined as

$$NEP = \frac{\text{rms noise current}}{G\eta}, \tag{17.10}$$

then the peak-to-peak signal-to-noise ratio in optical disk recording is

$$SNR = \frac{P_{p-p}^2}{NEP^2}, \quad SNR(\text{dB}) = 10\log\frac{P_{p-p}^2}{NEP^2}. \tag{17.11}$$

Note that the photodetector current is proportional to optical power (intensity), so the signal or noise power is proportional to the optical power squared!

For example, the noise sources in an M-O system mainly include the following:

1. *Photon shot noise in photodetector.* This is due to the electron number fluctuation resulting from photon number fluctuation. The rms photon shot noise current is

$$I_{shot} = G\sqrt{2eI_{dc}\Delta f} = G\sqrt{2e\eta P_{dc}\Delta f}, \tag{17.12}$$

where Δf is the amplifier bandwidth, $e = 1.6 \times 10^{-19}$ C, P_{dc} is the average laser power at the detector, and I_{dc} is the average detector current before gain.

2. *Electronics noise in amplifier.* The noise current due to electronics noise can be represented by noise equivalent current spectral density. If it is ~ 3 pA/$\sqrt{\text{Hz}}$, for example, then rms noise current due to electronics noise is

$$I_c = 3 \times 10^{-12}\frac{A}{\sqrt{\text{Hz}}} \times \sqrt{\Delta f}. \tag{17.13}$$

3. *Johnson thermal noise due to feedback resistance R.* The corresponding noise current is

$$I_e = \sqrt{4kT\Delta f/R}, \tag{17.14}$$

where k is the Boltzmann's constant, and T is the temperature.

4. *Laser noise.* It can be minimized by good optics design and reducing optical feedback, so is often ignored.

5. *Medium noise.* It is mainly due to substrate defects, M-O film imperfection, scatters in the shape and position of recorded transitions, etc. Medium noise and shot noise tend to be the most serious noise sources in M-O recording.

A quantity called the *carrier-to-noise ratio* (CNR) is often cited to characterize optical disk medium noise. It is typically defined as the ratio between the rms signal power (carrier) and the noise power *at a nearby frequency*, where power refers to voltage power measured by spectrum analyzer, not optical power. Similar to SNR, CNR is measured by a spectrum analyzer when a fixed frequency signal is sent through the data channel. By convention, the resolution bandwidth of the spectrum analyzer is set at 30 kHz. Usually, the noise level is measured at a frequency within a few hundred kHz range of the carrier signal.

If the noise is approximately white across the channel bandwidth, and the channel response is flat, then the noise power spectral density (in V^2/Hz) is constant. In other words, the noise power over the whole signal bandwidth should be multiplied by the ratio of signal bandwidth over resolution bandwidth. At high densities, the recording signal wave is approximately sinusoidal, so the rms signal amplitude is equivalent to the carrier power measured by spectrum analyzer. Therefore, the rms signal-to-medium noise ratio is

$$SNR_{rms} \equiv \frac{\text{rms signal power}}{\text{noise power}} = CNR \cdot \frac{\text{resolution bandwidth}}{\text{signal bandwidth}}, \quad (17.15)$$

where

$$CNR \equiv \frac{\text{carrier power}}{\text{noise power}}. \quad (17.16)$$

For example, if the signal bandwidth is 10 MHz, then SNR_{rms} (dB) = CNR (dB) − 25 dB. To reach a SNR_{rms} of the 25 dB, the required medium CNR is >50 dB.

17.1.6 RLL codes in optical disk recording

Similar to magnetic recording, RLL codes such as MFM code are widely used in optical recording channel. As shown in Fig 17.11, we can write channel bits according to the NRZ rule that all the marks represent the 1s and the spaces between them represent the 0s. This is called *mark-*

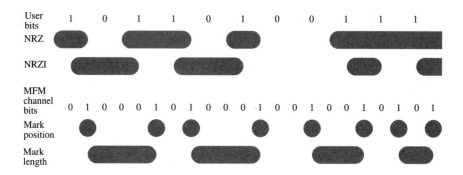

FIGURE 17.11.   User bits and RLL channel bits in optical recording. The marks can be thought as the bumps in CD-ROM, amorphous spot in phase-change disks, or magnetic domains in M-O disks.

position encoding or *peak-to-peak modulation* (PPM). Alternatively, we can write channel bits according to the NRZI implementation that the mark boundaries represent the 1s and the marks or spaces represent 0s. This is called *mark-length* encoding or *pulse width modulation* (PWM).

For simple NRZ or NRZI coding, the minimum mark size is the same. For MFM coding, however, the minimum mark size for PPM is twice that for PWM. In other words, mark-position encoding leads to a reduction in linear user data density if the minimum mark size if fixed. On the other hand, the uniform size of marks in PPM makes decoding relatively immune to defocus, sensitivity variations, and other system problems. The mark-position encoding was used in many first-generation optical disk systems, but mark-length encoding are used in more advanced products.

The DC content of a long sequence of channel bits, defined as the fraction of time the signal is high minus the fraction of time the signal is low, is an important parameter in designing RLL codes. For example, the maximum DC content of NRZ and NRZI is 100%. In magnetic recording the readback signals consist of alternating peaks with a constant baseline which is not affected by the DC content of RLL code. On contrary, the readback signals in optical disk recording have a drifting baseline depending on the DC content of RLL code. A large DC content in readback signal can cause drifts in automatic gain circuit (AGC) and high-pass filter, resulting in a higher bit error rate (BER). In addition, low frequency signals are often used by the optical head servomechanism, which will be interfered by the DC content of readback signals. Therefore, DC-free codes must be used in optical disk recording.[17]

A commonly used RLL code in optical disk recording is the *eight-to-fourteen modulation* (EFM) code, in which eight user bits are represented by fourteen channel bits plus three margin bits. The fourteen channel bits satisfy the run length constraint of $d = 2$ and $k = 10$. The three margin bits are used to eliminate the DC content and to avoid the run length violation between 14-bit channel code sequences. For example, a user data sequence of "11001010" can be represented by a channel data sequence of "10010010010010xxx", where "xxx" are the margin bits. The comparison between EFM and MFM is listed in Table 17.5. EFM code has a higher density ratio and a smaller DC content than MFM code. EFM code gives a 41.2% density gain over MFM code. Therefore, EFM code is widely used in CD-ROM and other optical disk drives. A slightly modified EFM code, called the EFM+ code, is employed in advanced products such as DVD. EFM+ code gives a 6% density gain over EFM code.

17.1.7 Servomechanism of optical head

Like magnetic hard disk drives, optical disk drives are also mechanical devices in which optical head (or heads) must be continuously controlled with a servomechanism. Both drives have a *tracking* servo to keep head on the right track. An optical head is usually not mounted on a flying slider like in a magnetic hard drive, thus requires an additional *focusing* servo to keep laser beam focused on recording surface.[18]

To keep the laser beam perfectly focused on a recording surface, the distance between the objective lens and the recording surface must be maintained at a constant focal length. *Focusing* of the optical head can be accomplished by the optical elements shown in Fig. 17.12. A four-element photodetector (quad-detector) is aligned at 45° to the axes of an astigmatic lens and positioned where the reflected beam is circular if the objective lens is at perfect focus. Any shift in focus causes the beam shape on the quad-detector to be distorted into an ellipse which illuminates one pair

TABLE 17.5. EFM Code vs. MFM Code

Code	Code rate m/n	d	k	Density ratio $(d + 1)m/n$	Maximum DC
MFM	1/2	1	3	1	33%
EFM	8/17	2	10	1.412	0
EFM+	8/16	2	10	1.5	0

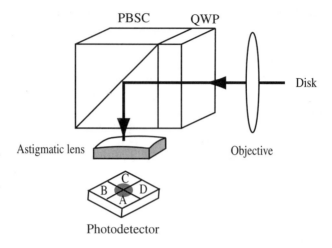

FIGURE 17.12. Focusing of optical head.

of detectors (*AD* or *BC*) more than the other. The ellipticity of the beam spot can be quantified as the focus error signal:

$$\text{focus error signal} = A + D - B - C, \text{ or } AD - BC, \quad (17.17)$$

where *A*, *B*, *C*, and *D* stand for the optical powers received by the corresponding detectors.

For optical head to function properly, the read/write laser beam must be focused not only at the right distance, but also at the right data track. The most common *tracking* method in optical recording uses tracking grooves formed into the recording surface as schematically shown in Fig. 17.13. When the laser beam is incident on a grooved surface, which is similar to a one-dimensional reflection grating, the laser is diffracted. Since the incident laser beam is focused (converging), the diffracted waves of each order form a diverging cone. The 0th order reflected beam passes the center of the numerical aperture, while the +1st and −1st diffraction orders of the reflected beam returns to the edges of the numerical aperture, respectively. The optics is aligned such that the ±1st spots are symmetrical with respect to the 0th spot. The light from the 0th spot is used to sense the focus error and the data signal. The signals from the ±1st spots as detected by split detectors (E and F) are fed into separate detection circuits, any difference between the signals indicates that the optical head is off track. This tracking scheme is also known as push-pull.[18]

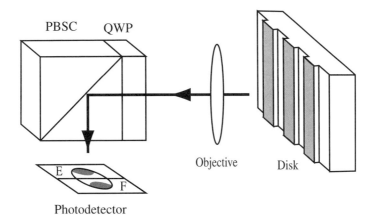

FIGURE 17.13. Tracking schemes using grooves. The shaded area between grooves is known as land.

In optical disk drives accurate focusing and tracking are achieved by closed-loop servos, which compensate for the unpredictable position errors, including both focus and tracking errors, as shown in Fig. 17.14. For example, a focusing sensor samples the light reflected from the medium surface and generates a signal proportional to the focus error. The error signal is amplified to a level that can drive a focusing motor that holds the objective lens. The motor moves the lens in the direction that reduces the focus error signal, thereby improving the focus.

Very similar servo control is done for tracking. In the push-pull tracking scheme,[18] the servo system maximizes the read signal from the main spot and makes the optical head stay on track by adjusting the beam position to balance the signal output of the two split detectors.

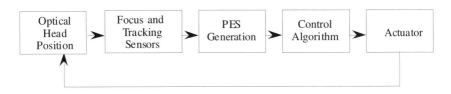

FIGURE 17.14. Focusing and tracking servos in optical disk drives. (PES = position error signal.)

17.2 HOLOGRAPHIC RECORDING

Holography was first proposed by Dennis Gabor in 1948[19] for which he later won a Nobel Prize. Since the discovery of off-axis holography in the early 1960s, 3D holographic recording has been heavily investigated for data storage.[20] In spite of its great potential, holographic data storage has not yet made it to the marketplace. However, the research activities and progresses in this area have accelerated recently. Here we briefly examine the basic principle, technique, state of the art, and the future prospect of holographic recording.

17.2.1 Basic principle of holography

Conventional optics usually records and detects light intensity only, but holography records and detects both amplitude and phase information of the object. Holograms can be classified into *amplitude hologram* and *phase hologram*. The former results from the diffraction that causes the modulation in the amplitude of illuminating light, while the latter results from the diffraction that causes the modulation in the phase.

For example, an amplitude hologram can be constructed as shown in Fig. 17.15. If O represents the electric field of a monochromatic wave from the object to be recorded and R the electric field of a reference wave coherent with O, the total field at the holographic recording medium is $O + R$. A square-law recording medium responds to the intensity $|O + R|^2$. After processing, the holographic recording material has a complex amplitude transmittance $T_a(x)$ which is proportional to the *exposure*, defined as the product of exposure time and exposure light intensity. The amplitude transmittance can be expressed as

$$T_a(x) = \beta t |O + R|^2 \qquad (17.18)$$
$$= \beta t (|O|^2 + |R|^2 + OR^* + O^*R),$$

where t is the exposure time, and β is a constant. When the hologram recorded is illuminated with the reference wave R, the transmitted wave is

$$E(x) = RT_a(x) = \beta t [(|O|^2 + |R|^2)R + |R|^2 O + R^2 O^*], \qquad (17.19)$$

where the first two terms represent the zero-order wave identical to the reference wave, the third term represents a *reconstructed wave* identical to the original object wave, and the fourth term is called the "conjugate" wave. When the reference beam R is off-axis, these three waves are spa-

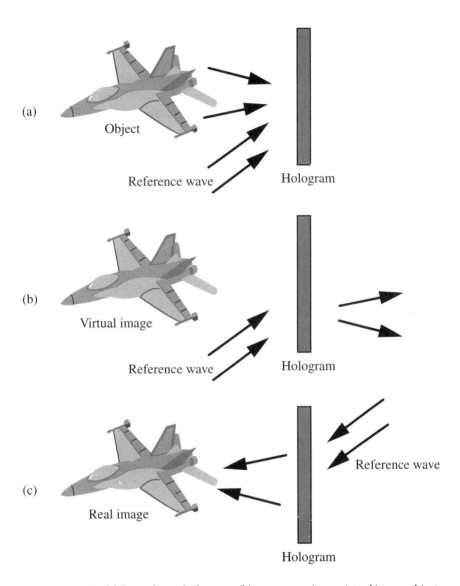

FIGURE 17.15.   (a) Recording a hologram, (b) reconstructing a virtual image object, (c) reconstructing a real image object.

tially separated from one another, i.e., their wave vectors are in different directions. Therefore, the reconstructed wave can be detected from an appropriate direction, and a *virtual image* of the original object is obtained.

If the hologram is illuminated with the conjugate of the reference beam, R^*, that is the reverse of the reference beam, then the transmitted wave is

$$E(x) = R^*T_a(x) = \beta t[(|O|^2 + |R|^2)R^* + |R|^2O^* + (R^*)^2O], \quad (17.20)$$

where the third term $|R|^2O^*$ is identical to the reverse of the original object wave, resulting in a *real image* of the object.

Holographic media are classified as *thin* or *thick* depending on their thickness d. We can define a Q parameter for a hologram consisting of a single sinusoidal grating with a grating normal to the surface of the emulsion:

$$Q \equiv \frac{2\pi\lambda d}{\Lambda^2}, \quad (17.21)$$

where λ is the light wavelength inside the holographic medium, and Λ is the period of the sinusoidal grating. The medium is considered thick if $Q > 2\pi$, and thin if otherwise. For 3D volume holographic recording, thick holographic media must be used.

17.2.2 Photorefractive effect

Many holographic recording schemes have been proposed, one of the most promising approaches is based on ferroelectric (FE) materials whose refractive index can be modulated by illumination. The so-called *photorefractive effect* originates from electric charge carriers generated by an interference pattern of light. Migration of these carriers toward areas of low optical illumination, and subsequent capture in localized traps, produce patterns of net space charge field. The space charge field and the strong *electro-optic effects* of FE materials lead to a pattern of variations in the refractive index, which will then result in a phase hologram because it will modulate the phase of illuminating beam.

An electric analog of magneto-optic effects, electro-optic effects refer to the modification of the refractive index ellipsoid of a material by an electric field. Noncentrosymmetric crystals (without centers of symmetry) can have a linear electro-optic effect called the Pockels effect, while centrosymmetric crystals (with centers of symmetry) can have a quadratic electro-optic effect called the Kerr effect. Both result in a net change in refractive index proportional to the space charge field.[21]

The origin of charge transport can be understood based on the energy band model. Electron or holes are excited from occupied levels (located in the band gap, but near conduction band edge) to the conduction band. Once excited, the charge carriers drift and diffuse from the bright area to the dark area, eventually relax to unoccupied levels. Since the charges redistribute according to the light interference pattern, a space charge field pattern develops. For example, one of the commonly used photorefractive crystals is $LaNbO_3$:Fe (doped with Fe^{2+}/Fe^{3+}), where Fe^{2+} act as electron donors and Fe^{3+} act as electron acceptors.

The band model of charge transport indicates that the read process of a hologram also excites the charge carriers that can redistribute and erase the recorded hologram. For data storage applications, *hologram fixing* techniques must be employed:

1. *Electrical fixing.* If a simple sinusoidal grating generated by free carrier transport is established as the principal grating, and subsequently an electric field comparable to the coercive field is applied, then the resulting FE domains (collective ionic displacements) tend to screen the principal space-charge field. When the field is removed, the ferroelectric domains remain, resulting in a fixed complementary ion grating that is the actual recorded data pattern. The ion grating may be revealed upon uniform illumination to remove free carrier pattern.

2. *Thermal fixing.* There are many nonphotorefractive ions in as grown crystals like $LiNbO_3$. Their mobility is strongly temperature-dependent, and tends to be very low at room temperature. If a recorded grating is heated to 100–150°C, ion relaxation occurs, forming a complementary ion grating. Upon cooling, the recorded electron grating may be erased by uniform illumination while the complementary ion grating is fixed.

3. *Two-photon recording.* In this method, light of one wavelength uniformly illuminates the recording material while light of the other wavelength records the electronic grating. During the write process, the combination of two photons of the two wavelengths excites the carriers to the conduction band. This requires either an intermediate state within the band gap or very strong laser intensities. During the read process, only the light of the second wavelength is used, so carriers cannot be excited. Consequently, the electronic grating can be read many times without erasure.

17.2.3 Holographic recording system

A fully digital storage system using volume holographic recording was demonstrated in 1994.[3] The recording medium was a $LiNbO_3$:Fe crystal measured 2 cm $\times$ 1 cm $\times$ 1 cm. The schematic of this benchmark demonstration system is shown in Fig. 17.16. During the write process, the signal beam passes through a spatial light modulator (SLM), where digital information is displayed. The reference beam is expanded to ensure sufficient overlap between the reference and signal beams. The angle of the reference beam is adjusted as a new page of information is displayed by the SLM, resulting in a new page being stored in the crystal at the same spatial location but with different wave vectors. During the read process, only the reference beam illuminates the crystal. As the reference beam scans the different recording angles, the diffracted waves containing digital information from corresponding pages are captured by a CCD (Charge-Coupled Device) camera. The output signals from the CCD camera are then decoded to retrieve the stored information.

The above method of recording different pages is called the *angular multiplexing*. The nearly orthogonal configuration of the reference beam and the signal beam offers the highest angular selectivity and lowest medium scattering noise. Other multiplexing methods include *phase-encoded multiplexing* (using a phase SLM to control the reference beam), *wavelength multiplexing* (using reference waves of different wavelengths but incident at the same angle), and *spatial multiplexing* (dividing the recording crystal volume into a number of regions and recording one stack of pages per region).

In the first digital demonstration, a user data capacity of 163 KB was recorded at a BER of 10^{-6}, giving a user data volume density of 650

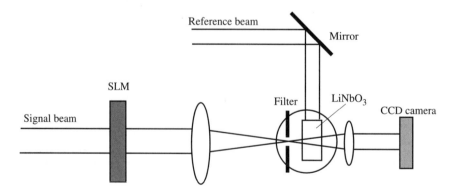

FIGURE 17.16.   Top view of a digital holographic recording system.

bits/mm^3. The data transfer rate was ~50 Mb/s, and recorded information was not fixed. The write speed was even slower—it took ~1 hour to write 163 KB. An improved version of the same system was reported in 1997 to have a data capacity of 16 MB, a data transfer rate of 250 Mb/s, and a write speed of 250 Kb/s. It suffices to say that the overall performance of holographic recording system is being improved rapidly, but is not yet competitive with magnetic hard disk drives.

17.2.4 Theoretical capacity and practical limitations of holographic recording

What is the ultimate capacity of holographic recording? Consider the 3D hologram as an information channel with W independent levels and a resolving power of at least $1/\lambda$ cycles/mm, where λ is the laser wavelength. The information capacity per unit volume is

$$\rho = \frac{\log_2 W}{\lambda^3}. \tag{17.22}$$

If $W = 2^{10} = 1024$, $\lambda = 0.5$ μm, then $\rho = 8 \times 10^{10}$ bits/mm^3. However, note that here we only counted the volume of the recording medium, but not the volume of other optical and electronic parts.

It is difficult to realize the theoretical capacity because noise sources such as medium scattering noise, interpage cross talk, CCD camera noise, and electronics noise tend to reduce the usable capacity long before approaching the theoretical limit. Furthermore, the *diffraction efficiency* of holograms (fraction of the reference beam that is diffracted into the reconstructed data beam) decreases as $1/N^2$, where N is the total number of pages stored in a crystal, thus the readback signal drops accordingly. If $\lambda = 500$ nm, a practical volume density limit is estimated[20-21] to be ~10^6 bits/mm^3.

A significant advantage of holographic recording is its intrinsic nature of parallel access of information, which is especially attractive for rapid access archival storage where frequent writing is not required. The major challenges for holographic recording include the following:

1. Recording media which satisfy these requirements: (a) the recorded gratings can be fixed (>10 years), (b) fast photorefractive response for fast write, (c) optical quality for high SNR, (d) large diffraction efficiency, (e) low cost.
2. Availability of compact blue-green lasers, spatial light modulators, and high-speed CCD cameras.
3. Efficient system integration and error correction approaches.

17.3 FLASH MEMORY

Semiconductor memories are essential building blocks of the data storage hierarchy in modern computers. Due to their fast access time, semiconductor memories are usually used as the main memory of central processing unit (CPU). The nonvolatile flash memory has been viewed as a possible replacement for magnetic hard disk drives particularly for portable computer applications. Therefore, it is interesting to examine the history, operation principle, and trends of flash memory.

17.3.1 Why flash?

As listed in Table 17.6, there are a variety of semiconductor memories being used extensively today, and their operation can be found in many VLSI textbooks.[22] The operation principles of DRAM, SRAM, ROM, and PROM are skipped here.

The first EPROM cell was developed with a poly-Si as a floating gate, as shown in Fig. 17.17a. To *program* (or *write*) a cell, one applies a high drain

TABLE 17.6.   Semiconductor Memories

Volatile {	Dynamic random access memory (DRAM)
	Static random access memory (SRAM)
Nonvolatile {	Read-only memory (ROM)
	Programmable read-only memory (PROM)
	Erasable programmable read-only memory (EPROM)
	Electrically erasable programmable read-only memory (E^{2}PROM)
	Flash E^{2}PROM or flash EPROM (flash memory)

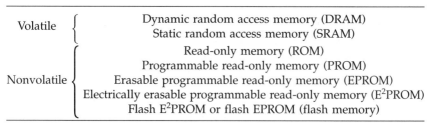
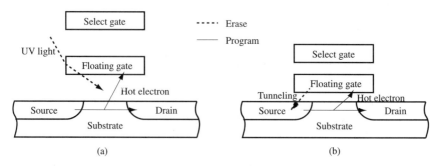

FIGURE 17.17.   Cell structures of (a) EPROM, and (b) E^{2}PROM (or flash) memory.

voltage to cause an avalanche process near the drain. The hot electrons generated are energetic enough to tunnel through the gate oxide and are eventually stored in the floating gate. The stored electrons increase the transistor turn-on threshold voltage (V_T) such that an applied voltage (V_G) at the select gate (or called the control gate) is not sufficient to turn it on. The absence of current results in a "0" at the memory output. To *erase* a cell, one uses UV radiation to excite electrons back to the bulk semiconductor. The resulted transistor can be turned on with the same applied voltage (V_G) at the select gate, producing a "1" at the memory output.

Although EPROM is erasable, UV radiation is cumbersome and shortens the transistor endurance. Therefore, an E^2PROM cell shown in Fig. 17.17b was invented to electrically remove the electrons from the floating gate. E^2PROM cells are *programmed* by hot electron injection like EPROM, but *erased* by applying a large voltage to the source which results in cold electron tunneling (Fowler-Nordheim tunneling) from the floating gate to the source. It should be noted that hot electron injection is only possible from channel to gate, and requires more voltage than Fowler-Nordheim tunneling, but it is possible to use Fowler-Nordheim tunneling mechanism for both programming and erasure. Tunneling is intrinsically slower than hot electron injection, though.

E^2PROM has many advantages over EPROM. For example it has more endurance by avoiding UV erasure which is harmful to transistors. However, E^2PROM high-voltage erasure procedure produces a wide scatter of cell gate thresholds, so a separate select transistor per cell is needed for E^2PROM to work properly. As a result, E^2PROM cell area is two to three times greater than that of EPROM. The basic idea of flash memory is to combine the advantages of both E^2PROM and EPROM. One way to accomplish this is that we keep most of the features of E^2PROM but give up the selectivity to each cell. By eliminating select transistor, flash memory cell can be made as small as that of EPROM. The cells can still be programmed or read individually, but they must be *erased one entire block at a time* (or several blocks at a time, but separately). All cells tunnel-erase uniformly to produce a very tight gate threshold distribution. This type of memory is called either flash E^2PROM or flash EPROM, and more often simply flash memories. The phrase "flash" did not mean "superfast", but was used to describe the erase operation of the chip. In a typical flash memory, a bit can be accessed in ~60–100 ns, but it takes ~1 ms to erase and ~5 μs to reprogram. In other words, flash memory can be almost instantly accessed, but it is much slower to write than magnetic hard disk drives.

17.3.2 Program/erase endurance

Like other semiconductor memories, flash memory is intrinsically much faster than magnetic hard disk drives in terms of access time. However, the program/erase mechanisms limit the endurance of flash memory. As shown in Fig. 17.18a, hot electron injection requires large voltage bias between the source and drain. The drawback of this process is that it may be destructive to the oxide barrier. Over a course of many programming cycles, the energetic traveling electrons can generate traps and fixed charges in the silicon/oxide interface, gradually degrading the turn-on characteristics of the transistor. In contrast, Fowler-Nordheim tunneling requires much less voltage bias, as shown in Fig. 17.18b, so it is less destructive to oxide barrier. Unfortunately, the thin oxide barrier layer (~10–20 nm) does degrade over time, and eventually leads to memory failure. In spite of many variations of flash memory cell structures which can be found in the literature, the product life of flash memory is largely limited to ~100,000 program/erasure cycles.

17.3.3 Flash memory and its future

The general architecture of flash memory is very similar to that of DRAM, as schematically shown in Fig. 17.19. The horizontal lines (rows) connecting all the select (control) gates are called the *word lines*. The vertical lines

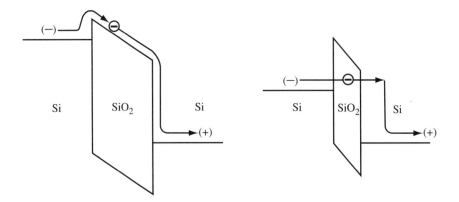

(a) Hot electron: over the energy barrier   (b) Tunneling: through the energy barrier

FIGURE 17.18.   (a) Hot electron injection. (b) Fowler-Nordheim tunneling under a strong electric field. If SiO_2 thickness is very small, direct tunneling becomes important.

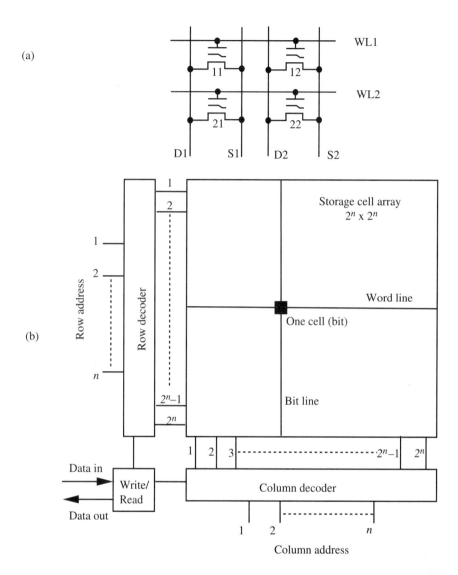

FIGURE 17.19. (a) Cell connection. (b) Block diagram of flash memory, here a bit line corresponds to a pair of D and S lines connecting drains and sources.

(columns) connecting all the drains and sources, respectively, are called the *bit lines*. In Fig. 17.19a, to *read* cell 11, the word line WL1 is selected (e.g., +5 V), then D1 = 1 V and S1 = 0 V are applied. If the cell is in "0", no current is sensed between S1 and D1. If the cell is in "1", the absence of charges in the floating gate allows the transistor to turn on, resulting in a current between S1 and D1. To *erase* cell 11 (turn "0" into "1"), WL1 = −11 V, D1 = float, and S1 = 5 V are applied such that electrons will tunnel from the floating gate to the bulk semiconductor. To *program* cell 11 (turn "1" into "0"), WL1 = 5 V, D1 = 20 V, and S1 = 0 V are applied such that hot electrons inject to the floating gate. Note that the voltages given are just for examples, the real values depend on the manufacturers and cell structures and are being scaled down as the chip density increases.

As mentioned previously, the flash memory cells are erased "in blocks." To read or program, the location of a memory cell is calculated by the row/column decoders. Data are first stored in shift registers, then programmed into each selected cell one by one. Typically a sequence of data are written in the same row, this is why the horizontal lines are called the word lines. In the read process, the cell is selected by the row decoder, the column decoder scans the entire row with a given clock cycle, and the data are read out to an external buffer. In this sense, the program/read operation of flash memory is very similar to that of DRAM.

The main advantages of flash memory are its ruggedness and short access time, currently around 60–100 ns, but it is at least one order of magnitude more expensive than magnetic hard disk drives in cost per megabytes ($/MB). For example, the state-of-art 16-MB flash chip costs ~$10/MB in 1997. The amount of technology required to lower its cost to a comparable level of magnetic hard disk drive is almost prohibitive, except for very small factor disk drives (<1.8 in.). In fact, current trends indicate that disk drive prices drop even more rapidly than those of flash memory and DRAM. Therefore, it is expected that flash memory will continue to play a dominant role in niche markets where highly rugged data storage with very fast access is required.

17.4 MAGNETIC RAM

Semiconductor memories such as DRAM are essential building blocks of modern computing machines, and flash memory is playing an increasing role in portable nonvolatile storage applications. The volatility of DRAM and the slow write time of flash memory render them not desirable for many information storage applications. The magnetic RAM (MRAM) technology based on giant magnetoresistance (GMR) materials and spin-dependent tunneling (SDT) junctions (or magnetic tunnel junction) is now considered a promising nonvolatile storage technology which could significantly alter the storage hierarchy of modern computers and other information systems. These storage devices are intrinsically nonvolatile, and can be rewritten an almost infinite number of times. In addition, they are radiation hard, which is especially attractive for space and military applications.

Magnetic RAM can be traced back to magnetic core memory that used to be the main memory for computers before being replaced by DRAM.[23,24] After the advent of more superior semiconductor memories, the subject of magnetic RAM had been largely abandoned for a long period until the late 1980s when MRAM based on anisotropic magnetoresistance (AMR) was proposed.[25] However, limited signal levels of the AMR-based MRAM lead to its slow access and small data capacity. Consequently, only Honeywell produced such MRAM chips at small quantities.

Magnetic RAM caught greater interests with the rapid development of giant magnetoresistive multilayers and spin valves, and more recently SDT junctions.[26-28] It is believed that larger signal levels attainable in these MRAM designs will enable much more attractive access times and greater data capacities. There are a variety of designs for MRAM storage cells in the literature. For example, an MRAM design based on spin valves is shown in Fig. 17.20. The storage cell comprises of two FM layers (F1 and F2) separated by a nonmagnetic metal (NM). By design F2 is the pinned layer and F1 is the free layer. Both F1 and F2 are soft ferromagnetic layers which are weakly (and generally ferromagnetically) coupled across the nonmagnetic spacer. Since magnetization of F1 is relatively free to respond to a relatively small applied field, the relative orientation of

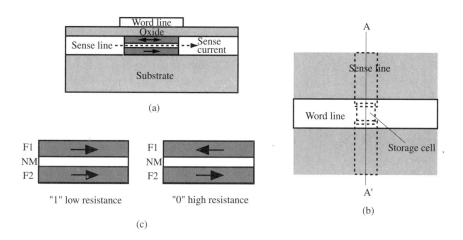

FIGURE 17.20. (a) Cross section of spin valve MRAM storage cell along AA' in (b). (b) Top view of spin valve MRAM storage cell. (c) Magnetization states of bit "1" and "0".

the two magnetizations (from parallel to antiparallel alignment) can be controlled by the field, resulting in a GMR effect.

Each storage cell is connected to a vertical (column) sense line and is crossed over by a horizontal (row) word line. In write operation, a word line current pulse combined with a sense line current pulse switches the magnetization of F1 to the right ("1") or to the left ("0"). Only the cell crossed by both the word line and the sense line will be switched. Upon turning off the currents, the F1 magnetization remains unchanged because of shape and intrinsic anisotropy.

In read operation, the total voltage drop on all the storage cells connected by a sense line is measured with a sense current pulse. Within the sense current pulse, a word line current pulse (interrogation current) sufficient to switch F1 to the right is applied. The sense voltage should remain high if the bit is "1" because F1 stays toward the right as in Fig. 17.20c. In contrast, the sense voltage will have a jump from low to high if the bit is originally "0" because F1 will switch from the left to the right. Only the cell crossed by the word line and the sense line may change the state. This interrogation process is destructive for bit "0", which must be rewritten after detection. The write/read operations of spin-valve MRAM are illustrated in Fig. 17.21. Note that the vertical scale is arbitrary.

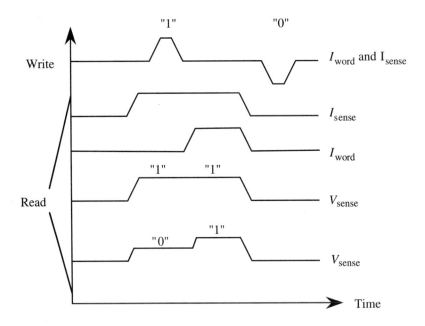

FIGURE 17.21.   Write/read current and sense voltage pulses of spin valve MRAM.

The important goals of MRAM design and fabrication include efficient utilization of the chip area, low power consumption, uniform switching thresholds, low BER, and short access time. If the magnetic layers in the storage cells are in single domain states as designed, they can switch very fast (~1 ns) from state "1" to "0", or vice versa, by a coherent domain rotation mode. Therefore, the access time of MRAM will be mainly limited by the SNR consideration, by the cell impedance, and by interconnects and complementary metal oxide semiconductor (CMOS) circuitry. At this writing, high-yield manufacturing processes based on either spin valves or SDT junctions are being demonstrated.

References

1. T. Suzuki, "High density optical storage materials," *Jpn. J. Appl. Phys.,* **64**(3), 208, 1995.

2. E. Betzig, J. R. Trantman, R. Wolfe, E. M. Gyorgy, P. L. Flinn, M. H. Kryder, and C. H. Chang, "Near-field magneto-optics and high density data storage," *Appl. Phys. Lett.,* **61**(2), 142, 1992.

3. J. F. Heanue, M. C. Bashaw, and L. Hesselink, "Volume holographic storage and retrieval of digital data," *Science,* **265**, 749, August, 1994.

4. A. G. Barre, "Flash memory: magnetic disk replacement?," *IEEE Trans., Magn.,* **29**, 4104, 1993.

5. J. M. Daughton, "Magnetoresistive memory technology," *Thin Solid Films,* **216**, 162, 1992.

6. B. D. Terris, H. P. Martin, M. E. Best, D. Rugar, and S. A. Rishton, "Nanoscale replication for scanning probe data storage," *Appl. Phys. Lett.,* **69**, 4262, 1996.

7. J. Lindmayer, P. Goldsmith, K. Gross, "Electron trapping for mass data storage memory," *Proc. SPIE,* **1401**, 103, 1990; A. D. McAulay and J. Wang, "Methods of addressing electron trapping optical memory materials," *Proc. SPIE,* **2754**, 210, 1996.

8. J. Watkinson, *Coding for Digital Recording,* (Kent, England: Focal Press, 1990), p. 26.

9. J. C. Briggs, "Enabling technologies for a 100 MB 3.5-in floppy (zip) disk drive," *Proc. SPIE,* **2604**, 1995.

10. H. Mimura, "DVD-Video format," *Proc. IEEE COMPCON 97*, San Jose, CA, 1997.

11. M. Elphick, "DVD technology: the new paradigm in optical storage," *Data Storage,* **25**, 25, January 1997. Note the access times and data rates of CD-ROM and DVD-5 were obtained from the web pages of Toshiba CD-ROM model XM-5701 and DVD-ROM model SD-C2002 Slim.

12. Taken from Toshiba DVD-RAM model SD-W1001, IBM hard disk drive model DTCA-24090 (Travelstar 4GT), and Seagate hard disk drive model ST19101 (Cheetah 9).

13. T. Ohta, K. Inoue, T. Akiyama, and K. Yoshida, "Effect of laser pulse width on overwrite cycle characteristics of phase change disk media," *Proc. SPIE,* **1663**, 436, 1992.

14. H.-P. D. Shieh and M. H. Kryder, "Operating margins for M-O recording materials with direct overwrite capability," *IEEE Trans. Magn.,* **23**, 171, 1987.

15. K. Tsutsumi and T. Fukami, "Direct overwrite in magneto-optic recording," *J. Magn. Magn. Mater.,* **118**, 231, 1993.

16. M. H. Kryder, "Magnetic storage of information," *Encyclopedia of Applied Physics,* 1994.

17. J. Watkinson, *Coding for Digital Recording,* (Kent, England: Focal Press, 1990), p. 57.

18. A. B. Marchant, *Optical Recording,* Addison-Wesley, Reading, MA, 1990; M. Mansuripur, *The Physical Principles of Magneto-Optical Recording,* Cambridge University Press, Cambridge, UK, 1995.

19. D. Gabor, "A new microscopic principle," *Nature,* **161**, 777, 1948.

20. H. M. Smith, *Holographic Recording Materials* (Berlin, Germany: Springer-Verlag, 1977); J. W. Goodman, *Introduction to Fourier Optics,* (New York, NY: McGraw-Hill, 1996).

21. L. Hesselink and M. C. Bashaw, "Optical memories implemented with photorefractive media," *Optical and Quantum Electronics,* **25**, S611, 1993.

22. For example, B. Prince and G. Due-Gundersen, *Semiconductor Memory,* (New York, NY: John Wiley, 1983); C. Hu, *Nonvolatile Semiconductor Memories,* (Piscataway, NJ: IEEE Press, 1991).

23. S. Chikazumi and S. H. Charap, *Physics of Magnetism,* (Malabar, FL: Krieger Publishing, 1964), p. 508.

24. J. Smit, *Magnetic Properties of Materials,* (New York, NY: McGraw-Hill, 1971), pp. 205, 271.

25. A. V. Pohm, C. S. Comstock, and J. M. Daughton, "Analysis of M-R elements for 10^8 bit/cm^2 arrays," *IEEE Trans. Magn.,* **25**, 4266, 1989.

26. K. T. M. Ranmuthu, A. V. Pohm, J. M. Daughton, and C. S. Comstock, "New low current memory modes with giant magnetoresistive materials," *IEEE Trans. Magn.,* **29**, 2593, 1993.

27. D. D. Tang, P. K. Wang, V. S. Speriosu, S. Lo, and K. K. Kung, "Spin-valve RAM cell," *IEEE Trans. Magn.,* **31**, 3206, 1995.

28. J. M. Daughton, "Magnetic tunneling applied to memory," *J. Appl. Phys.,* **81**, 3758, 1997.

Index